Contributions to
Analysis and Geometry

CONTRIBUTIONS TO ANALYSIS AND GEOMETRY

Edited by
D. N. Clark
G. Pecelli
R. Sacksteder

SUPPLEMENT TO THE AMERICAN JOURNAL OF
MATHEMATICS

The Johns Hopkins University Press
Baltimore and London

The Johns Hopkins University Press, Baltimore, Maryland 21218
The Johns Hopkins University Press Ltd., London

Library of Congress Catalog Card Number 81-48192
ISBN 0-8018-2779-5

It is with gratitude, admiration, and affection that we dedicate this volume to Professor Philip Hartman at his retirement from The Johns Hopkins University. Whether as teacher, colleague, or friend, his influence on all of us has been profound. We, along with the rest of the mathematical community, eagerly look forward to the continuation of his outstanding contributions to mathematics.

CONTENTS

FOREWORD

A conference was held at The Johns Hopkins University on April 24 and 25, 1980 to honor the distinguished tradition of analysis and geometry at the Hopkins Mathematics Department and the contributions of its members, students, and friends to these fields. We wish to express our gratitude to Professor J. A. Shalika for his skillful handling of the arrangements and to the National Science Foundation for its support.

This volume contains most of the invited addresses, which were given by François Trèves, Charles C. Pugh, Louis Nirenberg, Richard Sacksteder, Clifford Truesdell, Jack Hale, and Nelson Max, as well as papers by some of the more than 70 mathematicians attending the conference and by some who were unable to attend. The organizing committee is especially grateful to Joseph H. Sampson, Jun-Ichi Igusa, and the other editors of the *American Journal of Mathematics* for their gracious invitation to publish this volume of conference proceedings.

THE ORGANIZING COMMITTEE:
D. N. Clark
G. Pecelli
R. Sacksteder

COMPARISON THEOREMS FOR SELF-ADJOINT SECOND ORDER SYSTEMS AND UNIQUENESS OF EIGENVALUES OF SCALAR BOUNDARY VALUE PROBLEMS*

By Philip Hartman

Introduction. The problem motivating this paper is the generalization of the results of [5] on singular and nonsingular boundary value problems for scalar equations of the form

$$(1) \qquad x'' + q(t,\lambda)x = 0$$

to more general self-adjoint equations

$$(2) \qquad (p(t,\lambda)x')' + q(t,\lambda)x = 0.$$

It was pointed out to me by Professor Ahlbrandt that these results of [5] and those in Part III below answer some questions of uniqueness raised by Reid [8]. (This paper is not concerned with the main results of Part II in [5], involving complete monotonicity and infinite divisibility.)

The results of [5] depended on comparison theorems which were of interest in themselves. These comparison theorems, involving three equations

$$(3) \qquad x'' + q_i(t)x = 0 \qquad \text{for} \quad i = 1, 2, 3,$$

concerned either the comparison of the logarithmic derivatives of principal solutions of disconjugate equations (Theorems 3.1, $N = 2$) or the comparison of zeros of certain solutions (Corollary 6.1). By using a generalization of the Picone identity, Mingarelli [7] has given a simple proof of the latter type of comparison theorem for

$$(4) \qquad (p_i(t)x_i')' + q_i(t)x = 0 \qquad \text{for} \quad i = 1, 2, 3.$$

It turns out that Mingarelli's form of the Picone identity can be used to obtain both types of comparison theorems needed to extend the results of [5] from (1) to (2).

---

*Research partially supported by NSF Grant MCS78-01101.

1

In Part I, we further generalize the Picone identity to involve self-adjoint systems of linear second order equations,

(5) $(P(t)Y') + Q(t)Y = 0,$

in order to obtain comparison theorems for (5) and a scalar equation, using techniques of [6]. The results, when specialized to diagonal systems (for two scalar equations) give the desired comparison theorems (including that of Mingarelli [7]) for (4). The first result of Part I is an analogue of the first comparison theorem of Sturm (involving certain initial conditions) and generalizes some results of [6] which are analogues of the separation theorem in Sturm (involving the initial condition $Y = 0$). The second type of result of Part I is a comparison of principal solutions (and is an analogue of a theorem of Hartman and Wintner for scalar disconjugate equations; cf. Corollary 6.5, [2], pp. 358–359).

In Part II, we specialize the results of Part I to diagonal systems (5) (i.e., to sets of scalar equations as, for example, (4)).

In Part III, the results of Part II are applied to questions of uniqueness of eigenvalues of certain singular and nonsingular boundary value problems for (2).

The main results are stated in the first section of each part; see Sections 1, 4 and 6. The proofs for the basic comparison theorems (Parts I and II) are quite different from those of [5], while the proofs on the theorems about eigenvalues (Part III) follow those of [5].

It will remain an open question whether or not the oscillation result contained in Corollary 9.1 of [5] and the results of Part II of [5] dealing with the complete monotonicity, with respect to λ, of the logarithmic derivatives of principal solutions of disconjugate equations (1) have analogous generalizations.

I should like to acknowledge with thanks a useful correspondence with A. B. Mingarelli about [7].

Remark. See the Remark preceding Theorem 6.2 below for a correction to [5].

Part I.   Comparison theorems for self-adjoint systems.

1. Statement of results.   Let S be the linear space of real $n \times n$ symmetric matrices,

(1.1) $S = \{A : A = A^*, A \text{ is real } n \times n \text{ matrix}\}.$

For a t-interval T to be specified, let

(1.2) $P(t) = P^*(t) > 0, Q(t) = Q^*(t) \in C^0(T; S)$,

so that if y is an n-vector, the equation

(1.3) $(P(t)y')' + Q(t)y = 0$

is formally self-adjoint. We shall compare (1.3) with a scalar equation

(1.4) $(p_0(t)u')' + q_0(t)u = 0$,

where

(1.5) $p_0(t) > 0, q(t) \in C^0(T; \mathbb{R})$.

Etgen and Lewis [1] used real-valued linear functionals on the set of $n \times n$ real matrices to obtain comparison theorems for (1.3). Their results were generalized in [6] by the use of nonlinear real-valued functionals

(1.6) $p, q : S \to \mathbb{R}$.

Some of the properties of p, q required in [6] are the following, in which $A, B \in S$ and $0 \leqq \lambda \in \mathbb{R}$:

(1.7) $q(A + B) \geqq q(A) + q(B)$ (superadditive),

(1.8) $q(\lambda A) \geqq \lambda q(A)$ (superhomogeneous),

(1.9) $q(A) > 0$ if $A > 0$ and $q(I) = 1$

(positive and normalized),

(1.10) $p(A + B) \leqq p(A) + p(B)$ (subadditive),

(1.11) $p(\lambda A) \leqq \lambda p(A)$ (subhomogeneous),

(1.12) $q(A) \leqq p(A)$ for $A \geqq 0$,

(1.13) $q(A) > 0$ if $A \geqq 0$ and rank $A > n - k$ $(k - \text{positive})$.

Conditions (1.7)–(1.8) and (1.10)–(1.11) imply that q is a concave and p is a convex functional, and are therefore continuous. Below the functional p need only be defined on the cone of nonnegative $A \in S$. Note that "positivity" of q (cf. (1.9)) is equivalent to "1-positivity" (cf. (1.13)).

First, let $T = [a, b]$. Let M, N be constant $n \times n$ matrices such that

(1.14) $\operatorname{rank}(M, N) = n,$

(1.15) $N^* P(a) M = M^* P(a) N \in S.$

Thus (1.3) and the boundary conditions

(1.16) $My(a) - Ny'(a) = 0,$

(1.17) $y(b) = 0$

constitute a self-adjoint boundary value problem.

THEOREM 1.1.   *Assume (1.2) with $T = [a, b]$, (1.6)–(1.12), (1.14)–(1.15), and let $Y(t)$ be the $n \times n$ matrix solution of the initial value problem*

(1.18) $(P(t) Y')' + Q(t) Y = 0,$

(1.19) $Y(a) = N, Y'(a) = M.$

Assume (1.5) and

(1.20) $p_0(t) \geqq p(P(t)), \quad q_0(t) \leqq q(Q(t)),$

and let $u(t)$ be a solution of (1.4) satisfying

(1.21) $\displaystyle \limsup_{\alpha \to a+0} \left\{ p_o(\alpha) u'(\alpha) u(\alpha) + q\left(-u^2(\alpha) P(\alpha) Y'(\alpha) Y^{-1}(\alpha) \right) \right\} \geqq 0.$

(i) *Then, on any interval $(a, \beta) \subset [a, b]$ on which $u(t) \neq 0$ and $\det Y(t) \neq 0$,*

(1.22) $p_0 u' u \geqq q(u^2 P Y' Y^{-1}), \quad \text{hence} \quad p_0 u'/u \geqq q(P Y' Y^{-1}).$

(ii) *If "equality" holds in the first part of (1.22) for some $t = t_o \in (a, \beta)$, then "equality" holds in (1.21); also*

(1.23) $q(u^2 P Y' Y^{-1}) \leqq p_0 u' u \leqq -q(-u^2 P Y' Y^{-1}),$

$q_o(t) - q(Q(t)) \equiv 0 \quad \text{and} \quad u'^2(t)[p_0(t) - p(P(t))] \equiv 0$

on $(a, t_0]$ (so that $p_0 - p(P) \equiv 0$ on any interval on which $u' \neq 0$; e.g., on any interval on which $q_0 \neq 0$). Furthermore, if (1.13) holds, then there exist two k-dimensional linear manifolds $\mathbb{R}_0^k, \mathbb{R}_1^k \subset \mathbb{R}^n$ such that, on $[a, t_0]$,

$$(1.24) \qquad Y(t) y_0 \equiv u(t) y_1 \qquad \text{for all} \quad y_0 \in \mathbb{R}_0^k$$

and some $y_1 = y_1(y_0) \in \mathbb{R}_1^k$. (iii) If $u(t)$ satisfies

$$(1.25) \qquad u(t) > 0 \qquad \text{on} \quad (a, b) \qquad \text{and} \quad u(b) = 0,$$

then

$$(1.26) \qquad \det Y(t_0) = 0 \qquad \text{for some} \quad t_0 \in (a, b].$$

(iv) If (1.25) holds, $\det Y(t) \neq 0$ on (a, b), and (1.13) holds, then (1.23) and (1.24) hold on (a, b), so that rank $Y(b) \leq n - k$.
    Condition (1.21) always holds if

$$(1.27) \qquad u(a) = 0;$$

cf. Lemma 2.2 below. When $N = I$, so that (1.19) becomes

$$(1.28) \qquad Y(a) = I \qquad \text{and} \qquad Y'(a) = M,$$

then (1.21) is implied by

$$(1.29) \qquad p_0(a) u'(a) / u(a) \geq -q(-P(a) M),$$

where the left side is considered to be $+\infty$ if $u(a) = 0$. When $u(a) = 0$, assertion (iii) is contained in Theorem 1.1 of [6].
    Corresponding to Theorem 1.1, we can also obtain a comparison theorem for principal solutions if $T = [b, \omega)$, $b < \omega \leq \infty$, and (1.3) is disconjugate. (For general information about principal solutions of (1.4) and (1.3), see [2], pp. 350–361 and pp. 384–396.)

    THEOREM 1.2. Let $T = [b, \omega)$, $b < \omega \leq \infty$. Let P, Q, p, q and p_0, q_0 satisfy the assumptions of Theorem 1.1. Let (1.18) be disconjugate on $[b, \omega)$ and $Y = Y_0(t)$ a principal solution. (i) Then (1.4) is disconjugate on $[b, \omega)$ and if $u = u_0(t) > 0$ is a principal solution of (1.4), then

$$
(1.30) \quad
\begin{aligned}
p_0 u_0' u_0 &\leq -q(-u_0^2 P Y_0' Y^{-1}), \qquad \text{hence} \\
p_0 u_0' / u_0 &\leq -q(-P Y_0' Y_0^{-1}).
\end{aligned}
$$

(ii) *If* (1.13) *holds and if equality holds in* (1.30) *for some* $t = t_0 \in [b, \omega)$, *then there exist k-dimensional linear manifolds* $\mathbb{R}_0^k, \mathbb{R}_1^k \subset \mathbb{R}^n$ *such that*

$$(1.31) \qquad Y_0(t)y_0 \equiv u_0(t)y_1 \qquad for \quad t \in [t_0, \omega) \qquad and \quad y_0 \in \mathbb{R}_0^k,$$

and some $y_1 = y_1(y_0) \in \mathbb{R}_1^k$. *Furthermore,*

$$q\big(u_0^2 P Y_0' Y_0^{-1}\big) \leqq p_0 u_0' u_0 \leqq - q\big(- u_0^2 P Y_0' Y_0^{-1}\big),$$

$$(1.32)$$

$$q_0(t) - q(Q(t)) \equiv 0 \qquad and \quad u_0'^2(t)\big[\, p_0(t) - p(P(t))\big] \equiv 0$$

hold on $[t_0, \omega)$.

2. Proof of Theorem 1.1. On (i). Most of the proofs in this paper depend on the following lemma which can be considered as a "generalized Picone relation". For the relationship between (2.3), the standard Picone identity, and Mingarelli's generalization [7], see Section 4.

LEMMA 2.1. *Let* $Y(t)$ *be a prepared solution of* (1.18) *in the sense of* [3], *i.e.,*

$$(2.1) \qquad Y^* P Y' = (Y^* P Y')^* \in S \qquad for \quad t \in T$$

(e.g., let (2.1) *hold at* $t = a$*). Put*

$$(2.2) \qquad V = P Y' Y^{-1}, \qquad so \ that \quad V = V^* \in S,$$

wherever Y *is nonsingular. Then, for any scalar function* $u(t) \in C^1(T)$,

$$(2.3) \qquad (u^2 V)' + H^* H + u^2 Q - u'^2 P = 0,$$

where $H = H(t)$ *is the matrix*

$$(2.4) \qquad H = u' P^{1/2} - u P^{-1/2} V = P^{1/2} Y (Y^{-1} u)'.$$

(Also, if $U(t)$ *is a rectangular* $n \times k$ *matrix function of class* $C^1(T)$, *then*

$$(2.3') \qquad (U^* V U)' + H^* H + U^* Q U - U'^* P U = 0,$$

where $H = P^{1/2} U' - P^{-1/2} V U = P^{1/2} Y (Y^{-1} U)'$.)

In fact, V satisfies the Riccati equation

(2.5) $V' + VP^{-1}V + Q = 0,$

so that

(2.6) $(u^2V)' + u^2Q + u^2VP^{-1}V - 2uu'V = 0,$

which is equivalent to (2.3). (The result (2.3'), which will not be used below, follows similarly.)

In Theorem 1.1, $Y(t)$ is a prepared solution by (1.15) and (1.18)–(1.19). On any interval $(a, \beta) \subset [a,b]$, on which $Y(t)$ is nonsingular, we have the identity (2.3). Integrate over an interval $[\alpha, t] \subset (a, \beta)$,

(2.7) $\int_\alpha^t (u'^2P - u^2Q)\,ds = \int_\alpha^t H^*(s)H(s)\,ds + u^2(t)V(t) - u^2(\alpha)V(\alpha).$

Apply the functional q to this (symmetric) matrix equation and use $q(A - B) \leqq q(A) - q(B),$

$$q\left(\int_\alpha^t u'^2P\,ds\right) - q\left(\int_\alpha^t u^2Q\,ds\right) \geqq q\left(\int_\alpha^t H^*H\,ds\right) + q\left(u^2(t)V(t)\right)$$

$$+ q\left(-u^2(\alpha)V(\alpha)\right).$$

We argue now as in [6] to replace the first q by p (by virtue of (1.12)) and use convexity, concavity of p, q (i.e., Jensen's inequality) to obtain

$$\int_\alpha^t \left[u'^2p(P) - u^2q(Q)\right]\,ds \geqq q\left(\int_\alpha^t H^*H\,ds\right) + q\left(u^2(t)V(t)\right)$$

$$+ q\left(-u^2(\alpha)V(\alpha)\right).$$

Thus, by (1.20),

(2.8) $\int_\alpha^t \left[u'^2p_0 - u^2q_0\right]\,ds \geqq I_1(\alpha, t) + I_2(\alpha, t)$

$$+ q\left(u^2(t)V(t)\right) + q\left(-u^2(\alpha)V(\alpha)\right),$$

where

$$I_1(\alpha, t) = q\left(\int_\alpha^t H^* H \, ds\right),$$

$$I_2(\alpha, t) = \int_\alpha^t \left\{ u'^2[p_0 - p(P)] + u^2[q(Q) - q_0] \right\} ds \geqq 0.$$

If u is a solution of (1.4), the left side of (2.8) is $p_0(t)u'(t)u(t) - p_0(\alpha)u'(\alpha)u(\alpha)$. Hence

(2.9) $p_0(t)u'(t)u(t) - q\left(u^2(t)V(t)\right) \geqq I_1(\alpha, t) + I_2(\alpha, t) + I_3(\alpha),$

where

$$I_3(\alpha) = p_0(\alpha)u'(\alpha)u(\alpha) + q\left(-u^2(\alpha)V(\alpha)\right).$$

Letting $\alpha \to a + 0$, we obtain from (1.21),

(2.10) $p_0(t)u(t)u'(t) - q\left(u^2 V(t)\right) \geqq I_1(a, t) + I_2(a, t) \geqq 0.$

This proves (1.22).

   On (ii). We omit the proof of (ii) as it is similar to that of (iv) below.
   On (iii). Assume (1.25), but that $\det Y(t) \neq 0$ on $(a, b]$. Then (2.10) at $t = b$ implies that

(2.11) $I_1(a, b) = q\left(\int_a^b H^* H \, ds\right) = 0.$

Since q is a positive functional, $\int H^* H \, ds$ is not positive definite. Let $y_1 \neq 0$ be an n-vector such that

(2.12) $0 = \left(\int_a^b H^* H \, ds\right) y_1 \cdot y_1 = \int_a^b |H(s)y_1|^2 \, ds;$

so that $H(t)y_1 \equiv 0$ on $(a, b]$. From the definition of H in (2.4), $(uY^{-1})'y_1 = 0$; so that $uY^{-1}y_1 = y_0 \neq 0$ is a constant vector, i.e., $Y(t)y_0 \equiv u(t)y_1$. Since $u(b) = 0$ and $y_0 \neq 0$, this contradicts $\det Y(b) \neq 0$.
   On (1.24) in (iv). Suppose that

(2.13) $\det Y \neq 0$ on (a, b) and $u^2(t)V(t) \to 0$ as $t \to b$.

Then again we have (2.11). By the assumption (1.13), q is k-positive and (2.12) holds for all y_1 on a k-dimensional linear submanifold $\mathbb{R}_1^k$ of $\mathbb{R}^n$. The arguments of the last paragraph give (1.24). The last part of (2.13) is a consequence of Lemma 2.2 below.

On (1.23) in (iv). The left side of (2.10) is 0 at $t = b$, by (1.25) and the last part of (2.13). Hence $I_2(a,b) = 0$, and so the last part of (1.23) holds on $[a,b]$. Also, at $t = b$, (2.9) gives $0 \geqq I_3(\alpha)$ for $a < \alpha < b$. This is equivalent to the second inequality in the first part of (1.23), while the first inequality is contained in (1.22). This completes the proof.

LEMMA 2.2. *Let $T = [0,b]$ and $Y(t)$ be a prepared solution (cf. (2.1)) of (1.18) such that $\det Y(t) \neq 0$ on $(0,b]$. Then*

(2.14) $tY^{-1}(t)$ is bounded as $t \to +0.$

The proof will proceed as in the proof of the algebraic Lemma, [3], p. 31.

Proof of Lemma 2.2. Suppose (2.14) is false. Then there exist sequences $t_1 > t_2 > \ldots$ and $z_1, z_2 \ldots \in \mathbb{R}^n$ such that $t_n \to 0$ as $n \to \infty$, $z_n \neq 0$,

$$|t_n Y^{-1}(t_n)|/|z_n| \to \infty \qquad \text{as} \quad n \to \infty.$$

If $y_n = Y^{-1}(t_n)z_n$, then $Y(t_n)y_n/|y_n| = o(t_n)$ as $n \to \infty$. On replacing y_n by $y_n/|y_n|$, we can suppose

$$Y(t_n)y_n = o(t_n), \quad n \to \infty \qquad \text{and} \quad |y_n| = 1.$$

Since $Y \in C^1[0,b]$,

$$Y(t_n)y_n - Y(0)y_n - t_n Y'(0)y_n = o(t_n), \qquad n \to \infty.$$

Hence $Y(0)y_n + t_n Y'(0)y_n = o(t_n)$ or

$$P^{1/2}(0)Y'(0)y_n = -t_n^{-1}P^{1/2}(0)Y(0)y_n + o(1).$$

On choosing a subsequence, we can suppose that $y_0 = \lim y_n, n \to \infty$, exists and $|y_0| = 1$. It follows that

(2.15) $Y(0)y_0 = 0$ and $P^{1/2}(0)Y'(0)y_0 \in \text{range}\{P^{1/2}(0)Y(0)\},$

since the range of $P^{1/2}(0)Y(0)$ is closed. Thus

$$0 = Y^{*\prime}(0)P(0)Y(0)y_0 \cdot y \qquad \text{for all} \quad y \in \mathbb{R}^n.$$

By (2.1),

$$0 = Y^*(0)P(0)Y'(0)y_0 \cdot y = P^{1/2}(0)Y'(0)y_0 \cdot P^{1/2}(0)Y(0)y$$

for all y. Thus $P^{1/2}(0)Y'(0)y_0$ is orthogonal to the range of $P^{1/2}(0)Y(0)$, so that $Y'(0)y_0 = 0$ by (2.15). Since $Y(0)y_0 = Y'(0)y_0 = 0$ is impossible for $y_0 \neq 0$, we have a contradiction. This proves Lemma 2.2.

3. Proof of Theorem 1.2. On (i). Under the assumptions of Theorem 1.2, (1.4) is disconjugate on (b, ω) by Theorem 1.1. Let u be the solution of (1.4) satisfying for fixed t and τ, $b < t < \tau < \omega$, the boundary conditions

$$(3.1) \qquad u(t) = u_0(t) \qquad \text{and} \quad u(\tau) = 0.$$

Then, we obtain the analogue of (2.9),

$$(3.2) \qquad \int_t^\tau \left[u'^2 p_0(s) - u^2 q_0(s) \right] ds \geq q\left(\int_t^\tau H^* H \, ds \right) + q\left(-u^2(t)V_0(t) \right),$$

where $V_0 = PY_0'Y_0^{-1}$ and $H = P^{1/2}Y_0(Y_0^{-1}u)'$. The left side is $-p_0(t)u'(t)u(t)$, so that

$$(3.3) \qquad -p_0(t)u'(t)u(t) \geq q\left(-u^2(t)V_0(t) \right).$$

If t is fixed and $\tau \to \omega$, then $u(\cdot) \to u_0(\cdot)$ in $C^1[b', b'']$ for all (b', b''), $b < b' < b'' < \omega$; cf. e.g., [2], Exercise 6.5, pp. 357 and 571. This gives (1.30).

On (ii). Assume that equality holds in (1.30) at $t = t_0 \in (b, \omega)$. Let $t < \sigma < \tau$. From (3.2),

$$-p_0(t)u'(t)u(t) - q\left(-u^2(t)V_0(t) \right) \geq q\left(\int_t^\sigma H^* H \, ds \right) \geq 0.$$

If $t = t_0$ and $\tau \to \omega$, then we see that

$$(3.4) \qquad q\left(\int_{t_0}^\sigma H_0^* H_0 \, ds \right) = 0 \qquad \text{for} \quad t_0 \leq \sigma < \omega,$$

where $H_0 = P^{1/2}Y_0(Y_0^{-1}u_0)'$. Theorem 1.2 (ii) now follows from the arguments in the proofs of Theorem 1.1 (iii)–(iv) above. In verifying the first part of (1.32), use an analogue of (2.9), where (u, Y, α) are replaced by (u_0, Y_0, t_0) and, correspondingly, $I_1(\alpha, t) = I_2(\alpha, t) = I_3(\alpha) = 0$ for $(\alpha =)t_0 \leqq t < \omega$.

Part II. Comparison theorems for scalar equations.

4. Statement of results. Because the results seem of interest in themselves and for applications in Part III, we shall specialize the results of Part I to the case of diagonal systems. The theorems involve $n + 1$ scalar equations

(4.1) $(p_i(t)x')' + q_i(t)x = 0 \qquad \text{for} \quad i = 0, \ldots, n,$

where

(4.2) $p_i(t) > 0, q_i(t) \in C^0(T; \mathbb{R}) \qquad \text{for} \quad i = 0, \ldots, n.$

THEOREM 4.1. *Assume* (4.2) *with* $T = [a, b]$ *and let* $c_0, \ldots, c_n$ *be positive constants. Suppose that*

(4.3) $c_0 q_0 \leqq \sum_{i=1}^{n} c_i q_i, \qquad c_0 p_0 \geqq \sum_{i=1}^{n} c_i p_i.$

Let $x = x_i(t)$ *be a solution of* (4.1) *such that* $x_i(t) > 0$ *for small* $t - a > 0$ *and*

(4.4) $\limsup_{\alpha \to a + 0} \left\{ c_0 p_0(\alpha)x_0'(\alpha)x_0(a) - x_0^2(\alpha) \sum_{i=1}^{n} c_i p_i(\alpha)x_i'(\alpha)/x_i(\alpha) \right\} \geqq 0.$

(i) *Then, on any interval* $(a, \beta) \subset [a, b]$ *on which all* $x_i(t) > 0$,

(4.5) $c_0 p_0 x_0'/x_0 \geqq \sum_{i=1}^{n} c_i p_i x_i'/x_i.$

(ii) *If "equality" holds in* (4.5) *for some* $t_0 \in (a, \beta)$, *then "equality" holds in* (4.5) *for* $a < t < t_0$ *and in* (4.4). *Furthermore*

(4.6) $c_0 q_0 - \sum_{i=1}^{n} c_i q_i \equiv 0 \qquad \text{and} \qquad x_0'^2 \left[c_0 p_0 - \sum_{i=1}^{n} c_i p_i \right] \equiv 0$

on $[a, t_0]$, and there exist constants $\alpha_1, \ldots, \alpha_n$ such that

(4.7) $x_i(t) = \alpha_i x_0(t)$ *for* $i = 1, \ldots, n,$

on $[a, t_0]$.
(iii) *If*

(4.8) $x_0(t) > 0$ *on* (a, b) *and* $x_0(b) = 0,$

then there exists an i, $0 < i \leqq n$, and a $t_0 \in (a, b]$ such that $x_i(t_0) = 0$.
(iv) *If $x_i(t) > 0$ on (a, b) for $0 \leqq i \leqq n$ and (4.8) holds, then (4.6) and*
(4.7), with constant $(\alpha_1, \ldots, \alpha_n)$, hold on $[a, b]$ (and, in particular, $x_i(b)$
$= 0$ for $0 \leqq i \leqq n$); and "equality" holds in (4.5) on (a, b).

Theorem 4.1, (i) and (iii) for $n = 2$, is essentially Theorem 2 of
Mingarelli [7]. This generalizes Corollary 6.1 in [5] in which $n = 2$,
$p_0 \equiv p_1 \equiv p_2 \equiv 1$, $c_0 = 2$, $c_1 = c_2 = 1$, (4.4) is replaced by

(4.9) $x_i(a) = 1$ *and* $x_i'(a) = r_i,$

where $r_0, \ldots, r_n$ are numbers satisfying

$$(4.10) \quad c_0 p_0(a) r_0 \geqq \sum_{i=1}^{n} c_i p_i(a) r_i,$$

and, as observed by Mingarelli, there is the unneeded assumption
$q_1(t) \leqq q_0(t) \leqq q_2(t)$.

THEOREM 4.2. *Assume (4.2) with $T = [b, \omega)$, $b < \omega \leqq \infty$, and let*
$c_0, \ldots, c_n$ be positive constants. Suppose that (4.3) holds, and that (4.1) is
disconjugate on T for $1 \leqq i \leqq n$. Then (4.1), for $i = 0$, is disconjugate. If
$x = v_i(t)$ is a principal solution of (4.1), then

$$(4.11) \quad c_0 p_0(t) v_0'(t) / v_0(t) \leqq \sum_{i=1}^{n} c_i p_i(t) v_i'(t) / v_i(t).$$

If "equality" holds in (4.11) for some $t = t_0 \in [b, \omega)$, then

$$(4.12) \quad v_i(t) / v_i(t_0) \equiv v_0(t) / v_0(t_0)$$

$$\text{for} \quad t_0 \leqq t < \omega \quad \text{and} \quad i = 1, \ldots, n,$$

"equality" holds in (4.11) *on* $[t_0, \omega)$, *and*

(4.13) $\quad c_0 q_0 - \sum_{i=1}^{n} c_i q_i \equiv 0 \quad$ and $\quad v_0'^2 \left[c_0 p_0 - \sum_{i=1}^{n} c_i p_i \right] \equiv 0$

$$\text{for} \quad t_0 \leqq t < \omega.$$

This generalizes the part $N = 2$ of Theorem 3.1 of [5] in which $p_0 \equiv p_1 \equiv p_2$, $c_0 = 2$, $c_1 = c_2 = 1$, and there is the unneeded assumption $q_1(t) \leqq q_0(t) \leqq q_2(t)$.

Of course, for $n = 1$, Theorem 4.1 is the first comparison theorem of Sturm (cf. [2], p. 334), and Theorem 4.2 is the comparison theorem of Hartman and Wintner (cf. [2], pp. 358–359).

Remark. It may be useful to explain our use of the term "generalized Picone relation" for (2.3) in Lemma 2.2. The relation (2.3) for the scalar equation (4.1) is

(4.14) $\quad (u^2 p_i x_i' / x_i)' + p_i (u x_i' / x_i - u')^2 + q_i u^2 - p_i u'^2 = 0.$

For $i = 0$ or $1 \leqq i \leqq n$, multiply this relation by c_0 or $-c_i$ and add, with $u = x_0$,

(4.15) $\quad \left(x_0^2 \left[c_0 p_0 x_0' / x_0 - \sum_{i=1}^{n} c_i p_i x_i' / x_i \right] \right)'$

$$= \sum_{i=1}^{n} c_i p_i (x_0 x_i' / x_i - x_0')^2 - x_0^2 \left(c_0 q_0 - \sum_{i=1}^{n} c_i q_i \right)$$

$$+ x_0'^2 \left(c_0 p_0 - \sum_{i=1}^{n} c_i p_i \right).$$

We obtain the Mingarelli identity [7] by integrating (4.15) over a t-interval, and the result reduces to the standard Picone identity if $n = 1$ and $c_0 = c_1 = 1$. (Incidentally, a symmetric form of the relation (7.5) in [5] is the same as Mingarelli's identity for $n = 2$, $c_0 = 2$ and $c_1 = c_2 = 1$ and $p_i(t) \equiv 1$.) A direct proof of Theorem 4.1 can be obtained from (4.15) (cf. Mingarelli [7]), but we deduce it as a special case of Theorem 1.1. An analogous remark applies to Theorem 4.2.

5. Proofs of Theorems 4.1 and 4.2.   It can be supposed that

(5.1) $c_0 = 1$ and $c_1 = \cdots = c_n = 1/n$.

For otherwise, multiply equations (4.1) by c_0 if $i = 0$ and by nc_i if $1 \leq i \leq n$, and rename $c_0 p_0, c_0 q_0$ and $nc_i p_i$, $nc_i p_i$ to p_0, q_0 and p_i, q_i for $1 \leq i \leq n$. This does not affect the solutions $x_i(t)$, but (4.3)–(4.4) now hold in the case (5.1).

In order to obtain Theorem 4.1, apply Theorem 1.1, where $P = \mathrm{diag}(p_1, \ldots, p_n)$, $Q = \mathrm{diag}(q_1, \ldots, q_n)$, $p(A) = q(A) = \mathrm{trace}\,A/n$, $M = \mathrm{diag}(x_1'(a), \ldots, x_n'(a))$, $N = \mathrm{diag}(x_1(a), \ldots, x_n(a))$ and $Y(t) = \mathrm{diag}(x_1(t), \ldots, x_n(t))$. Assertions (ii) and (iv) depend on the fact that $q(A) = \mathrm{trace}\,A/n$ is an n-positive functional (cf.(1.13)). Similarly, Theorem 4.2 follows from Theorem 1.2.

Part III. Boundary value problems.

6. Statement of results.   The first theorem deals with a real scalar nonsingular boundary value problem

(6.1) $(p(t,\lambda)u')' + q(t,\lambda)u = 0,$

(6.2) $r(\lambda)u(a) - u'(a) = 0,$ $s(\lambda)u(b) - u'(b) = 0,$

in which the eigenvalue parameter occurs nonlinearly in the differential equation (6.1) and in the (separated) boundary conditions (6.2). It is not assumed that the coefficients p, q are monotone functions if λ. Instead, we make convexity-type assumptions.

Let $u_\lambda(t)$ and $u^\lambda(t)$ denote the solutions of (6.1) satisfying the initial conditions

(6.3) $u_\lambda(a) = 1,$ $u_\lambda'(a) = r(\lambda)$ and $u^\lambda(b) = 1,$ $u^{\lambda\prime}(b) = s(\lambda).$

Correspondingly, let

(6.4)
$$N(\lambda) = \#\{\text{zeros of } u_\lambda \text{ on } [a,b]\},$$
$$\nu(\lambda) = \#\{\text{zeros of } u^\lambda \text{ on } [a,b]\},$$

so that $|N(\lambda) - \nu(\lambda)| \leq 1$.

THEOREM 6.1.   *Let* $p(t,\lambda) > 0$, $q(t,\lambda) \in C^0\{[a,b] \times [0,\infty)\}$; $r(\lambda)$, $s(\lambda) \in C^0[0,\infty)$ *such that, for fixed* t,

(6.5) $q(t,\cdot)$ *is a convex function of* $\lambda \geqq 0$,

(6.6) $p(t,\cdot)$ *is a concave function of* $\lambda \geqq 0$,

(6.7) $r(\cdot)p(a,\cdot)$ *is a concave function of* $\lambda \geqq 0$,

(6.8) $s(\cdot)p(b,\cdot)$ *is a convex function of* $\lambda \geqq 0$.

Suppose that

(6.9) $N(\mu_0) = 0$ *and* $\nu(\mu^0) = 0$ *for some* $\mu_0, \mu^0 \geqq 0$.

(i)   *Then* $N(\lambda)$, $\lambda \geqq \mu_0$, *and* $\nu(\lambda)$, $\lambda \geqq \mu^0$, *are nondecreasing.*
(ii)   *If, for some integer* $k > 0$,

(6.10) $\infty \geqq \lim\limits_{\lambda \to \infty} N(\lambda) \geqq k + 1$ *or* $\infty \geqq \lim\limits_{\lambda \to \infty} \nu(\lambda) \geqq k + 1$,

then, on $\lambda > \max(\mu_0, \mu^0)$, *there exist* k *unique* λ-*values* $\lambda_1 < \ldots < \lambda_k$ *such that* (6.1)–(6.2) *has a solution* $(\not\equiv 0)$ *for* $\lambda = \lambda_j$ *with* $N(\lambda_j) = \nu(\lambda_j) = j$ *for* $j = 1, \ldots, k$.

The other theorems of this section deal with a real scalar singular boundary value problem on an open interval (α, ω) or half-open interval $[a, \omega)$. The differential equation is again (6.1), but the boundary conditions in the first case are

(6.11) u is an α- and an ω-principal solution.

Assume that

(6.12) $(-\infty \leqq)\alpha < a < b < \omega(\leqq \infty)$

and that (6.1) is disconjugate on $(\alpha, a]$ and on $[b, \omega)$ (e.g., that $q(t,\lambda) \leqq 0$ for $t \in (\alpha, a] \cup [b, \omega)$). Let (6.1) have solutions $u_\lambda(t)$, $u^\lambda(t)$ such that

u_λ is α- principal and $u_\lambda(a) = 1$,

(6.13)

u^λ is ω-principal and $u^\lambda(b) = 1$.

Remark. These solutions u_λ, u^λ exist if and only if (6.1) is disconjugate on open intervals (which may depend on λ) containing $(\alpha, a]$ and $[b, \omega)$. This assumption was not made explicit in some statements in [5].

THEOREM 6.2. *Assume* (6.12). *Let* $p(t, \lambda) > 0, q(t, \lambda) \in C^0\{(\alpha, \omega) \times [0, \infty)\}$ *satisfy* (6.5)–(6.6). *Suppose that* (6.1) *is disconjugate on* $(\alpha, a]$ *and on* $[b, \omega)$ *and has solutions* $u_\lambda(t), u^\lambda(t)$ *satisfying* (6.13), *and that* $N(\lambda)$ *and* $\nu(\lambda)$, *defined by* (6.4), *satisfy* (6.9). *Then assertions* (i) *and* (ii) *of Theorem* 6.1 *hold, with the boundary value problem* (6.1)–(6.2) *in* (ii) *replaced by* (6.1), (6.11).

Since an α-[or ω-] principal solution of (6.1) has no zero on $(\alpha, a]$ [or on (b, ω)], a solution ($\not\equiv 0$) of (6.1)–(6.11) for $\lambda = \lambda_j$ has exactly j zeros on (α, ω). A somewhat different statement from Theorem 6.2 is obtained if we count zeros on (α, ω), rather than on $[a, b]$; so let

(6.14) $N_0(\lambda) = \#\{\text{zeros of } u_\lambda(t) \text{ on } (\alpha, \omega)\}.$

THEOREM 6.3. *Assume the hypotheses of Theorem* 6.2 *with* (6.9) *replaced by*

(6.15) $N_0(\lambda) = 0$

(*the conditions* $u_\lambda(a) = 1, u^\lambda(b) = 1$ *can be omitted*). (i) *Then* $N_0(\lambda)$ *is nondecreasing and* $N_0(\lambda + 0) - N_0(\lambda) \leq 1$. (ii) *If, for some integer* $k \geq 0$,

(6.16) $\infty \geq \lim_{\lambda \to \infty} N_0(\lambda) \geq k + 1,$

then there exist $k + 1$ *nonnegative* λ-*values* $\lambda_0 < \ldots < \lambda_k$ *such that*

(6.17) $N_0(\lambda_j) = j, N_0(\lambda_j + 0) = j + 1$ *for* $j = 0, \ldots, k;$

(6.1), (6.11) *has a solution for* $\lambda = \lambda_j$ *with* j *zeros on* (α, ω), *but has no solution for* $\lambda \geq \lambda_0$ *unless* λ *is a discontinuity point of* $N_0(\lambda)$. (iii) *If* $\lambda_0 > 0$ (*so that* $N_0(\lambda) = 0$ *for* $0 \leq \lambda \leq \lambda_0$), *we have only the following possibilities*: (*a*) (6.1), (6.11) *has no solution* ($\not\equiv 0$) *for* $0 \leq \lambda < \lambda_0$; (*b*) (6.1), (6.11) *has a solution* ($\not\equiv 0$) *for* $\lambda = 0$ *but none for* $0 < \lambda < \lambda_0$; (*c*) (6.1, (6.11) *has a solution* ($\not\equiv 0$) *for every* λ *on* $0 \leq \lambda \leq \lambda_0$, *in which case,* $u_\lambda(t), u^\lambda(t)$, $q(t, \lambda), u_0'^2(t)p(t, \lambda)$ *are independent of* $\lambda \in [0, \lambda_0]$.

We can also obtain results on singular boundary value problems on half-open intervals $[a, \omega)$, where $-\infty < a < \omega \leqq \infty$, with boundary conditions of the type

(6.18) $u'(a) - r(\lambda)u(a) = 0$ u is ω − principal.

THEOREM 6.4.   *Let $p(t, \lambda), q(t, \lambda)$ satisfy the hypotheses of Theorem 6.2, ignoring conditions on $\alpha < t < a$. Let $r(\lambda) \in C^0[0, \infty)$ be such that (6.7) holds. Let u_λ be defined by (6.3) and u^λ by (6.13). Let $N_0(\lambda) = \#\{zeros\ of\ u_\lambda\ on\ [a, \omega)\}$ satisfy (6.15). Then the conclusions of Theorem 6.3 remain valid if (6.11) is replaced by (6.18) and (α, ω) by $[a, \omega)$.*

Theorems 6.1, 6.3, and 6.4 generalize Theorems 6.2, 1.2 and 1.3 of [5] in which $p(t, \lambda) \equiv 1$ and there is the unneeded assumption that $q(t, \lambda)$ is monotone with respect to λ (for fixed t).

7. Lemma 7.1.   The proof of Theorem 6.1 follows that of Theorem 6.2 of [4] and requires the following generalization of Theorem 6.1 there.

LEMMA 7.1.   *Assume the hypotheses of Theorem 6.1, except those applying to $u^\lambda(t)$ and $v(\lambda)$. Then $N(\lambda)$ is nondecreasing for $\lambda \geqq \mu_0$ and, if $\infty \geqq \lim N(\lambda) \geqq k$, as $\lambda \to \infty$, then there exist numbers $\mu_0 < \mu_1 < \cdots < \mu_k$ such that $N(\lambda) = j$ on $[\mu_j, \mu_{j+1})$ for $j = 0, \ldots, k - 1$ and $N(\lambda) \geqq k$ on $[\mu_k, \infty)$. If $(a <)t_1(\lambda) < t_2(\lambda) < \cdots$ are the ordered zeros of $u_\lambda(t)$, so that $t_j(\lambda)$ exists if and only if $\lambda \geqq \mu_j$ then, for $j = 1, \ldots, k$,*

(7.1) $t_j(\mu_j) = b,$

(7.2) $t_j(\lambda)$ *is strictly decreasing for $\lambda \geqq \mu_j$.*

Proof.   On $t_1(\lambda)$. Let $\lambda = \mu_1$ be the least value of $\lambda > \mu_0$ at which $N(\lambda) \geqq 1$. Then (7.1), $j = 1$, holds by continuity, since $u_\lambda(t_0) = 0$ implies that $u_\lambda'(t_0) \neq 0$.

We first show that $t_1(\lambda)$ exists (i.e., $N(\lambda) \geqq 1$) for $\lambda \geqq \mu_1$ and that $t_1(\lambda) < b$ for $\lambda > \mu_1$. Suppose, if possible, that there is a λ-value, say $\lambda = \lambda_2 > \mu_1$, such that either $t_1(\lambda_2)$ does not exist or that $t_1(\lambda_2) = b$. Put $\lambda_0 = \mu_1$ and let $\lambda_1 = \mu_0$. Choose $c_0 = 1, c_1 > 0, c_2 > 0$ so that $\lambda_0 = c_1\lambda_1 + c_2\lambda_2$ and $c_1 + c_2 = 1$. Put $p_i(t) = p(t, \lambda_i)$, $q_i = q(t, \lambda_i)$, $x_i(t) = u_{\lambda_i}(t)$ for $i = 0, 1, 2$. Then, by (6.5)–(6.7), it follows that assumptions (4.3)–(4.5) of Theorem 4.1 hold for $n = 2$ (cf. (4.8)–(4.9)). Since

$x_i(t) > 0$ on (a,b) for $i = 0,1,2$, it follows from Theorem 4.1 (iii) that $x_1(t), x_2(t)$ are constant multiples of $x_0(t)$. In particular, $x_1(t) \equiv u_0(t) = 0$ for $t = b$. This contradicts $N(\mu_0) = 0$. Hence $t_1(\lambda) < b$ exists for $\lambda > \mu_1$.

If $\mu_1 \leqq \lambda < \mu$, then the same argument, in which b is replaced by $t_1(\lambda)$, shows that $t_1(\mu) < t_1(\lambda)$, i.e., (7.2) holds for $j = 1$. This proves (7.1)–(7.2) for $j = 1$.

On $t_j(\lambda), j = 2, \ldots, k$. Let $k > 1$ and assume the existence of $0 = \mu_0 < \mu_1 < \ldots < \mu_{k-1}$ and (7.1)–(7.2) for $j = 1, \ldots, k - 1$. Let $\lambda = \mu_k$ be the least value of $\lambda > \mu_{k-1}$ for which $N(\lambda) \geqq k$. Then, as above, (7.1), $j = k$, holds by continuity.

We show that $t_k(\lambda)$ exists for $\lambda \geqq \mu_k$ and that $t_k(\lambda) < b$ for $\lambda > \mu_k$. Suppose that there is a λ-value, say $\lambda = \lambda_2 > \mu_k$, such that either $t_k(\lambda_2)$ does not exist or that $t_k(\lambda_2) = b$. Put $\lambda_0 = \mu_k$ and let $\lambda_1 = \mu_0$. Choose $c_0 = 1$, $c_1 > 0$ and $c_2 > 0$ so that $\lambda_0 = c_1\lambda_1 + c_2\lambda_2, c_1 + c_2 = 1$. Put $p_i(t) = p(t,\lambda_i), q_i = q(t,\lambda_i), x_0(t) = u_0(t)$ and $x_i(t) = (-1)^{k-1}u_{\lambda_i}(t)$ for $i = 1,2$. Then by (6.5)–(6.7), it follows that assumption (4.3) of Theorem 4.1 holds. If $[a,b]$ in Theorem 4.1 is replaced by $[t_{k-1}(\mu_k),b]$, then assumptions (4.4)–(4.5) hold. For $x_i(t) > 0$ for small $t - t_{k-1}(\mu_k) > 0$ and $u_0(t) > 0$ on $[a,b]$, $u_\lambda(t)$ changes signs at each of its zeroes, and $t_{k-1}(\lambda_2) < t_{k-1}(\mu_k)$ by (7.2) for $j = k - 1$. Also (4.4) holds if a is replaced by $t_{k-1}(\mu_k)$ since $x_0(t) = u_{\mu_k}(t)$ vanishes at $t = t_{k-1}(\mu_k)$.

Since $x_i(t) > 0$ on $(t_{k-1}(\mu_k),b)$ for $i = 0,1,2$, it follows from Theorem 4.1(iii) that $x_1(t)$, $x_2(t)$ are constant multiples of $x_0(t)$ and, hence, vanish at $t = b$. This contradicts $x_1(t) = u_0(t) > 0$ on $[a,b]$. Hence $t_k(\lambda) < b$ exists for $\lambda > \mu_k$.

As in the case $j = 1$, it follows that $t_j(\lambda)$ is strictly decreasing for $\lambda > \mu_k$. This completes the proof of Lemma 7.1.

8. Proof of Theorem 6.1. We can suppose that the second alternative in (6.10) holds, otherwise we replace t by $-t$ and interchange the roles of u_λ, u^λ. Since $|N(\lambda) - \nu(\lambda)| \leqq 1$, this alternative implies that $\lim N(\lambda) \geqq k$ as $\lambda \to \infty$.

Let $0 \leqq \mu_0 < \mu_1 < \ldots < \mu_k$ and $t_1(\lambda) < \ldots < t_k(\lambda)$ be as in Lemma 7.1. By an analogue of this lemma, there exist $0 \leqq \mu^0 < \mu_1 < \ldots < \mu^{k+1}$ such that $u^\lambda(t)$ has j zeroes if $\mu^j \leqq \lambda < \mu^{j+1}$ for $j = 0, \ldots, k$ and at least $k + 1$ zeroes if $\mu^{k+1} \leqq \lambda < \infty$. If $t^1(\lambda) > t^2(\lambda) > \ldots$ are the ordered zeroes of $u^\lambda(t)$, so that $t^j(\lambda)$ exists if and only if

$\lambda \geqq \mu^j$, then

(8.1) $t^j(\mu^j) = a$,

(8.2) $t^j(\lambda)$ is strictly increasing for $\lambda \geqq \mu^j$,

for $j = 1, \ldots, k + 1$.

By the Sturm separation theorem, $u_\lambda(t)$ has a zero on $[t^{j+1}(\lambda), t^j(\lambda)]$ for $\lambda = \mu^{j+1}$. In particular, $t_1(\lambda)$ exists and $t_1(\lambda) \leqq t^j(\lambda)$ for $\lambda = \mu^{j+1}$. It follows from (7.1)–(7.2) and (8.1)–(8.2), that there exists a unique λ-value $\lambda = \lambda_j$ such that $t_1(\lambda_j) = t^j(\lambda_j)$, $\mu^j \leqq \lambda_j \leqq \mu^{j+1}$. Thus, for $\lambda = \lambda_j$, $u_\lambda(t)$ and $u^\lambda(t)$ have a common zero and hence are linearly dependent. Consequently, either is a solution of (6.1)–(6.2) with $\lambda = \lambda_j$. Since $u_\lambda(a) = 1 \neq 0$, it follows that $\mu^j < \lambda_j < \mu^{j+1}$, so that $N(\lambda_j) = \nu(\lambda_j) = j$.

9. Proof of Theorem 6.2.   Theorem 4.2 implies the following:

LEMMA 9.1.   *Let* $p(t,\lambda) > 0$, $q(t,\lambda) \in C^0\{(b,\omega) \times [0,\infty)\}$ *such that* (6.5)–(6.6) *hold, and let* (6.1) *be disconjugate on* (b,ω) *for each* $\lambda \geqq 0$. *Let* $u^\lambda(t)$ *be an* ω-*principal solution of* (6.1). *Then*

(9.1) $s(t,\lambda) = p(t,\lambda)u^{\lambda\prime}(t)/u^\lambda(t)$

is a convex function of $\lambda = 0$, *for fixed* $t \in (b,\omega)$.

Lemma 9.1 generalizes the case $N = 2$ of Theorem 3.1 of [5], where it is assumed that $p(t,\lambda) \equiv 1$ and that $q(t,\cdot)$ is also nondecreasing for $\lambda \geqq 0$ (and fixed t).

Lemma 9.1 is a consequence of Theorem 4.2 for if $\lambda_0 = c_1\lambda_1 + c_2\lambda_2$, where $c_1 > 0$, $c_2 > 0$ and $c_0 = c_1 + c_2 = 1$, and if $p_i(t) = p(t,\lambda_i), q_i(t) = q(t,\lambda_i)$ then (4.3) holds and the conclusion (4.9) means that $s(t,\lambda)$ is a convex function of $\lambda \geqq 0$ (for fixed t).

Similarly, in Theorem 6.2,

(9.2) $r(t,\lambda) = p(t,\lambda)u_\lambda'(t)/u_\lambda(t)$

is a concave function of $\lambda \geqq 0$ for fixed $t \in (\alpha,a]$.

If we put

(9.3) $r(\lambda) = r(a,\lambda)$ and $s(\lambda) = s(b,\lambda)$, then the singular boundary value problem (6.1), (6.11) on (α, ω) is equivalent to the regular problem (6.1), (6.2) on $[a, b]$. Thus Theorem 6.2 is a corollary of Theorem 6.1.

10. Proof of Theorem 6.3. (We follow the proof of Theorem 1.2 in [5].) If $N_0(\lambda) \equiv 0$, Theorem 6.3 is trivially correct. Suppose, therefore, that there exists an integer $k \geqq 0$ such that

$$(10.1) \qquad \infty \geqq \lim_{\lambda \to \infty} \sup N_0(\lambda) \geqq k + 1.$$

By [4], every discontinuity point of $N_0(\lambda)$ is an eigenvalue for (6.1), (6.11) and (10.1) implies the existence of eigenvalues $(0 \leqq)\lambda_0 < \ldots < \lambda_k$ for which

$$(10.2) \qquad N_0(\lambda_j) = j \qquad \text{and} \quad \lim \sup N(\lambda) \geqq j + 1, \lambda \to \lambda_j + 0,$$

for $j = 0, \ldots, k$.

In particular, u_λ and u^λ are linearly dependent and do not vanish for $\alpha < t < \omega$ if $\lambda = \lambda_0$; so that $\nu(\lambda_0) = 0$. This together with (6.15) imply (6.9) and hence, by virtue of the proof of Lemma 7.1 and Theorem 6.2, that $N_0(\lambda)$ is nondecreasing for $\lambda \geqq \lambda_0$ and that, on $\lambda \geqq \lambda_0$, the eigenvalues $\lambda_0, \ldots, \lambda_k$ are uniquely determined by $N_0(\lambda_j) = j$.

It only remains to consider $0 \leqq \lambda \leqq \lambda_0$ if $\lambda_0 > 0$. Since $N_0(\lambda_0) = 0$, it is clear that $\lambda_0 < \mu_1$, where μ_1 is defined in Lemma 7.1.

The function (9.2) satisfies $r(b, 0) \geqq s(0)$, for otherwise $u_\lambda(t)$ has a zero on (b, ω), contradicting $N_0(0) = 0$; cf. [2], Theorem 6.4, p. 355.

For $\lambda = \mu_1, u_\lambda(t) = 0$ at $t = b$ in Lemma 7.1, so that $r(t, \mu_1) \to -\infty$ as $t \to b$. Also, the argument above leading to Lemma 9.1 shows that $r(t, \lambda)$ is a concave function of λ, $0 \leqq \lambda < \mu_1$, for fixed $t \in (\alpha, b]$; cf. Theorem 4.2. The problem (6.1), (6.11) has a solution $(\not\equiv 0)$ if and only if

$$(10.3) \qquad r(b, \lambda) = s(\lambda).$$

Since $r(b, 0) \geqq s(0), r(b, \lambda)$ is concave on $0 \leqq \lambda < \mu_1$, and $s(\lambda)$ is convex by Lemma 9.1, we have the following possibilities:

(a) $r(b, 0) > s(0)$, i.e., $\lambda = 0$ is not an eigenvalue, hence (10.1) holds for at most one λ-value, say, $\lambda = \lambda_0, 0 < \lambda_0 < \mu_1$.

(b) $r(b,0) = s(0)$ (i.e., $\lambda = 0$ is an eigenvalue), hence (10.3) holds on some interval $0 \leqq \lambda \leqq \lambda_0$ or for at most one positive λ-value λ_0, $0 < \lambda_0 < \mu_1$.

On (a). By [4], there exists at least one eigenvalue λ with $N(\lambda) = 0$. Thus, in case (a), there exists exactly one eigenvalue $\lambda_0 \in (0, \mu_1)$. Also, $r(b,\lambda) > s(\lambda)$ on $[0,\lambda_0)$ and $r(b,\lambda) < s(\lambda)$ on (λ_0, μ_1). Thus $N_0(\lambda) = 0$ on $[0,\lambda_0]$ and $N(\lambda) = 1$ on (λ_0, μ_1), [2], Theorem 6.4, p. 355. Since every discontinuity point of $N_0(\lambda)$ is an eigenvalue by [4], it follows from the uniqueness of $\lambda_j, j \geqq 1$, that $N_0(\lambda)$ is nondecreasing on $\lambda \geqq 0$.

On (b), *first alternative*, $\lambda_0 > 0$. Suppose that the first alternative in (b) holds with $\lambda_0 > 0$. Then (10.3) holds for $0 \leqq \lambda \leqq \lambda_0$, so that $r(b,\lambda)$, $s(\lambda)$ are constant on this λ-interval. By the last part of Theorem 4.2, $u^\lambda(t), q(t,\lambda), u^{0,2}(t)p(t,\lambda)$ are independent of $\lambda \in [0,\lambda_0]$ for $b \leqq t < \omega$. Since we can replace t by $-t$ and, thus interchange α and ω in this argument, we see that $u_\lambda(t), q(t,\lambda), u_0'^2(t)p(t,\lambda)$ are independent of $\lambda \in [0,\lambda_0]$ for $\alpha < t \leqq a$.

Since $r(b,\lambda) = s(\lambda)$ and $s(a,\lambda) = r(\lambda)$ are constants on $0 \leqq \lambda \leqq \lambda_0$, Theorem 4.1 (ii) implies that $u_\lambda(t), u^\lambda(t)$, $q(t,\lambda), u_0'^2(t)p(t,\lambda)$ are independent of λ for $a \leqq t \leqq b$.

Furthermore, $r(b,\lambda) < s(\lambda)$ on $\lambda_0, \mu_1)$, so that $N_0(\lambda) = 1$ on $(\lambda_0 \mu_1)$ by [2], Theorem 6.4, p. 355. Thus, the assertions concerning $N_0(\lambda)$ in the paragraph "On (a)" above are valid. This completes the proof of Theorem 6.3.

THE JOHNS HOPKINS UNIVERSITY

REFERENCES

[1] G.J. Etgen and R.T. Lewis, Positive functionals and oscillation criteria for differential systems, *Optimal Control and Differential Equations* (ed. A.B. Schwarzkopf, W.G. Kelley, S.B. Eliason), Academic Press 1978, 245-275.

[2] P. Hartman, Ordinary Differential Equations, S.M. Hartman, Baltimore, 1973.

[3] P. Hartman, Self-adjoint, nonoscillatory systems of ordinary second order linear differential equations, *Duke Math. J.* **24** (1957), 25-36.

[4] P. Hartman, Boundary value problems for second order, ordinary differential equations involving a parameter, *J. Differential Equations* **12** (1972), 194-212.

[5] P. Hartman, Uniqueness of principal values, complete monotonicity of logarithmic derivatives of principal solutions, and oscillation theorems, *Math. Annalen* **241** (1979), 257-281.

[6] P. Hartman, Oscillation criteria for self-adjoint second order differential systems and "principal sectional curvatures," *J. Differential Equations* **34** (1979), 326-338.

[7] A.B. Mingarelli, Some extensions of the Sturm-Picone theorem, *C.R. Math. Rep. Acad. Sci.* Canada **1** (1979), 223-226.

[8] W.T. Reid, Boundary problems of Sturmian type on an infinite interval, *SIAM J. Math. Anal.* **4** (1973), 185-197.

REGULARIZING EFFECTS OF HOMOGENEOUS EVOLUTION EQUATIONS*

By Philippe Bénilan and Michael G. Crandall

Introduction.   Each of the three evolution equations

$$(1)_\alpha \qquad \frac{\partial u}{\partial t} = \frac{\partial^2}{\partial x^2}\left(|u|^{\alpha-1}u\right) \qquad t > 0, x \in \mathbb{R},$$

$$(2)_\alpha \qquad \frac{\partial u}{\partial t} = \frac{\partial}{\partial x}\left(|u|^{\alpha-1}u\right) \qquad t > 0, x \in \mathbb{R},$$

and

$$(3)_\alpha \qquad \frac{\partial u}{\partial t} = \frac{\partial}{\partial x}\left(\left|\frac{\partial u}{\partial x}\right|^{\alpha-1}\frac{\partial u}{\partial x}\right) \qquad t > 0, x \in \mathbb{R},$$

may, if $\alpha > 0$, be solved subject to the initial condition

$$(IC) \qquad u(x,0) = u_0(x), \qquad x \in \mathbb{R},$$

where $u_0 \in L^1(\mathbb{R})$ for a solution $u(x,t)$ in such a way that $t \to u(\cdot,t)$, which we call $u(t)$, is a continuous curve in $L^1(\mathbb{R})$ and if $S(t)$ is defined by $S(t)u_0 = u(t)$ then it is a nonexpansive self-map of $L^1(\mathbb{R})$ for each $t \geqslant 0$. (This is discussed further in Section 2, but the reader need know nothing of these matters beforehand.) A main result of this note, as applied to these three problems, establishes that

$$(4) \qquad \lim_{h\downarrow 0} \frac{\|u(t+h) - u(t)\|}{h} = \frac{2\|u_0\|}{|\alpha - 1|t},$$

with $\| \ \|$ the norm of $L^1(\mathbb{R})$, provided $\alpha \neq 1$. Indeed, it is shown that (4) holds for any evolution equation $u'(t) = B(u(t))$ in which B is homogeneous of degree $\alpha > 0$, $\alpha \neq 1$, and for which there is an associated nonexpansive family $S(t)$. In addition, each of the problems $(1)_\alpha$, $(2)_\alpha$,

---

*Sponsored by the United States Army under Contract No. DAAG29-75-C-0024 and DAAG29-80-C-0041.

$(3)_\alpha$ give rise to operators $S(t)$ which respect the natural order on $L^1(\mathbb{R})$. It will follow from another abstract result of this paper that therefore the pointwise estimate

$$(5) \qquad (\alpha - 1)\frac{\partial u}{\partial t} \geqslant -\frac{1}{t}u \qquad t > 0, x \in \mathbb{R},$$

holds for nonnegative solutions of any of the above problems ((5) being understood in the sense of distributions).

Estimates of the form (4), (5) are types of "regularizing effects" in that the quantities estimated for $t > 0$ need not be sensible at $t = 0$. We comment on the rather subtle implications of the estimates (4) and (5) in particular cases at some length in Section 2, and there we pay due respect to the difference between the assertion (4), which is an estimate of the "speed", and the stronger assertion that the "velocity" $u'(t)$ exists and admits the corresponding estimate.

This note is divided into two sections. Section 1 presents the abstract results concerning solutions of the equation $u' = B(u)$ and its perturbations under various assumptions (always including that B is homogeneous). These results are elementary estimates on the difference quotients $h^{-1}(u(t + h) - u(t))$. Section 2 discusses the interaction of the abstract results with particular problems, including those listed above, and it is partly expository.

Estimates in evolution problems of velocities $u'(t)$ by expressions involving $1/t$ are familiar in several contexts. Perhaps the closest in spirit to those given here occur when B is the (linear) infinitesimal generator of a strongly continuous semigroup, in which case an estimate of $h^{-1}\|u(t + h) - u(t)\|$ in the form $C\|u(0)\|/t$ is essentially equivalent to B being the generator of a holomorphic semigroup (see, e.g., [17], [29]). (This is the case for the linear problems $(1)_1 = (3)_1$ in a variety of spaces.) Another known case is the result of Brezis [7, chp. III] which applies if $B = \partial\Phi$ is the subdifferential of a convex function on a Hilbert space. Brezis' estimates apply to variants of $(1)_\alpha$, $(3)_\alpha$ to give L^2 based results like (4) which do not use the homogeneity of the right-hand sides. See [3, pg. 200]. Other regularizing effects are to be found in [5], [16], [26].

Section 1. Let $\| \ \|$ be a semi norm on the vector space X and $B : D(B) \subseteq X \to X$ be an operator in X which is homogeneous of degree

$\alpha > 0$. That is

(H) $B(rx) = r^{\alpha}B(x)$ for $r \geqslant 0$ and $x \in D(B)$,

where it is understood in (H) that $rD(B) \subseteq D(B)$ for $r \geqslant 0$. We are interested in the evolution problem

(E) $\begin{cases} \dfrac{du}{dt} = B(u) \\ u(0) = x, \end{cases}$

but rather than deal with (E) directly we shall work here only with its solutions. These solutions are assumed to be presented to us by some theory or construction in the form $u(t) = S(t)x$ where each $S(t)$, $t \geqslant 0$, is a mapping $S(t): C \to X$ and C is subset of X. The property (H) of B is taken to be reflected in S by the identities

(HS) $\lambda^{1/(\alpha-1)}S(\lambda t)x = S(t)(\lambda^{1/(\alpha-1)}x)$ for $t, \lambda \geqslant 0$ and $x \in C$.

This is arrived at in the following way: If $u(t)$ is a classical solution of (E) and (H) holds and $\lambda > 0$, then $v(t) = \lambda^{1/(\alpha-1)}u(\lambda t)$ satisfies (dv/dt)
$\cdot (t) = \lambda^{1/(\alpha-1)}\lambda u'(\lambda t) = \lambda^{\alpha/(\alpha-1)}B(u(\lambda t)) = B(\lambda^{1/(\alpha-1)}u(\lambda t))$
$= B(v(t))$ so that v is again a classical solution of (E) and further satisfies $v(0) = \lambda^{1/(\alpha-1)}u(0)$. The corresponding property of the notion of solution of (E) provided by S is what is requested by (HS). It is understood in (HS) that $rC \subseteq C$ for $r \geqslant 0$. The other major requirement we place upon S is the Lipschitz condition

(L) $\|S(t)x - S(t)\hat{x}\| \leqslant L\|x - \hat{x}\|$ for $t \geqslant 0, x, \hat{x} \in C$,

where $\| \ \|$ is the norm in X.

THEOREM 1. *Let* $C \subseteq X$ *and* $S(t): C \to X$ *for* $t \geqslant 0$ *and satisfy* (HS) *with* $\alpha > 0, \alpha \neq 1$, (L) *and* $S(t)0 \equiv 0$. *Then for* $x \in C$ *and* $t, h > 0$,

(6) $\begin{cases} \|S(t + h)x - S(t)x\| \leqslant 2L\|x\| \left|1 - \left(1 + \dfrac{h}{t}\right)^{1/(1-\alpha)}\right|, \\ \text{and, in particular,} \\ \limsup\limits_{h \downarrow 0} \dfrac{\|S(t + h)x - S(t)x\|}{h} \leqslant \left(\dfrac{2L\|x\|}{|\alpha - 1|}\right)\dfrac{1}{t}. \end{cases}$

Proof. We use (HS) with

(7) $\lambda = \left(1 + \dfrac{h}{t}\right)$

to compute

(8) $S(t + h)x - S(t)x = S(\lambda t)x - S(t)x$

$$= \lambda^{1/(1-\alpha)}S(t)(\lambda^{1/(\alpha-1)}x) - S(t)x$$

$$= \lambda^{1/(1-\alpha)}\big(S(t)(\lambda^{1/(\alpha-1)}x) - S(t)x\big)$$

$$+ (\lambda^{1/(1-\alpha)} - 1)S(t)x.$$

Now (8), (L) and $S(t)0 \equiv 0$ imply

$$\|S(t + h)x - S(t)x\| \leqslant \lambda^{1/(1-\alpha)}L|\lambda^{1/(\alpha-1)} - 1|\,\|x\|$$

$$+ |\lambda^{1/(1-\alpha)} - 1|L\|x\|$$

$$= 2L\|x\|L|1 - \lambda^{1/(1-\alpha)}|.$$

The estimates (6) follow from this and (7).

For the next result we assume that X is equipped with a relation $\geqslant$ under which it is an ordered vector space and that $S(t)$ respects this order.

THEOREM 2. *Let X be an ordered vector space with the order relation denoted by $\geqslant$. Let $S(t)$ satisfy*

(O) $S(t)x \geqslant S(t)y$ *if* $x, y \in C$ *and* $x \geqslant y,$

and satisfy (HS) *with* $\alpha > 0, \alpha \neq 1$. *If* $x \in C, x \geqslant 0$ *and* $t, h > 0$ *then*

(9) $(\alpha - 1)(S(t + h)x - S(t)x)$

$$\geqslant (\alpha - 1)\left(\left(1 + \dfrac{h}{t}\right)^{1/(1-\alpha)} - 1\right)S(t)x.$$

Proof. There are two cases. If $\alpha > 1$, we return to (8) and observe that $\lambda = (1 + (h/t)) > 1$ and $\lambda^{1/(\alpha-1)} > 1$. Thus $\lambda^{1/(\alpha-1)}x \geqslant x$ and so $\lambda^{-1/(\alpha-1)}(S(t)(\lambda^{1/(\alpha-1)}x) - S(t)x) \geqslant 0$ by (O). The inequality (9) is obtained by dropping this nonnegative term from the right hand side of (8) and multiplying by $\alpha - 1 > 0$. The parallel reasoning if $0 < \alpha < 1$ shows that the term just dropped is now nonpositive, so we have the opposite inequality than above coming from (8), which becomes (9) again upon multiplication by $\alpha - 1 < 0$.

In applications of Theorem 2 there is sometimes a nonnegative linear functional $\Lambda : X \to \mathbb{R}$ which is preserved by $S(t)$, i.e.

$$(10) \qquad \Lambda S(t)x = \Lambda x \qquad \text{for} \quad t \geqslant 0, x \in C, x \geqslant 0,$$

and X is a lattice. The notation $x^+ = \sup\{x, 0\}, x^- = -\inf\{x, 0\}$ will be used.

COROLLARY 3. *In addition to the conditions of Theorem 2 assume that X is a vector lattice, Λ is a nonnegative linear functional on X and (10) holds. Let $x \in C, x \geqslant 0$ and $u(t) = S(t)x$. The following estimate is valid for $t, h > 0$ and $\nu \in \{+, -\}$:*

$$(11) \qquad \Lambda\big((u(t + h) - u(t))^\nu\big) \leqslant \Big|1 - \Big(1 + \frac{h}{t}\Big)^{1/(1 - \alpha)}\Big|\Lambda x.$$

Proof. From (10) we have

$$(12) \qquad \Lambda\big((u(t + h) - u(t))^+\big) - \Lambda\big((u(t + h) - u(t))^-\big)$$
$$= \Lambda u(t + h) - \Lambda u(t) = 0.$$

From (9) and $S(t)x \geqslant 0$ we also have

$$(13) \qquad \begin{cases} (u(t + h) - u(t))^- \leqslant \Big(1 - \Big(1 + \dfrac{h}{t}\Big)^{1/(1 - \alpha)}\Big)S(t)x \\ \qquad\qquad\qquad \text{if} \quad \alpha > 1, \\ (u(t + h) - u(t))^+ \leqslant \Big(\Big(1 + \dfrac{h}{t}\Big)^{1/(1 - \alpha)} - 1\Big)S(t)x \\ \qquad\qquad\qquad \text{if} \quad 0 < \alpha < 1. \end{cases}$$

Applying Λ to the inequalities (13) and using (12) implies (11).

Remarks. If Λ is as above and if (10) holds, then (O) is essentially equivalent to the property

$$(14) \qquad \Lambda\big((S(t)x - S(t)y)^+\big) \leqslant \Lambda(x - y)^+.$$

See [14]. Hence (11) represents a slight refinement of (6) with $\|x\| = \Lambda x^+$. Also the proof shows (11) is valid for $\alpha > 1$ if (10) is weakened to $\Lambda S(t)x \leqslant \Lambda x$.

We turn now to the "forced" problem

$$
\text{(FE)} \qquad
\begin{cases}
\dfrac{du}{dt} = B(u) + f(t) \\
u(0) = x
\end{cases}
$$

where $f:[0, T] \to X$ for some $T > 0$. Again, it is most efficient to assume that solutions of (FE) are presented to us in the form $u(t) = S(t, x, f)$ and lay our conditions directly upon S. Computing the equation satisfied by $v(t) = \lambda^{1/(\alpha-1)}u(\lambda t)$ if u is a classical solution of (FE) and B is homogeneous of degree α leads to

$$
\text{(15)} \qquad v'(t) = B(v(t)) + \lambda^{\alpha/(1-\alpha)}f(\lambda t).
$$

In order to minimize bookkeeping problems we will assume simply that X is a Banach space and $C \subset X \times L^1_{\text{loc}}(0, \infty : X)$ is given together with

$$
S : [0, \infty) \times C \to X
$$

such that $u(t) = S(t, x, f)$ is the solution of (FE) of interest for $(x, f) \in C$. We let

$$
\text{(16)} \qquad f_\lambda(t) = f(\lambda t)
$$

and assume $(x, f) \in C \Rightarrow (x, \lambda^{\alpha/(1-\alpha)}f_\lambda) \in C$ for $\lambda \geqslant 0$. The equation (15) satisfied by $\lambda^{1/1-\alpha}u(\lambda t)$ is to be reflected in S by

$$
\text{(FH}_S\text{)} \qquad \lambda^{1/(\alpha-1)}S(\lambda t, x, f) = S(t, \lambda^{1/(\alpha-1)}x, \lambda^{\alpha/(\alpha-1)}f_\lambda).
$$

Motivated by known existence theories (see Section 2) the Lipschitz condition (L) is generalized to

$$
\text{(FL)} \qquad \|S(t, x, f) - S(t, \hat{x}, \hat{f})\| \leqslant L\left(\|x - \hat{x}\| + \int_0^t \|f(\tau) - \hat{f}(\tau)\| d\tau\right)
$$

when the arguments lie in the domain of S.

THEOREM 4. *Let S satisfy* (FH$_S$), (FL) *and* $S(t,0,0) \equiv 0$. *If* $t, h > 0, (x, f) \in C, \alpha > 0, \alpha \neq 1$ *and* $u(t) = S(t, x, f)$, *then*

(17) $\|u(t+h) - u(t)\|$

$$\leqslant L\left(\left|1 - \left(1 + \frac{h}{t}\right)^{1/(1-\alpha)}\right|\left(2\|x\| + \int_0^t \|f(\tau)\| \, d\tau\right)\right.$$

$$+ \left|\left(1 + \frac{h}{t}\right) - \left(1 + \frac{h}{t}\right)^{1/(1-\alpha)}\right|\int_0^t \left\|f\left(\tau + \frac{h}{t}\tau\right)\right\| d\tau$$

$$\left. + \left(1 + \frac{h}{t}\right)^{1/(1-\alpha)}\int_0^t \left\|f\left(\tau + \frac{h}{t}\tau\right) - f(\tau)\right\| d\tau\right).$$

In particular, if

(18) $$V(t, f) = \limsup_{\xi \downarrow 0} \int_0^t \frac{\|f(\tau + \xi\tau) - f(\tau)\|}{\xi} \, d\xi$$

then

(19) $$\limsup_{h \downarrow 0} \frac{\|u(t+h) - u(t)\|}{h}$$

$$\leqslant \frac{L}{t}\left(\frac{2\|x\| + (1 + \alpha)\int_0^t \|f(\tau)\| \, d\tau}{|\alpha - 1|} + V(t, f)\right)$$

and u is Lipschitz continuous on each compact subset of $(0, T]$ if $V(T, f) < \infty$.

Proof. Of course the argument is just as before. The relation (FH$_S$) yields, with $\lambda = (1 + (h/t))$,

(20) $u(t+h) - u(t) = S(\lambda t, x, f) - S(t, x, f)$

$$= \lambda^{1/(1-\alpha)}\left[\left(S(t, \lambda^{1/(\alpha-1)}x, \lambda^{\alpha/(\alpha-1)}f_\lambda)\right.\right.$$

$$\left. - S(t, x, f_\lambda)\right)$$

$$+ \left(S(t, x, f_\lambda) - S(t, x, f)\right)\bigg]$$

$$+ (\lambda^{1/(1-\alpha)} - 1)S(t, x, f).$$

Using (LH) in conjunction with (20) proves (17) and (19) follows by taking the indicated limit.

Remark 5. If f is absolutely continuous and differentiable almost everywhere on each compact subset of $(0, T]$, then

$$(21) \qquad V(t, f) = \int_0^t \tau \|f'(\tau)\| \, d\tau.$$

In general $V(T, f) < \infty$ is equivalent to $t \to tf(t)$ being of (essentially) finite variation on $[0, T]$.

Remark 6. It is quite interesting that Theorem 2 has a forced analogue. If $S(t, x, f)$ is nondecreasing in x and f (where $f \geqslant g$ means $f(t) \geqslant g(t)$ a.e.) and $t \to (\alpha - 1)(t^{\alpha/(\alpha-1)}f(t))$ is nondecreasing in t, then (20) implies (9) with $u(t) = S(t, x, f)$ in place of $S(t)x$.

The final abstract case we consider is (E) perturbed by a Lipschitz continuous function $p : D(p) \subseteq X \to X$. That is

$$(PE) \qquad \begin{cases} \dfrac{du}{dt} = B(u) + p(u) \\ u(0) = x \end{cases}$$

where p satisfies

$$(22) \qquad \|p(x) - p(y)\| \leqslant M\|x - y\| \qquad \text{for } x, y \in D(p)$$

and some $M \geqslant 0$. We regard (PE) as a special case of (FE) in the sense that we assume solutions $S(t, x, f)$ of (FE) are known and understand a solution u of (PE) to be a function u with values in $D(p)$ such that $(x, p(u)) \in C$ and $u(t) = S(t, x, p(u))$. The results will be a modulus of continuity of any solution u of (PE).

THEOREM 7. *Let S satisfy the assumptions of Theorem* 4 *with* $\alpha > 0, \alpha \neq 1$. *Let* $p : D(p) \subseteq X \to X$ *satisfy* (22), $u \in C([0, \infty) : D(p))$, $(x, p(u)) \in C$ *and* $u = S(t, x, p(u))$. *Then for each* $T > 0$

$$\sup_{\substack{0 \leqslant t \leqslant T \\ 0 \leqslant h \leqslant t}} \frac{t}{h} \|u(t + h) - u(t)\| \leqslant c(T, \alpha, \|x\|, L, M)$$

where the right hand side above depends only on the indicated quantities. In particular, u is Lipschitz continuous on compact subsets of $(0, T]$ for each $T > 0$.

Proof. The Lipschitz condition (22) implies

(23) $\| p(u(t)) \| \leqslant a + M \| u(t) \|$

for some a. Using (23), (LF) and $S(t, 0, 0) \equiv 0$ one deduces that

$$\| u(t) \| = \| S(t, x, p(u)) - S(t, 0, 0) \|$$

$$\leqslant L\left(\| x \| + at + M \int_0^t \| u(\tau) \| \, d\tau \right)$$

from which flows the estimate

(24) $\| u(t) \| \leqslant L(\| x \| + aT) e^{\mathrm{LMT}}$ for $0 \leqslant t \leqslant T$.

Next we use (17) with $f(t) = p(u(t))$ and the estimates (23), (24) to conclude that for $T > 0$ and $0 \leqslant t \leqslant t + h \leqslant T$ there is a constant $C = C(T, \alpha, \| x \|, L, M)$ for which

(25) $\dfrac{t}{h} \| u(t + h) - u(t) \|$

$$\leqslant C\left[\frac{t}{h} \left| 1 - \left(1 + \frac{h}{t} \right)^{1/(1-\alpha)} \right| \right.$$

$$+ \frac{t}{h} \left| \left(1 + \frac{h}{t} \right) - \left(1 + \frac{h}{t} \right)^{1/(1-\alpha)} \right|$$

$$\left. + \frac{t}{h} \left(1 + \frac{h}{t} \right)^{1/(1-\alpha)} \int_0^t M \left\| u\left(\tau + \frac{h}{t} \tau \right) - u(\tau) \right\| d\tau \right].$$

Set $\xi = h/t$ above and

(26) $g(t, \xi) = \dfrac{\| u(t(1 + \xi)) - u(t) \|}{\xi}$.

Then (25) implies

$$(27) \qquad g(t,\xi) \leqslant C\left[1 + \int_0^t g(\tau,\xi)\,d\tau\right]$$

for some new constant C and $0 \leqslant t \leqslant T/(1 + \beta)$ $0 \leqslant \xi \leqslant \beta \leqslant 1$, where β is chosen in $(0, 1]$. The estimate (27) gives a new estimate

$$(28) \qquad g(t,\xi) \leqslant \hat{C} \qquad \text{for} \quad 0 \leqslant t \leqslant T/(1 + \beta), 0 \leqslant \xi \leqslant \beta \leqslant 1$$

where $\hat{C}$ is yet another constant, whose precise structure we leave to the reader, but depends only on allowed quantities. T being arbitrary the proof is complete.

Section 2. Examples and Applications.

We begin by reviewing one abstract theory for (E) which guarantees that (FHS) and (FL) hold whenever B satisfies (H) and one additional condition. The theory encompasses the three classes of examples $(1)_\alpha$, $(2)_\alpha$, $(3)_\alpha$ and generalizations of them as well as the equation

$$(26)_\alpha \qquad \frac{\partial u}{\partial t} = \sum_{i=1}^{N} |\frac{\partial u}{\partial x_i}|^{\alpha-1}\frac{\partial u}{\partial x_i} \qquad t > 0, x \in \mathbb{R}^N$$

and a host of other possibilities.

Following this we discuss briefly the two classes of examples $(1)_\alpha$, $(2)_\alpha$ in their simplest setting to make various points and orient the reader. We make no attempt to write down the new results which obviously flow from the estimates of Section 1 even as applied to the examples mentioned here.

Given a Banach space X and $B: D(B) \subseteq X \to X, T > 0$, and $f \in L^1(0, T: X)$ we call $u \in C([0, T]: X)$ a *mild solution* of

$$(EF)' \qquad u' = B(u) + f$$

on $[0, T]$ provided for every $\epsilon > 0$ we can find a partition $\{0 = t_0 < t_1 < \cdots < t_n\}$ of $[0, t_n]$ and finite sequences $\{x_i\}_{i=0}^n, \{f_i\}_{i=1}^n$ in X such that

$$(27) \quad \begin{cases} \text{(i)} & \dfrac{x_{i+1} - x_i}{t_{i+1} - t_i} - B(x_{i+1}) = f_{i+1}, \quad i = 0, 1, \ldots, n - 1 \\[2mm] \text{(ii)} & t_{i+1} - t_i < \epsilon \qquad\qquad\qquad i = 1, \ldots, n - 1 \\[2mm] \text{(iii)} & 0 \leqslant T - t_n < \epsilon \\[2mm] \text{(iv)} & \displaystyle\sum_{i=1}^{n-1} \int_{t_i}^{t_{i+1}} \| f_i - f(s) \| \, ds < \epsilon, \end{cases}$$

and

$$(28) \qquad \| u_\epsilon(t) - u(t) \| \leqslant \epsilon \qquad \text{on} \quad [0, t_n)$$

where

$$(29) \qquad u_\epsilon(t) = x_i \qquad \text{for} \quad t_i \leqslant t < t_{i+1}, i = 1, \ldots, n - 1.$$

A function u_ϵ, piecewise constant as in (29), is called an ϵ-*approximate solution* of (EF)′ when the various conditions of (27) are satisfied. Roughly, (27) defines a simple implicit Euler approximation of (EF)′ and we are *defining* solutions of (EF)′ to be the uniform limits of solutions of these difference approximations. We have:

PROPOSITION 8. *Let B be homogeneous of degree $\alpha > 0, \alpha \neq 1$. Let $T > 0, \lambda > 0$ and $f \in L^1(0, T : X)$. If $u \in C([0, T] : X)$ is a mild solution of* (EF), *then $v(t) = \lambda^{1/(\alpha-1)}u(\lambda t)$ is a mild solution of* (EF) *on $[0, T/\lambda]$ with f replaced by $\lambda^{\alpha/(\alpha-1)}f(\lambda t)$.*

The proof is left to the reader. If B is also dissipative (equivalently, $- B$ is accretive—see, e.g., [3], [11], [16]) one has:

PROPOSITION 9. *Let B be dissipative. Let $x \in closure(D(B)), T > 0$ and $f \in L^1(0, T : X)$. If for each $\epsilon > 0$ there is an ϵ-approximate solution u_ϵ of* (EF)′ *satisfying $\| u_\epsilon(0) - x \| < \epsilon$ then* (EF)′ *has a mild solution u on $[0, T]$. Moreover, if $f, \hat{f} \in L^1(0, T : X)$ and $u, \hat{u}$ are mild solutions of $u' = B(u) + f, \hat{u}' = B(\hat{u}) + \hat{f}$ respectively, then*

$$\| u(t) - \hat{u}(t) \| \leqslant \| u(0) - \hat{u}(0) \| + \int_0^t \| f(s) - \hat{f}(s) \| \, ds$$

for $0 \leqslant t \leqslant T$.

This is proved in [13], although the definition of "mild solution" is not given there. See also [18].

It follows from Propositions 8 and 9 that letting $u(t) = S(t, u(0), f)$ when u is a mild solution of (EF)' and B is dissipative and homogeneous of degree α defines an operator S with the desired properties (FHS) and (FL) with $L = 1$. (The special case $f = 0$ gives (HS) and (L) with $L = 1$.)

Associated with each of the problems $(1)_\alpha$, $(2)_\alpha$, $(3)_\alpha$, $\alpha > 0$, is a densely defined m-dissipative operator B_α in $L^1(\mathbb{R})$, that is B_α is dissipative and the range of $(I - \lambda B_\alpha)$ is $L^1(\mathbb{R})$ for $\lambda > 0$. This in conjunction with Propositions 8 and 9, guarantees the existence of mild solutions. Moreover, each operator S so obtained is order preserving with respect to the natural order on L^1. This provides one precise sense in which these problems fall under the scope of this paper. (One may, of course, treat these problems by any other suitable method which provides the information (FHS) and (FL), etc.) Some references are: (i) [6] which shows how to make precise the m-dissipative operator in $L^1(\mathbb{R}^N)$ associated with equations $u_t = \Delta\varphi(u)$ for more general nonlinearities than in $(1)_\alpha$ and in any number of dimensions N, (ii) [4] and [9] which contain results defining m-dissipative operators associated with initial-boundary value problems for $u_t = \Delta\varphi(u)$, (iii) [2], [23] which contain results defining m-dissipative operators in L^p spaces, $1 \leqslant p \leqslant \infty$, associated with variants of $(3)_\alpha$ (which must be modified for the pure initial value problem), (iv) [10], [4] which establish m-dissipative operators for generalizations of $(2)_\alpha$. The equation $(26)_\alpha$, $1 < \alpha \leqslant 2$, corresponds to an m-dissipative operator in the space of uniformly continuous functions on $\mathbb{R}^N$, as is proved in [25].

Of course, there is a huge literature concerning other approaches and results for these problems. We continue in this section by choosing $(1)_\alpha$ and $(2)_\alpha$ for further discussion to illustrate the significance of the results of Section 1 in applications and something of the relationship with known results.

The problem $(1)_\alpha$ for $\alpha = 1$ is the initial-value for the linear heat equation which is solved by

$$(30) \qquad u(x,t) = \frac{1}{\sqrt{4\pi t}} \int_{-\infty}^{\infty} e^{-|x-y|^2/4t} u_0(y)\, dy.$$

If X is any one of the Banach spaces $L^p(\mathbb{R})$, $1 \leqslant p < \infty$, or $BU(\mathbb{R})$ (the bounded uniformly continuous functions on $\mathbb{R}$) equipped with the usual norm and $u_0 \in X$, then $t \to u(\cdot, t)$ with u given by (30) is a continuous curve in X for $t > 0$. Moreover $u(t) = u(\cdot, t) \to u_0$ in X as $t \downarrow 0$. In each space X define an operator B by

(31) $D(B) = \{ v \in X : v'' \in X \}$ and $Bv = v''$ for $v \in D(B)$,

where differentiation is in the sense of distributions. Then it is very well-known that B is m-dissipative and the mild solution of $u' = Bu, u(0) = u_0$ is $S(t)u_0 = u(t)$. Direct examination of (30) shows that if $u_0 \in X$, then $u(t)$ is differentiable, $u(t) \in D(B)$ and $u'(t) = Bu(t)$ for $t > 0$. Moreover

(32) $\left\| \dfrac{du}{dt}(t) \right\| = \| Bu(t) \| \leqslant \dfrac{c}{t} \| u(0) \|$ for $t > 0$.

Thus one has very explicit regularizing here, most convincingly illustrated by the formula (30). The estimate (32) implies that $S(t)$ is an analytic semigroup in X (see, e.g., [17], [29]). We note that while (32) has much the same character as the estimate (6) of Theorem 1, Theorem 1 does not apply here for $\alpha = 1$.

If $\alpha > 0$ and $\alpha \neq 1$, we do not know a formula for the solution of (1). However, the operator B_α in $X = L^1(\mathbb{R})$ given by

(33) $\begin{cases} D(B_\alpha) = \left\{ v \in L^1(\mathbb{R}) : |v|^{\alpha - 1} v \in L^1_{\text{loc}}(\mathbb{R}) \right. \\ \qquad \text{and} \quad \left. \left(|v|^{\alpha - 1} v \right)'' \in L^1(\mathbb{R}) \right\} \\ B_\alpha v = \left(|v|^{\alpha - 1} v \right)'' \quad \text{for} \quad v \in D(B_\alpha) \end{cases}$

is m-accretive in $L^1(\mathbb{R})$ ([6]). The mild solutions provided by this B_α are uniquely characterized as solutions of $(1)_\alpha$ in the sense of distributions (see [8]). Thus for $\alpha > 0$, $\alpha \neq 1$ Theorems 1 and 2 apply with these choices and we conclude that the solution u of $(1)_\alpha$ satisfies

(34) $\limsup\limits_{h \downarrow 0} \displaystyle\int_{\mathbb{R}} \dfrac{|u(x, t+h) - u(x,t)|}{h} \, dx \leqslant \dfrac{1}{t} \dfrac{2}{|\alpha - 1|} \int_{\mathbb{R}} |u_0(x)| \, dx,$

and also

(35) $\dfrac{\partial u}{\partial t} \geqslant -\dfrac{1}{(\alpha - 1)t} u$

provided $u_0 \geqslant 0$ so $u \geqslant 0$. The analogue of (34) in $L^p(\mathbb{R})$ or $BU(\mathbb{R})$ does not hold in view of explicit examples. See, e.g. [24]. The relation (35) follows from Theorem 2 applied to this example by dividing (9) by h and letting $h\downarrow 0$. (The limit of $(u(t + h, x) - u(t, x))/h$ is taken in the sense of distributions.) The curve $t \rightarrow u(t)$ in $L^1(\mathbb{R})$ solving $(1)_\alpha$ thus has a "speed" bounded in the form c/t for $t > 0$, as was true in the linear case, but we cannot so easily assert here the existence of the velocity $\lim_{h\rightarrow 0} h^{-1}(u(t + h) - u(t)) = u'(t)$ in $L^1(\mathbb{R})$ or that $u(t) \in D(B_\alpha)$ for almost all $t > 0$. If $u_0 = u(0) \geqslant 0$, these desirable properties hold true—see [1]. The current proof of this is a long story beginning with (35). (Ongoing work of various investigators indicates that the results of [1] extend to u_0 not necessarily of fixed sign and to more general nonlinearities.) However, it is known that if $S(t)$ is constructed from an (abstract) m-dissipative B as explained above, then $\lim_{h\downarrow 0} \| S(t + h)x - S(t)x\|/h \leqslant M < \infty$ exactly when there is a sequence $\{x_n\} \subset D(B)$ with $x_n \rightarrow S(t)x$ and $\limsup_{n\rightarrow\infty} \| Bx_n\| \leqslant M$. (See, e.g., [12], [14]). Thus (34) itself and the explicit nature (33) of B_α imply that $(|u(t)|^{\alpha - 1}u(t))''$ is a measure on $\mathbb{R}$ of variation at most $2/t|\alpha - 1|$.

With respect to other literature about $(1)_\alpha$ and variants, we mention in particular that the L^1-non expansiveness is noted in [27], that [3], [22], [24] are of interest and the references listed therein provide access to the large literature, that the estimate (34) is not new if $u_0 \geqslant 0$ (see [1]), and that the result of our paper applied to $(1)_\alpha$ and generalizations of it with u_0 not of fixed sign and the equation either perturbed or forced seem to be new.

The distinction between finite speed and possessing a velocity is clearly illustrated by the case of problems $(2)_\alpha$. The linear problem $\alpha = 1$ is explicitly solved by $u(x, t) = u_0(x + t)$. If $u_0 \in X$ and X is one of the spaces $L^p(\mathbb{R})$, $1 \leqslant p < \infty$ or $BU(\mathbb{R})$, then $u(t) = u_0(- + t)$ is a continuous curve in X. The velocity $u'(t)$ exists at some t if and only if it exists for $t = 0$ if and only if $u_0 \in D(B) = \{v \in X : v' \in X\}$. The speed $\lim_{h\downarrow 0} h^{-1}\|u(t + h) - u(t)\|$ is independent of t and is finite if and only if

$$(36) \quad \begin{cases} \text{(i)} & u_0 \in D(B) \text{ when } X = L^p(\mathbb{R}), 1 < p < \infty, \\ \text{(ii)} & u_0 \text{ is Lipschitz continuous when } X = BU(\mathbb{R}), \\ \text{(iii)} & u_0 \text{ is of essentially bounded variation on } \mathbb{R} \\ & \quad \text{when } X = L^1(\mathbb{R}). \end{cases}$$

Moreover, the speed is $\| Bu_0\| = \|u_0'\|$ in case (i), the least Lipschitz constant in case (ii) and the variation of u_0 in case (iii). There is no regularizing in this example. The differentiability and speed of u are independent of t. An estimate on the speed does imply some regularity in x as described above, but it does not imply $u(t) \in D(B)$.

The nonlinear problems $(2)_\alpha$, $\alpha \neq 1$, corresponds to the m-dissipative operators

$$
(37) \quad \begin{cases} D(B_\alpha) = \left\{ v \in L^\infty(\mathbb{R}) : (|v|^{\alpha-1}v)' \in L^1(\mathbb{R}) \right\}, \\ B_\alpha v = (|v|^{\alpha-1}v)' \end{cases}
$$

in $L^1(\mathbb{R})$ and the operators S_α to which they give rise respect the order of $L^1(\mathbb{R})$ ([10], [4]). It is not true here that solutions of $(1)_\alpha$ in the sense of distributions are unique and extra conditions must be laid upon solutions—so called entropy conditions. See [20], [21], [28], which further explain other approaches to $(1)_\alpha$. The entropy solutions of $(1)_\alpha$ are given by S_α ([10]). Simple analysis by the method of characteristics shows that even if u_0 is smooth and compactly supported, the solution of $(2)_\alpha$ must become discontinuous as t increases—i.e. "shocks form." This is reflected in the S_α and in general $S_\alpha(t)u_0 \notin D(B_\alpha)$ large t and $u_0 \neq 0$. Here we have "regularizing" in that Theorem 1 estimates the speed of a solution $u(t)$ in the form c/t and additional considerations explained above then estimate the variation in x of $|u(t)|^\alpha$ sign $u(t)$ by the same quantity, but we also have "roughing" in that $u(t)$ need not be smooth in x (or even lie in $D(B_\alpha)$) even if u_0 is smooth.

Estimates on the variation of solutions of $\partial u/\partial t + \partial f(u)/\partial x = 0$ which decay like c/t are classical for convex functions f ([21]). Our estimates for, e.g., $\partial u/\partial t + \partial u^5/\partial x = 0$ are perhaps new, as are the pointwise estimates (35) for nonnegative solutions and the estimates for the perturbed and forced equations. See also [15]. Concerning generalizations of $(26)_\alpha$ see [19].

A final point of interest here is that Theorem 1 did not capture the regularizing present in the linear heat equation $(1)_1$ and, indeed, it could not for Theorem 1 uses only properties shared by $(2)_\alpha$ for which there is no regularizing.

UNIVERSITY OF BESANÇON
UNIVERSITY OF WISCONSIN, MADISON

REFERENCES

[1] Aronson, D. G., and Ph. Bénilan, Régularité des solutions de l'équation des milieux poreux dans $\mathbb{R}^N$, *C. R. Acad. Sci. Paris Ser. A-B* **288** (1979), 103–105.

[2] Attouch, H. and A. Damlamian, Application des méthodes de convexité et monotonie a l'étude de certaines équations quasi-linéaires, *Proc. Royal Soc. Edinburgh*, to appear.

[3] Barbu, V., "Nonlinear Semigroups and Differential Equations in Banach Spaces," Nordhoff International Publishing Co., Leyden (1976).

[4] Bénilan, Ph., Équations d'évolution dans un espace de Banach quelconque et applications, thesis, Orsay 1972.

[5] Bénilan, Ph., Opérateurs accrétifs et semi-groupes dans les espaces $L^p(1 \leqslant p \leqslant \infty)$, France-Japan Seminar, Tokyo, 1976.

[6] Benilan, Ph., H. Brezis and M. G. Crandall, A semilinear elliptic equation in $L^1(\mathbb{R}^N)$, *Ann. Scuola Norm. Sup.* Pisa, Serie IV—Vol. II (1975), 523–555.

[7] Brezis, H., Opérateurs maximaux monotones et semi-groupes de contractions dans les espace de Hilbert, Amsterdam, North-Holland, 1977.

[8] Brezis, H. and M. G. Crandall, Uniqueness of solutions of the initial-value problem for $u_t - \Delta\varphi(u) = 0$, *J. Math. Pures et Appl.* **58** (1979), 153–163.

[9] Brezis, H. and W. Strauss, Semilinear elliptic equations in L^1, *J. Math. Soc. Japan*, **25** (1973), 15–26.

[10] Crandall, M. G., The semigroup approach to first order quasilinear equations in several space variables, *Israel J. Math.* **12** (1972), 108–132.

[11] ———, An introduction to evolution governed by accretive operators, Dynamical Systems—An Internation Symposium, L. Cesari, J. Hale, J. LaSalle eds., Academic Press, New York, 1976, 131–165.

[12] ———, A generalized domain for semigroup generators, *Proc. Amer. Math. Soc.* **37** (1973), 434–440.

[13] Crandall, M. G. and L. C. Evans, On the relation of the operator $\partial/\partial s + \partial/\partial\tau$ to evolution governed by accretive operators, *Israel J. Math.* **21** (1975), 261–278.

[14] Crandall, M. G. and L. Tartar, Some relations between nonexpansive and order preserving mappings, *Proc. Amer. Math. Soc.* **78** (1980), 385–390.

[15] Dafermos, C., Asymptotic behaviour of solutions of hyperbolic balance laws, in *Bifurcation Phenomena in Mathematical Physics and Related Topics*, C. Bardos and D. Bessis eds., D. Riedel Publishing Co. NATO Advanced Study Institute Series, Series C, Mathematics and Physical Sciences, 54, Reidel, Dordrecht, 1980.

[16] Evans, L. C., Application of nonlinear semigroup theory to certain partial differential equations, *Nonlinear Evolution Equations*, M. G. Crandall, ed., Academic Press, New York, 1978.

[17] Friedman, A., *Partial Differential Equations*, Holt, Rhinehart and Winston, Inc., Chicago, 1969.

[18] Kobayashi, Y., Difference approximation of Cauchy problems for quasi-dissipative operators and generation of nonlinear semigroups, *J. Math. Soc. Japan* **27** (1975), 640–665.

[19] Kružkov, S. N., Generalized solutions of first order nonlinear equations in several independent variables, I. *Mat. Sb.* **70** (112), (1966), 394–415, II. *Mat. Sb.* (*N.S.*) **72** (114) 1967, 108–134.

[20] Kružkov, S. N., First order quasilinear equations with several space variables, *Math. USSR Sb.* **10** (1970), 217–243.

[21] Lax, P. D., "Hyperbolic Systems of Conservation Laws and the Mathematical Theory of Shock Waves", *SIAM Regional Conference Series in Applied Mathematics* #11.

[22] Lions, J. L., Quelgues Méthodes de Résolution des Problèmes aux Limites Non Linéairies, Dunod, Gauthier-Villars, Paris 1969.

[23] Lê Chaû-Hoàn, Etude de la classe des opérateurs m-accrétifs de $L^1(\Omega)$ et accrétifs dans $L^\infty(\Omega)$, Thèse de 3e Cycle, Université de Paris VI.

[24] Peletier, L. A., The porous media equation, Notes, University of Leiden, Netherlands.

[25] Tamburro, M. B., The evolution operator approach to the Hamilton-Jacobi equation, *Israel J. Math.* **26** (1977), 232–264.

[26] Veron, L., Coercité et propriétés régularisantes des semi-groupes non linéaires dans les espaces de Banach, *Publ. Math. Fac. Sci. Besançon*, **3** (1977).

[27] Vol'pert, A. I. and S. I. Hudjaev, Cauchy's problem for degenerate second order quasi-linear parabolic equations, *Math. USSR Sb.*, **7** (1969), 365–387.

[28] Vol'pert, A. I., The spaces BV and quasilinear equations, *Math. USSR Sb.*, **2** (1967), 225–267.

[29] Yosida, K., Functional Analysis, Fifth Edition, Grundlehren der Mathematischen Wissenschaften 123, Springer-Verlag, New York 1978.

COMPACT HANKEL OPERATORS

By Jeffrey R. Butz

---

1. Introduction. During the last 25 years, a significant connection has developed between classical function theory on the unit circle and the study of bounded linear operators on Hilbert space. A clear illustration of this is given by recent investigations concerning Hankel operators. In the present, largely expository, paper, we give a brief synopsis of these developments and suggest what the future may hold.

In what follows, the complex L^p space of normalized Lebesgue measure on the unit circle in the complex plane will be denoted simply by L^p. H^p will denote the Hardy subspace of L^p consisting of those functions whose Fourier coefficients vanish on the negative integers. Functions that differ only on zero sets will be considered equal.

Given a function $\varphi \in L^\infty$, we define the corresponding *Hankel operator* $H_\varphi : H^2 \to (H^2)^\perp$ by $H_\varphi f = (1 - P_+)\varphi f$, where P_+ is the orthogonal projection of L^2 onto H^2. A related class is that of the *Toeplitz operators* $T_\varphi : H^2 \to H^2$ given by $T_\varphi f = P_+ \varphi f$. In each case, we refer to the function φ as the corresponding *symbol*.

Clearly, two functions φ_1, φ_2 whose difference is in H^∞ induce the same Hankel operator. Also, relative to the usual orthonormal basis for H^2 (the one consisting of the functions $e^{in\theta}$, $n \geqslant 0$) and the corresponding orthonormal basis for $(H^2)^\perp$, we can express the matrix representations for H_φ and T_φ in terms of the Fourier coefficients of φ. Namely, if φ has the Fourier series $\sum_{-\infty}^{\infty} c_n e^{in\theta}$, then

$$H_\varphi = (c_{-j-k+1}), \qquad T_\varphi = (c_{j-k}).$$

Thus, Hankel matrices are constant on the cross diagonals, and Toeplitz matrices are constant on the main diagonals. An important special case occurs when $\varphi(e^{i\theta}) = i(\theta - \pi)$ $(0 < \theta < 2\pi)$; then $c_{-n} = 1/n$, and the Hankel matrix for φ is the famous Hilbert matrix.

Both Hankel and, especially, Toeplitz operators have been the object of extensive study in the last twenty years, and a vast literature exists concerning them. In particular, it is well known and not difficult to show that the operators T_φ, with $\varphi \in L^\infty$, are characterized by the

41

property that $U_+^* T_\varphi U_+ = T_\varphi$, where U_+ is the unilateral shift, given on H^2 by multiplication by $e^{i\theta}$, and that $\|T_\varphi\| = \|\varphi\|_\infty$. Likewise, a Hankel operator H_φ, when given a definition different but equivalent to the one we have adopted, can be characterized in terms of U_+. But, by contrast, H_φ may exist as a bounded operator even though φ is not in L^∞. Since, as noted above, the values of H_φ are independent of the part of φ in H^2, this is not surprising. An explicit criterion for the boundedness of H_φ is given in the following famous theorem of Z. Nehari.

NEHARI'S THEOREM [16]. *A Hankel operator H_φ is bounded if, and only if, there exists a function $\psi \in H^\infty$ such that $\varphi + \psi \in L^\infty$. In such a case, the function ψ can be chosen so that $\|\varphi + \psi\|_\infty = \|H_\varphi\|$.*

In 1958, one year after Nehari's theorem appeared, Philip Hartman obtained a striking characterization of the compact Hankel operators.

HARTMAN'S THEOREM [11]. *A Hankel operator H_φ is compact if, and only if, there exists a function $\psi \in H^2$ such that $\varphi + \psi$ is continuous.*

Thus, a Hankel operator is bounded iff the symbol φ can be perturbed so as to be a bounded function; a Hankel operator is compact ("completely continuous") iff the symbol φ can be perturbed so as to be a continuous function. By contrast, non-zero Toeplitz operators are never compact.

The proof of this theorem, as given by Hartman, is quite brief and involves showing that, when H_φ is compact, it can be uniformly approximated by finite Hankel matrices. Subsequently, several alternative proofs have been obtained. Some of these are discussed below; here we mention [4] and [17].

In the twelve years since it appeared, Hartman's theorem has been the starting point for several deep and diverse investigations. In the following, we describe these developments as they have occurred in function theory on the unit circle, in several complex variables, and in the harmonic analysis of linear operators on Hilbert space.

2. Function theory on the unit circle. Nehari's theorem can be restated as the formula

$$\|H_\varphi\| = \mathrm{dist}(\varphi, H^\infty).$$

In a similar way, we can rephrase Hartman's theorem. To do this, recall that the essential norm of an operator is, by definition, its distance from the set of compact operators. We denote the essential norm by $\| \cdot \|_e$. For a self-adjoint Hankel operator, the essential norm was first studied by Clark, a student of Hartman's, in his thesis [6]. Subsequently, the general case was described.

THEOREM 2.1. $\|H_\varphi\|_e = \text{dist}(\varphi, H^\infty + C)$.

Proof [3]. If K is a compact operator and n is a positive integer, then

$$\|H_\varphi - K\| \geq \|(H_\varphi - K)U_+^n\| \geq \|H_\varphi U_+^n\| - \|KU_+^n\|.$$

Since $U_+^{*n} \to 0$ strongly, we have $\|KU_+^n\| \to 0$ as $n \to \infty$, and thus $\lim \|H_\varphi U_+^n\| \leq \|H_\varphi\|_e$. In fact, equality holds since $H_\varphi U_+^n$ is a rank n perturbation of H_φ and so has norm greater than or equal to $\|H_\varphi\|_e$. Now, since $H_\varphi U_+^n = H_{z^n \varphi}$, we can use Nehari's theorem to get

$$\|H_\varphi\|_e = \lim_{n \to \infty} \text{dist}(z^n \varphi, H^\infty)$$
$$= \lim_{n \to \infty} \text{dist}(\varphi, \bar{z}^n H^\infty) = \text{dist}(\varphi, H^\infty + C),$$

where the last equality follows from the fact that continuous functions can be uniformly approximated by trigonometric polynomials. This proves the result.

In [20], Sarason proves that $H^\infty + C$ is closed in L^∞, and therefore Hartman's result follows immediately from this theorem. We mention in passing that the approach adopted by Sarason in [20] can itself be simplified via Hankel matrices. The results in that paper are based upon a study of operators that are multiplication by an H^∞ function followed by projection onto a subspace of H^2 invariant for U_+^*. Such an operator can, however, be written as the product of a unitary operator and a Hankel matrix, and, from this, it is easy to conclude that $H^\infty + C$ is closed. This alternative approach to Sarason's results was first observed by Clark [7].

If ϕ is any function in L^∞, then Nehari's theorem (or a standard normal families argument) shows that $\text{dist}(\phi, H^\infty)$ is attained for some function $\psi \in H^\infty$. The question of when such a "minifunction" ψ is unique is difficult. A condition in terms of H_ϕ has been given by Adamyan, Arov, and Krein [1]: ϕ has a unique best approximation in H^∞ if, and only if, the constant function 1 does not lie in the range of the operator $(H_\phi^* H_\varphi - \|H_\varphi\|^2)^{1/2}$. As one would suspect, this condition is difficult to check in practice; it would be much preferable to have a structural condition on φ itself.

In view of the above, it is natural to ask whether every function in L^∞ has a best approximation in $H^\infty + C$. Theorem 2.1 states that the distance of a function φ from $H^\infty + C$ equals the distance of H_φ from the set of compact operators. It is known [12] that this latter distance is always attained. Thus, the question becomes: Does every Hankel operator have a best compact approximation that is also a Hankel operator? The recent answer to this question—affirmative—and its proof provide a beautiful application of the interplay between operator theory and classical analysis:

THEOREM 2.2. [3]   *Every function in L^∞ has a best approximation in $H^\infty + C$. Moreover, if this function is not itself in $H^\infty + C$, then the best approximation is never unique.*

Recently, D. Leucking [15] has obtained an alternative proof to the above by showing that the compact Hankel operators form an M-ideal in the space of Hankel operators. It is a result, due to Alfsen and Effros [2], that such ideals in a Banach space always yield best approximations. Likewise, the non-uniqueness part of the theorem can be obtained via this route.

As a generalization, it is also natural to ask whether Theorem 2.2 remains true when $H^\infty + C$ is replaced by other subalgebras of L^∞ containing H^∞. This is an open question.

It is likewise of interest to know when other subclasses of Hankel operators can be characterized in a simple way in terms of their symbol function φ. Subclasses of particular importance are the Schatten ideals S^p. [A compact operator T is in S^p, $1 \leqslant p < \infty$, if $\sum_{n=0}^\infty |\lambda_n((T^*T)^{1/2})|^p < \infty$, where $\{\lambda_n((T^*T)^{1/2})\}$ is the sequence of eigenvalues of the operator $(T^*T)^{1/2}$.] M. Rosenblum [19] first considered this question for S^1 (the "trace class operators"), and J. Howland [13], using work of Rosenblum, and also Clark [8] later obtained conditions, some necessary and some sufficient, for a Hankel operator to be of trace class. Recently, V. Peller has obtained criteria that are both necessary and sufficient for H_φ to be in S^p. These criteria involve the Besov spaces $\wedge^p_{1/p}$, $1 \leqslant p < \infty$. $\wedge^1_1$ consists of those functions φ that are analytic on the unit disc $D = \{z : |z| < 1\}$, continuous on $T = \partial D$, and that satisfy

$$\int_{-\pi}^\pi \frac{\int_{-\pi}^\pi |\varphi(e^{ix+it}) + \varphi(e^{ix-it}) - 2\varphi(e^{ix})| \, dx}{|t|^2} \, dt < \infty.$$

Similarly, $\wedge^p_{1/p}$, $1 < p < \infty$, consists of such functions φ that satisfy

$$\int_{-\pi}^{\pi} \frac{\int_{-\pi}^{\pi} |\varphi(e^{ix+it}) - \varphi(e^{ix})|^p \, dx}{|t|^2} \, dt < \infty.$$

It is known [22] that $\varphi \in \wedge^1_1$ iff $\iint_D |\varphi''(z)| \, dz < \infty$ and that $\varphi \in \wedge^p_{1/p}$, $1 < p < \infty$, iff $\iint_D |\varphi'(re^{i\theta})|^p (1 - r)^{p-2} \, dr \, d\theta < \infty$.

THEOREM 2.3. [18]   *A Hankel operator H_φ is in the Schatten ideal S^p if, and only if, the function $\psi(z) = \overline{z\varphi(z)}$ is in the Besov space $\wedge^p_{1/p}$.*

3. Extensions to higher dimensions. During the last ten years, substantial progress has been made in extending function theory for the unit circle in $\mathbb{C}$ to that for the unit sphere in $\mathbb{C}^n$. Here, we briefly summarize that part of this work which relates to Hankel operators.

Let $B = B_n$ be the unit ball in $\mathbb{C}^n$, i.e., $B_n = \{z = (z_1, \ldots, z_n) : |z| < 1\}$, and let $\partial B = \partial B_n = \{z \in \mathbb{C}^n : |z| = 1\}$ denote the corresponding sphere. Lebesgue measure on ∂B_n is denoted by $d\sigma(z)$. The Hardy space $H^p(\partial B_n)$, $1 \leq p < \infty$, is then defined to be the space of functions F that are holomorphic on B_n and for which

$$\|F\|_p = \sup_{r<1} \left[\int_{\partial B_n} |F(rz)|^p \, d\sigma(z) \right]^{1/p} < \infty.$$

We will identify functions in $H^p(\partial B_n) = H^p$ with their boundary values on ∂B_n.

Fundamental to recent work on Hardy spaces has been the duality between H^1 and the space BMO. Since we will need to make reference to this duality in what follows, we indicate the relevant definitions. For a function $\varphi \in H^2(\partial B_n)$, define

$$(S_1\varphi)(z) = \sup_S \left[\frac{1}{|S|} \int_S |\varphi(z) - m_S(\varphi)| \, d\sigma(z) \right]$$

where the supremum is taken over all spheres S in B_n, $|S|$ denotes the measure of S, and

$$m_S(\varphi) = \frac{1}{|S|} \int_S \varphi(z) \, d\sigma(z).$$

A function φ is said to belong to BMO (Bounded Mean Oscillation) if $\|\varphi\|_{BMO} = \|S_1(\varphi)\|_\infty < \infty$.

In 1972, Fefferman and Stein [10] proved that, for Euclidean space, BMO consists of those functions that can be written as the sum of an L^∞ function and the harmonic conjugate of an L^∞ function, i.e., BMO $= L^\infty + \tilde{L}^\infty$. Since $L^\infty + \tilde{L}^\infty$ can be naturally identified with the dual of H_1 (those functions whose real and imaginary parts are the real parts of H^1 functions), the above shows that, essentially, BMO is the dual of H^1. In fact, Coifman, Rochberg, and Weiss [9] have recently explicitly proved that the Fefferman-Stein result for Euclidean space carries over to ∂B_n and that this result is equivalent to the fact that an H^2 function φ is in BMO if, and only if, it is the projection (defined below) of an L^∞ function.

We now define Hankel operators for $H^2(\partial B_n)$. Given $f \in L^1(\partial B_n)$, define

$$(Pf)(z) = f(z) + P.V. \int_{\partial B_n} \frac{f(\xi)}{(1 - \bar{\xi} \cdot z)^n} \, d\sigma(\xi),$$

where $\bar{\xi} = (\bar{\xi}_1, \ldots, \bar{\xi}_n)$. Koranyi and Vagi [14] have shown that for $1 < p < \infty$, P is a bounded projection of $L^p(\partial B_n)$ onto $H^p(\partial B_n)$. Now, for $\varphi \in L^2(\partial B_n)$, define the corresponding Hankel operator $H_\varphi : H^2(\partial B_n) \to (H^2(\partial B_n))^\perp$ as before:

$$H_\varphi(f) = (I - P)\varphi f.$$

Coifman, Rochberg, and Weiss, using their result cited above, have proved the following generalization of the Nehari theorem.

THEOREM 3.1. [9]   For $\varphi \in H^2(\partial B_n)$, the following are equivalent:

(a) H_φ is a bounded map of $H^2(\partial B_n)$ into $(H^2(\partial B_n))^\perp$.
(b) $\varphi = (I - P)\psi$ for some $\psi \in L^\infty(\partial B_n)$.
(c) $\varphi \in$ BMO.

Moreover, if any of these conditions hold, then ψ can be chosen so that the norms $\|\psi\|_\infty$, $\|\phi\|_{\mathrm{BMO}}$, and $\|H_\phi\|$ are equivalent.

Likewise, in that same paper, an extension of the Hartman theorem is obtained. To state this result, we first define a subspace of BMO. For $\varphi \in$ BMO, let

$$M_r(\varphi) = \sup_{|S| < r} \frac{1}{|S|} \int_S |\varphi(z) - m_S(\varphi)| \, d\sigma(z),$$

where S denotes a sphere on ∂B_n with respect to the metric $d(z, \zeta)$ $= |1 - \bar{\zeta} \cdot z|^{1/2}$ and where $|S|$ denotes the Lebesgue measure of S. We say that φ is in the space VMO (Vanishing Mean Oscillation) if $\lim_{r \to 0} M_r(\varphi) = 0$.

THEOREM 3.2. [9] *For $\varphi \in H^2(\partial B_n)$, the following are equivalent:*

(a) H_φ *is a compact operator from $H^2(\partial B_n)$ to $(H^2(\partial B_n))^\perp$.*
(b) $\varphi = (I - P)\psi$ *for some ψ continuous on ∂B_n.*
(c) $\tilde{\varphi}$ *is an analytic function, where $\tilde{\varphi}(z) = \overline{\varphi(\bar{z})}$.*

The equivalence of (b) and (c) in the one-dimensional case is a result of Sarason [21], who introduced and first studied VMO.

4. Extensions to arbitrary Hilbert space. In this final section, we consider yet another avenue of work that has been motivated by Hartman's compactness criterion for the classical Hankel matrices. The results to be described have been obtained during the last several years, and they suggest prospective future developments.

If we adopt the alternative (but equivalent) definition of a Hankel operator alluded to earlier, so that its range is H^2 rather than $(H^2)^\perp$ (viz., $H_\varphi f(e^{i\theta}) = P_+ \varphi(e^{i\theta}) f(e^{-i\theta}))$, then such an operator is characterized by the formula

$$U_+^* H_\varphi = H_\varphi U_+,$$

where, as before, U_+ is the unilateral shift on H^2.

Unilateral shifts can, however, be defined as operators on an arbitrary Hilbert space, and the above characterization of the Hankel operators enables us likewise to define them in this more general setting. To be specific, let W be a unitary operator on a Hilbert space $\mathcal{K}$, and suppose that there exists a subspace $\mathcal{L}$ of $\mathcal{K}$ such that

$$\mathcal{K} = \bigoplus_{n=-\infty}^{\infty} W^n \mathcal{L}.$$

If we let $\mathcal{K} = \mathcal{L} \oplus W\mathcal{L} \oplus W^2\mathcal{L} \oplus \cdots$ and $U = W|\mathcal{K}$, then U is said to be the *unilateral shift* on $\mathcal{K}$ corresponding to the *bilateral shift* W on $\mathcal{K}$, of multiplicity $\dim \mathcal{L}$. An operator $H \in B(\mathcal{K})$ is said to be *Hankel* with respect to U if $U^*H = HU$.

In a standard way, the Hilbert space $\mathcal{K}$ described above can be associated with the "vector-valued" Hardy space $H_\mathcal{L}^2$, consisting of the square-integrable $\mathfrak{L}$-valued functions on T whose Fourier coefficients vanish on the negative integers. Using this model, Page [17] extended Hartman's theorem to operators that are Hankel with respect to unilateral shifts of countable multiplicity. Subsequently, Bonsall and Power [4] obtained a model-free proof valid for unilateral shifts of arbitrary multiplicity. The analogue of $H^\infty + C$ in this context is the norm closure in $B(\mathcal{K})$ of the algebra

$$U\{ W^{-n}\mathcal{K} : n \geqslant 0 \},$$

where, by analogy with the classical function space representation,

$$\mathcal{K}^\infty = \{ A \in B(\mathcal{K}) : AW = WA \quad \text{and} \quad A\mathcal{K} \subset \mathcal{K} \}.$$

There is an alternative way of generalizing the Hankel operators $H_\varphi : H^2 \to H^2$ in which the domain remains H^2 but the range can be any Hilbert space. To obtain such a generalization, let T be a contraction on a Hilbert space $\mathcal{K}$ such that T has no reducing subspace $\mathfrak{N}$ for which the restriction $T|\mathfrak{N}$ is unitary. Such a contraction is said to be *completely non-unitary*, and, for such an operator, a result of Sz.-Nagy and Foias enables us to define a functional calculus for functions $u \in H^\infty$:

THEOREM 4.1 ([23], p. 114). *For a completely non-unitary contraction T on a Hilbert space $\mathcal{K}$, the mapping $u \mapsto u(T)$ of H^∞ into $B(\mathcal{K})$, defined by*

$$(4.1) \qquad u(T) = \lim_{r \uparrow 1} \sum_{k=0}^\infty r^k c_k T^k \quad \text{for} \quad u(e^{i\theta}) = \sum_{k=0}^\infty c_k e^{ik\theta},$$

is a norm-decreasing algebra homomorphism of H^∞ into $B(\mathcal{K})$.

Suppose now that T is a fixed completely non-unitary contraction on $\mathcal{K}$ and that φ is a fixed vector in $\mathcal{K}$. Using the above theorem we can then define the desired generalized Hankel operator on H^∞ by setting $H_\varphi u = u(T)\varphi$. If we now assume that φ has the property that there exists a constant C such that

$$\|u(T)\varphi\|_{\mathcal{K}} \leqslant C\|u\|_{H^2} \quad \text{for all} \quad u \in H^\infty,$$

then, in such a case, we can extend, via continuity, the definition of H_φ to obtain an operator, still called H_φ, with domain all of H^2. It is easy to see that, in such a framework, we obtain the classical Hankel operators $H_\varphi : H^2 \to H^2$ by choosing $\mathcal{K} = H^2$, $\varphi = \varphi(e^{i\theta})$, and $T = U_+^*$.

Recently, by using the notion of the minimal unitary dilation corresponding to a contraction T (cf. [23]), the author obtained an extension of Nehari's theorem for the generalized Hankel operators defined above.

THEOREM 4.2. [5] *Let T be a complete non-unitary contraction on a Hilbert space $\mathcal{K}$. Let $W : \mathcal{K} \to \mathcal{K}$ be the minimal unitary dilation of T. Let $\varphi \in \mathcal{K}$ define a generalized Hankel operator $H_\varphi : H^2 \to \mathcal{K}$ by $H_\varphi u = u(T)\varphi$. Then there exists an element $\psi \in \mathcal{K}$ such that $P_\mathcal{K}\psi = \varphi$ and the operator $J_\psi : L^2 \to \mathcal{K}$ defined by $J_\psi u = u(W)\psi$ satisfies*

$$P_\mathcal{K} J_\psi \,|\, H^2 = H_\varphi, \qquad J_\psi U = W J_\psi, \qquad and \qquad \|J_\psi\| = \|H_\varphi\|.$$

It is well known that a condition such as $J_\psi U = W J_\psi$, where U is the bilateral shift on L^2 and W is its adjoint, forces ψ to be in L^∞, and we do therefore obtain the classical Nehari criterion for boundedness from the above.

Obtaining a compactness criterion is, however, quite another matter. At present, there are no known necessary and sufficient conditions on the contaction $T \in B(\mathcal{K})$ and the symbol $\varphi \in \mathcal{K}$ so that the corresponding Hankel operator $H_\varphi : H^2 \to \mathcal{K}$ is compact. The initial step, is, of course, to define an appropriate analogue for the subalgebra $H^\infty + C$ of L^∞. What this should be is not clear. At any rate, obtaining a compactness criterion in this context seems to be both interesting and difficult. It is a question motivated, in common with virtually all of the results of this paper, by Hartman's now-famous 1958 *Proceedings* paper.

ACKNOWLEDGEMENT. The author would like to thank Professor Douglas Clark for several helpful suggestions in the preparation of this work.

FORDHAM UNIVERSITY

REFERENCES

[1] V. M. Adamyan, D. Z. Arov, and M. G. Krein, Infinite Hankel matrices and generalized problems of Caratheodory-Fejer and J. Schur, *Funkcional. Anal. i Prilozen.* (4) **2** (1968), 1–17; English translation: *Functional Anal. Appl.* **2** (1968), 269–281.

[2] E. Alfsen and E. Effros, Structure in real Banach spaces, *Ann. of Math.* **96** (1972), 98–173.

[3] S. Axler, J. D. Berg, N. P. Jewell, and A. Shields, Approximation by compact operators and the space $H^\infty + C$, *Ann. of Math.* **109** (1979), 601–612.

[4] F. F. Bonsall and S. C. Power, A proof of Hartman's theorem on compact Hankel operators, *Math. Proc. Camb. Phil. Soc.* **78** (1975), 447–450.

[5] J. R. Butz, Hankel operators with Hilbert space range, *Ill. J. Math.*, **24** (1980), 477–482.

[6] D. N. Clark, On the spectra of bounded, Hermitian, Hankel matrices, *Amer. J. Math.* **90** (1968), 627–656.

[7] ——, On matrices associated with generalized interpolation problems, *Pacific J. Math.* **27** (1968), 241–253.

[8] ——, On interpolating sequences and the theory of Hankel and Toeplitz matrices, *J. Func. Anal.* **5** (1970), 247–258.

[9] R. R. Coifman, R. Rochberg, and G. Weiss, Factorization theorems for Hardy spaces in several variables, *Ann. of Math.* (2) **103** (1976), 611–635.

[10] C. Fefferman and E. M. Stein, H^p spaces of several variables, *Acta Math.* **129** (1972), 137–193.

[11] P. Hartman, On completely continuous Hankel matrices, *Proc. Amer. Math. Soc.* **9** (1958), 862–866.

[12] R. B. Holmes and B. R. Kripke, Best approximation by compact operators, *Indiana Univ. Math. J.* **21** (1971), 255–263.

[13] J. S. Howland, Trace class Hankel operators, *Quart. J. Math Oxford* (2) **22** (1971), 147–159.

[14] A. Koranyi and S. Vagi, Singular integrals on homogeneous spaces and some problems of classical analysis, *Annali della Scada Normale superiore di Pisa Classe di Scienze* **25** (1971), 575–648.

[15] D. H. Leucking, The compact Hankel operators form an M-ideal in the space of Hankel operators, *Proc. Amer. Math. Soc.* **79** (1980), 222–224.

[16] Z. Nehari, On bounded bilinear forms, *Ann. of Math.* **65** (1957), 529–540.

[17] L. B. Page, Bounded and compact vectorial Hankel operators, *Trans. Amer. Math. Soc.* (2) **150** (1970), 529–540.

[18] V. V. Peller, Nuclearity of Hankel operators, *Leningrad Math. Institute Preprint*, 1979.

[19] M. Rosenblum, The absolute continuity of Toeplitz's matrices, *Pacific J. Math.* **10** (1960), 987–996.

[20] D. Sarason, Generalized interpolation in H, *Trans. Amer. Math. Soc.* **127** (1967), 179–203.

[21] ——, Functions of vanishing mean oscillation, *Trans. Amer. Math. Soc.* **207** (1975), 391–405.

[22] E. M. Stein, *Singular Integrals and Differentiability Properties of Functions*, Princeton, 1970.

[23] B. Sz.-Nagy and C. Foias, *Harmonic Analysis of Operators on Hilbert Space*, North Holland, Amsterdam, 1970.

PAIRS OF MAPS INTO COMPLEX PROJECTIVE SPACE*

By Kuo-Tsai Chen

Let N be an oriented compact C^∞ manifold. A pair of C^∞ maps

$$f_0, f_1 : N \to CP^n$$

are studied by constructing a smooth family of geodesics from $f_0(y)$ to $f_1(y)$. Of course, there may be a subset Γ of N consisting of points y, for which there is no unique shortest geodesic from $f_0(y)$ to $f_1(y)$. Let $P(CP^n)$ be the space of C^∞ paths of CP^n. Then there is a map

$$\alpha : N - \Gamma \to P(CP^n)$$

sending y to the geodesic from $f_0(y)$ to $f_1(y)$. For simplicity, it is assumed that Γ is a closed submanifold of N, whose connected components are $\Gamma_1, \ldots \Gamma_l$. Associated to each Γ_i, there is an integer ord $_\alpha\Gamma_i$ defined through the Kähler $(1,1)$-form w of CP^n. The results in this paper are applications of the formula

$$(0.1) \qquad \int_N (f_1^*w - f_0^*w) \wedge \sigma = \sum_i (\mathrm{ord}_\alpha \Gamma_i) \int_{\Gamma_i} \sigma$$

where σ is any closed p-form on N with $p + 2 = \dim N$. Theorem 2.1 is a generalized version of this formula.

First we apply this formula to the case where f_0 is a constant map and obtain a sufficient condition for the validity of another formula (Theorem 3.1):

$$\int_N (f_1^* w^q) \wedge \sigma = (-1)^q \sum_j \mu_j \int_{\Delta_j} \sigma$$

where $\{\Delta_j\}$ are the connected components of $f_1^{-1}(L)$ for a given subspace L of codimension q of CP^n, and $\{\mu_j\}$ are certain intersection numbers.

---

*Work supported in part by NSF Grant MCS 79-00321.

Remark. This formula has a holomorphic version, which can be taken as a variation of a special case of the First Main Theorem in the value distribution theory. Our version is similar to that of Stoll [5]. Our approach regarding CP^n is in the spirit of Chern [2] and Levine [4].

In the general case, it is known that Γ consists of those points y such that $f_0(y)$ and $f_1(y)$ are orthogonal in CP^n. Using the formula (0.1), we express the Lefschetz coincidence number as

$$L(f_0, f_1) = (n + 1)\deg f_1 + \sum (\operatorname{ord}_\alpha \Gamma_i) \int_{\Gamma_i} \sigma$$

where $\sigma = f_0^* w^{n-1} + 2f_0^* w^{n-2} \wedge f_1^* w + \ldots + nf_1^* w^{n-1}$ (Theorem 4.1).

In Section 1, an almost complete account of required material on path space differential forms is given. Theorem 2.1 is obviously applicable to spaces other than CP^n and can be further generalized by using path space differential forms of other types.

1. Path space differential forms. Let M be a C^∞ manifold. Denote by $P(M)$ the space of all C^∞ paths $\gamma : I \to M$. The path space $P(M)$ has a natural C^∞ structure, and the two end point maps

$$p_0, p_1 : P(M) \to M,$$

respectively given by $\gamma \mapsto \gamma(0)$ and $\gamma(1)$, are C^∞ maps. Thus, if w is a p-form on M, so are $p_0^* w$ and $p_1^* w$ on $P(M)$.

Denote by $\eta : P(M) \to P(M)$ the map sending every path γ to the constant path at the initial point $\gamma(0)$. Denote by γ^t the path which, up to the linear reparametrization, is the restriction of γ to the interval $[0, t]$, $t \in I$. Then the map

$$P(M) \times I \to P(M)$$

given by $(\gamma, t) \mapsto \gamma^t$ is a C^∞ homotopy from the map η to the identity map of $P(M)$. As in the case of manifolds, the homotopy induces a cochain homotopy of the de Rham complex of $P(M)$. If $\int$ denotes this cochain homotopy, then, for every p-form u on $P(M)$, $\int u$ is a $(p-1)$-form on $P(M)$ with

$$(1.1) \qquad \int du + d\int u = u - \eta^* u.$$

For a p-form w on M, we shall write

$$\int w = \int p_1^* w.$$

If w is closed, then $p_1^* w$ is also closed, and, owing to $p_1 \eta = p_0$,

$$d \int w = - \int dp_1^* w + p_1^* w - \eta^* p_1^* w = p_1^* w - p_0^* w.$$

The reader is referred to [1] for more detailed discussions of path space differential forms.

2. Order. Let w be a closed q-form on $M, q \geqq 1$. Let N be an oriented compact C^∞ manifold, and let Γ be a closed submanifold of codimension q and let

$$\alpha : N - \Gamma \to P(M)$$

be a C^∞ map such that both composite maps $p_0 \alpha$ and $p_1 \alpha : N - \Gamma \to M$ extend respectively to two C^∞ maps

$$f_0, f_1 : N \to M.$$

Choose a Riemann metric for N. For $y \in \Gamma$, the fiber of an ϵ-tubular neighborhood of Γ over y is a geodesic q-ball, whose boundary will be denoted by $F(\epsilon, y)$. Let $\{\Gamma_i\}$ be the connected components of Γ.

Definition. The order of Γ_i with respect to α is

$$\mathrm{ord}_\alpha \Gamma_i = \lim_{\epsilon \to 0} \int_{F(\epsilon, y)} \alpha^* \int w$$

where y is any point of Γ_i in the interior of N, and the orientation of $F(\epsilon, y)$ is chosen so that the r.h.s. is nonnegative.

In order to justify this definition and to describe further properties, we need to verify the following assertions:

(a) For each Γ_i, $\mathrm{ord}_\alpha \Gamma_i$ exists and does not depend on the choice of y.

(b) If w is replaced by another closed q-form w' in the same de Rham cohomology class, the value of $\mathrm{ord}_\alpha \Gamma_i$ does not change.

(c) The value of $\mathrm{ord}_\alpha \Gamma_i$ does not change under a smooth homotopy of the map $\alpha : N - \Gamma \to P(M)$.

In order to verify the first part of (a), choose $y \in \Gamma_i$ and let

$$D = \bigcup_{0 \leq t \leq 1} F((1 - t)\epsilon_0 + t\epsilon_1, y), \qquad \epsilon_0, \epsilon_1 \in (0, \epsilon).$$

Then $\partial D = F(\epsilon_1, y) - F(\epsilon_0, y)$ as a chain, and

$$\int_{\partial D} \alpha^* \int w = \int_D \alpha^*(p_1^* w - p_0^* w) = \int_D (f_1^* w - f_0^* w) = o(\epsilon).$$

Consequently $\mathrm{ord}_\alpha \Gamma_i$ exists for this choice of y. Now let y be a simple path from y_0 to y_1 in Γ_i and let

$$D' = \bigcup_{0 \leq t \leq 1} F(\epsilon, \gamma(t)).$$

Then $\partial D' = F(\epsilon, y_1) - F(\epsilon, y_0)$ and

$$\int_{\partial D'} \alpha^* \int w = \int_{D'} (f_1^* w - f_0^* w) = 0(\epsilon)$$

so that the value of $\mathrm{ord}_\alpha \Gamma_i$ does not depend on the choice of y.

In order to see (b), let $dv = w' - w$. Then

$$\lim_{\epsilon \to 0} \int_{F(\epsilon, y)} \alpha^* \int dv = \lim_{\epsilon \to 0} \int_{F(\epsilon, y)} \alpha^*\left(-d\int v + p_1^* v - p_0^* v\right)$$

$$= \lim_{\epsilon \to 0} \int_{F(\epsilon, y)} (f_1^* v - f_0^* v) = 0.$$

In order to prove (c), let $h : I \times (N - \Gamma) \to P(M)$ be a homotopy from α to α' such that $p_0 h(t, y) = f_0(y)$ and $p_1 h(t, y) = f_1(y)$ $\forall (t, y) \in I \times (N - \Gamma)$. Regard $\hat{h} = h | I \times F(\epsilon, y)$ as a q-chain. Then

$$\int_{F(\epsilon, y)} \left(\alpha'^* \int w - \alpha^* \int w\right) = \int_{\hat{h}} d \int w = \int_{\hat{h}} (p_1^* w - p_0^* w)$$

$$= \int_{p_1 \hat{h} - p_0 \hat{h}} w = 0$$

because both $p_0\hat{h}$ and $p_1\hat{h}$ represent degenerate q-chains given by

$$(p_i\hat{h})(t, y') = f_i(y'), \qquad i = 0, 1.$$

Remark.   Of course, $\text{ord}_\alpha \Gamma_i$ depends on the de Rham cohomology class of w.

THEOREM 2.1.   Let w be a closed q-form on a C^∞ manifold M, and let N be an oriented compact manifold (without boundary). Let

$$f_0, f_1 : N \to M$$

be two C^∞ maps. If there exists a closed submanifold Γ of N of codimension q which consists of connected components $\Gamma_1, \ldots, \Gamma_l$ and if there exists a C^∞ map

$$\alpha : N - \Gamma \to P(M)$$

such that $p_0\alpha$ and $p_1\alpha$ extend respectively to f_0 and f_1, then for any closed p-form σ on N with $p + q = \dim N$,

$$\int_N (f_1^*w - f_0^*w) \wedge \sigma = -\sum_i (\text{ord}_\alpha \Gamma_i) \int_{\Gamma_i} \sigma.$$

Remark.   The orientation of Γ_i is determined as follows: If $\text{ord}_\alpha \Gamma_i = 0$ (including the case of a nonorientable Γ_i), an orientation of Γ_i is not needed. If $\text{ord}_\alpha \Gamma_i > 0$, then $F(\epsilon, y)$, $y \in \Gamma_i$, has an orientation, which induces an orientation of geodesic q-ball $B(\epsilon, y)$ normal to Γ_i at y. The orientation of Γ_i is chosen such that the product orientation of $\Gamma_i \times B(\epsilon, y)$ is induced by the orientation of N.

Proof.   Observe that

$$\int_N (f_1^*w - f_0^*w) \wedge \sigma = \int_{N-\Gamma} d\left(\alpha^* \int w\right) \wedge \sigma$$

$$= -\lim_{\epsilon \to 0} \sum_i \int_{T_i(\epsilon)} d\left(\alpha^* \int w\right) \wedge \sigma$$

where $T_i(\epsilon)$ is the ϵ-tubular neighborhood of Γ_i. The integration over $T_i(\epsilon)$ as a q-ball bundle over Γ_i is first taken over fibers and then over Γ_i.

Since the p-form σ is defined everywhere on N, its contribution during integration around fibers can be ignored as ϵ tends to 0. In particular, if $\text{ord}_\alpha \Gamma_i = 0$, then $\lim_{\epsilon \to 0} \int_{T_i(\epsilon)} d(\alpha^* \int w) \wedge \sigma = 0$. If $\text{ord}_\alpha \Gamma_i > 0$, then

$$\lim_{\epsilon \to 0} \int_{T_i(\epsilon)} d\left(\alpha^* \int w\right) \wedge \sigma = \lim_{\epsilon \to 0} \int_{\partial T_i(\epsilon)} \alpha^* \int w \wedge \sigma = (\text{ord}_\alpha \Gamma_i) \int_{\Gamma_i} \sigma.$$

Hence the theorem is proved.

3. The case of a C^∞ map to CP^n. Hereafter M will be the complex projective space CP^n. Let w be a closed 2-form on CP^n representing the preferred generator of the integral cohomology group $H^2(CP^n; Z)$.

We choose a base point in CP^n and a hyperplane L_1 away from the base point. For every C^∞ map

$$f : N \to CP^n$$

take $f_1 = f$ and let $f_0 : N \to CP^n$ be the constant map to the base point. Given these data, we are going to construct an associated C^∞ map

$$\alpha = \alpha_f : N - \Gamma \to P(CP^n), \qquad \Gamma = f^{-1}(L_1),$$

such that $p_1 \alpha = f|N - \Gamma$ and $p_0 \alpha$ is the constant map to the base point. This is done as follows: Choose homogeneous coordinates $[z] = [z_0, \ldots, z_n]$ of CP^n such that the base point is $[1, 0, \ldots, 0]$ and L_1 is the hyperplane at infinity given by $z_0 = 0$. Then, $\forall y \in N - \Gamma, \alpha(y)$ is the path from the base point $[1, 0, \ldots, 0]$ to the point $f(y)$ given by

$$t \mapsto [f_0(y), tf_1(y), \ldots, tf_n(y)]_0, \quad f(y) = [f_0(y), \ldots, f_n(y)].$$

Assume that Γ is a closed submanifold of codimension 2 in N. Let $\Gamma_1, \ldots, \Gamma_l$ be its connected components. We are going to verify analytically that $\text{ord}_\alpha \Gamma_i$ is the intersection number of L_1 and the image of a fiber of a tubular neighborhood of Γ_i (i.e. the image of a disc bounded by $F(\epsilon, y), y \in \Gamma_i$).

For this purpose, we take w to be the Kähler $(1, 1)$-form on CP^n. Let $Z = (Z_1, \ldots, Z_n)$ be the coordinates of $C^n \subset CP^n$ so that Z_i

$= z_i/z_0$. Then w can be expressed as

(3.1) $\quad w|C^n = (2\pi i)^{-1}(1 + Z\overline{Z})^{-1}$

$$\times \sum_i \left\{ d\overline{Z}_i \wedge dZ_i - (1 + Z\overline{Z})^{-1} \sum Z_i d\overline{Z}_i \wedge \sum \overline{Z}_i dZ_i \right\}$$

where $Z\overline{Z} = \sum_i Z_i\overline{Z}_i$.

The inclusion $C^n \subset CP^n$ has an associated map

$$\tilde{\alpha} : C^n \to P(CP^n)$$

such that $\alpha(Z)$ is the path given by $t \mapsto [1, tZ]$. Let

$$\phi_{\tilde{\alpha}} : I \times C^n \to CP^n$$

be the map given by $(t, Z) \mapsto \tilde{\alpha}(Z)(t) = [1, tZ]$. Then

$$\phi_{\tilde{\alpha}}^* w = dt \wedge (2\pi i)^{-1}(1 + t^2 Z\overline{Z})^{-1} t \sum (\overline{Z}_i dZ_i - Z_i d\overline{Z}_i) + \ldots .$$

Set $\tau = \tilde{\alpha}^* \int w$. Then, according to [1],

$$\tau = \int_0^1 \left(\frac{\partial}{\partial t} \lrcorner \phi_{\tilde{\alpha}}^* w \right) dt = (4\pi i)^{-1}(1 + Z\overline{Z})^{-1} \sum (\overline{Z}_i dZ_i - Z_i d\overline{Z}_i)$$

so that

(3.2) $\quad \tau = (4\pi i)^{-1}(1 + Z\overline{Z})^{-1} \sum_i Z_i \overline{Z}_i d \log(Z_i/\overline{Z}_i).$

About $y \in \Gamma_i$, write $f = [f_0, \ldots, f_n]$ and $f\bar{f} = \sum_{0 \le i \le n} f_i \bar{f}_i$. Then $\alpha = f\tilde{\alpha}$ and

$$\alpha^* \int w = f^* \tilde{\alpha}^* \int w = (4\pi i)^{-1}(f\bar{f})^{-1}$$

$$\times \sum_{1 \le i \le n} f_i \bar{f}_i d\left(\log f_i/\bar{f}_i - \log f_0/\bar{f}_0 \right)$$

$$= (4\pi i)^{-1} \left[(f\bar{f})^{-1} \sum_{0 \le i \le n} f_i \bar{f}_i d \log(f_i/\bar{f}_i) + d \log(f_0/\bar{f}_0) \right].$$

Since $\lim_{\epsilon \to 0} \int_{F(\epsilon, y)} f_i \bar{f}_i \, d\log(f_i/\bar{f}_i) = 0$, we obtain

$$\text{ord}_\alpha \Gamma_i = (4\pi i)^{-1} \lim_{\epsilon \to 0} \int_{F(\epsilon, y)} d\log(f_0/\bar{f}_0).$$

About y in $N - \Gamma$, write $f_0 = re^{i\theta}$. Then

$$\text{ord}_\alpha \Gamma_i = (2\pi)^{-1} \int_{F(\epsilon, y)} d\theta,$$

which is the intersection number mentioned before.

With f and w as given, $\text{ord}_\alpha \Gamma_i$ is an integer and depends on the choice of the hyperplane L_1 but not on that of the base point.

Theorem 2.1 can now be applied to the map α, and the following formula is obtained:

$$(3.3) \qquad \int_N (f^*w) \wedge \sigma = -\sum (\text{ord}_\alpha \Gamma_i) \int_{\Gamma_i} \sigma.$$

Let $CP^n = L_0 \supset L_1 \supset \ldots \supset L_q$ be a sequence of projective subspaces of CP^n such that L_i is of codimension i in CP^n. Such a sequence is said to be regular with respect to a C^∞ map $f: N \to CP^n$ if, for each i, $f^{-1}(L_i)$ is a closed submanifold of (real) codimension $2i$ in N.

THEOREM 3.1. *Let N be an oriented compact manifold and let $f: N \to CP^n$ be a C^∞ map. Let L be a projective subspace of codimension q in CP^n such that $f^{-1}(L)$ consists of connected components $\Delta_1, \ldots, \Delta_l$. If there exists a regular sequence $CP^n = L_0 \supset L_1 \supset \ldots \supset L_q = L$ with respect to f, then, for any closed p-form σ with $p + 2q = \dim N$, the following formula holds:*

$$(3.4) \qquad \int_N f^*w^q \wedge \sigma = (-1)^q \sum_{1 \leq j \leq l} \mu_j \int_{\Delta_j} \sigma$$

where μ_j is the intersection number of L with the image of the fiber of a tubular neighborhood of Δ_j under f.

Remark. In a way, we may think that μ_j is the multiplicity of the connected component Δ_j of $f^{-1}(L)$. Then $\sum \mu_j \Delta_j$ represents $f^{-1}(L)$ as a p-cycle of N. Observe that the l.h.s. of (3.4) is independent of the choice of L. With this interpretation, we can conclude that the real homology class of $f^{-1}(L)$ is independent of the choice of L provided there exists a regular sequence with respect of f terminating at L. In other words, $(-1)^q f^{-1}(L)$ gives rise to the real Poincaré dual of the cohomology class of $f^* w^q$ under this condition.

Proof. Let $\{\Gamma_j^{(i)}\}$ be the connected components of $f^{-1}(L_i)$ that contains Δ_j. Applying the formula (3.3) repeatedly, we obtain

$$\int_N f^* w^q \wedge \sigma = (-1)^q \sum_j v_j^{(1)} \cdots v_j^{(q)} \int_{\Delta_j} \sigma$$

where $v_j^{(i)}$ is the order of $\Delta_j^{(i)}$ with respect to the restriction of f as a map $f^{-1}(L_{i-1}) \to L_{i-1}$ and the choice of L_i as the hyperplane in L_{i-1}. It remains to prove that

$$\mu_j = v_j^{(1)} \cdots v_j^{(q)}.$$

We use induction on q. The case of $q = 1$ has been shown. For $q > 1$, choose local coordinates $\eta = (\eta_1, \ldots, \eta_k)$ about $y \in \Delta_j$ with $\eta(y) = 0$ where $k = p + 2q = \dim N$. It is further required that, for each i, $\Gamma_j^{(i)}$ is locally given by $\eta_1 = \ldots = \eta_{2i} = 0$. For $\epsilon > 0$ sufficiently small, let

$$D = \left\{ \eta : \eta_1^2 + \eta_2^2 < \epsilon^2, \ldots, \eta_{2q-1}^2 + \eta_{2q}^2 < \epsilon^2, \right.$$

$$\eta_{2q+1} = \ldots = \eta_k = 0 \right\}$$

and

$$B = D \cap \{\eta; \eta_1 = \eta_2 = 0\} = D \cap \Gamma_j^{(1)}.$$

By the induction hypothesis, the intersection number of $f|B$ and L_q in L_1 is equal to $v_j^{(2)} \cdots v_j^{(q)}$.

Choose a perturbation g of $f|D : D \to CP^n$ so that

(a) $g^{-1}(L_1) = B$ and $g|B$ meets L_q transversally in L_1;

(b) for a preassigned $\delta > 0$, $|g(y') - f(y')| < \delta$, $\forall y' \in D$.

Then $g^{-1}(L_q)$ is a finite set contained in B. We need only to count the local intersection number of L_q and g at each point y_λ of $g^{-1}(L_q)$. At y_λ, let μ_λ' be the local intersection number of $g|B$ and L_q in L_1. The transversality implies that $\mu_\lambda' = \pm 1$. Moreover $\sum_\lambda \mu_\lambda' = v_j^{(2)} \cdots v_j^{(q)}$.

Set $D_\lambda = \{\eta \in D; \eta_i = \eta_i(y_\lambda), i = 3, 4, \ldots, k\}$, which is a 2-disc. The intersection number of $g|D_\lambda$ and L_1 is equal to that of $f|D_\lambda$ and L_1, which is $\nu_j^{(1)}$. Consequently, the local intersection number of g and L_q at y_λ is $\nu_j^{(1)}\mu_\lambda'$. The sum of these local intersection numbers is then equal to $\nu_j^{(1)} \ldots \nu_j^{(q)}$. Hence the theorem is proved.

4. Orthogonality and map pairs. Let $z = (z_0, \ldots, z_n)$ be the coordinates of the complex $(n + 1)$-space C^{n+1}, which is equipped with the hermitian metric so that two points z and z' are orthogonal if $\bar{z}z' = \sum \bar{z}_i z_i' = 0$. This notion of orthogonality is carried over to the CP^n through the canonical projection

$$C^{n+1} - \{0\} \to CP^n.$$

Thus two points $[z]$ and $[z']$ are orthogonal in CP^n if $\bar{z}z' = 0$.

Definition. The orthogonal locus of a pair of C^∞ maps

$$f_0, f_1 : N \to CP^n$$

is the set

$$\Gamma = \{y \in N; f_0(y) \text{ and } f_1(y) \text{ are orthogonal}\}.$$

For each pair of C^∞ maps f_0 and f_1 as above, we are going to construct a C^∞ map

$$\alpha : N - \Gamma \to P(CP^n)$$

with $p_0\alpha = f_0|N - \Gamma$ and $p_1\alpha = f_1|N - \Gamma$. Let N be an oriented compact C^∞ manifold, and assume that Γ is a closed submanifold of codimension 2, whose connected components are $\Gamma_1, \ldots, \Gamma_l$. Then Theorem 2.1 yields the formula

$$(4.1) \qquad \int_N (f_1^* w - f_0^* w) \wedge \sigma = -\sum_i (\mathrm{ord}_\alpha \Gamma_i) \int_{\Gamma_i} \sigma,$$

where σ is any closed p-form on N with $p + 2 = \dim N$.

Let K be the subset of $CP^n \times CP^n$ consisting of orthogonal pairs of points of CP^n. In order to construct α, it suffices to construct a C^∞ map

$$(4.2) \qquad \chi : CP^n \times CP^n - K \to P(CP^n).$$

Then α can be defined as the composite map χf, where

$$f = (f_0, f_1)|N - \Gamma.$$

Let $([z], [z']) \in CP^n \times CP^n - K$. With respect to the Fubini-Study mertric of CP^n, there is a unique geodesic from $[z]$ to $[z']$ of arc length s given by

$$\cos s = \frac{\bar{z}z'}{|z||z'|} .$$

Define $\chi([z], [z'])$ to be the geodesic with a linear reparametrization. Therefore χ and, consequently, α are well defined C^∞ maps.

THEOREM 4.1. *Let N be an oriented compact C^∞ $2n$-manifold and let*

$$f_0, f_1 : N \to CP^n$$

be a pair of C^∞ maps, whose Lefschetz coincidence number is $L(f_0, f_1)$. Assume that the orthogonal locus of the map pair (f_0, f_1) is a closed submanifold of codimension 2 consisting of connected components $\Gamma_1, \ldots, \Gamma_l$. Then

$$L(f_0, f_1) = (n + 1)\deg f_1 + \sum (\mathrm{ord}_\alpha \Gamma_i) \int_{\Gamma_i} \sigma$$

where

$$\sigma = f_0^* w^{n-1} + 2f_0^* w^{n-2} \wedge f_1^* w + \ldots + nf_1^* w^{n-1}.$$

Remark. Each integral $\int_{\Gamma_i} \sigma$ is geometrically meaningful. This will be explained through subsequent examples.

Proof. The Lefschetz coincidence number $L(f_0, f_1)$ can be given by the integral $\int_N \sum_{0 \leq i \leq n} f_0^* w^{n-i} \wedge f_1^* w^i$. (See p. 419 [3].) Thus

$$\int_N (f_1^* w - f_0^* w) \wedge \sigma = (n + 1) \int_N f_1^* w^n - L(f_0, f_1).$$

The theorem then follows from (5.1).

Example. When $n = 1$, Γ consists of a finite number of points $\{y_i\}$, and

$$L(f_0, f_1) = 2 \deg f_1 + \sum \operatorname{ord} y_i.$$

When $n = 2$, $\sigma = f_0^* w + 2f_1^* w$, and $\int_{\Gamma_i} f_\lambda^* w$, $\lambda = 0, 1$, is the intersection number of $f_\lambda | \Gamma_i$ and a line in CP^2. For higher n, the geometric significance can be interpreted through a repeated application of Theorem 3.1.

UNIVERSITY OF ILLINOIS, URBANA

REFERENCES

[1] K. T. Chen, Iterated path integrals, *Bull. Amer. Math. Soc.* **83** (77), 831-879.
[2] S. S. Chern, The integrated form of the first main theorem for complex analytic mappings in several complex variables, *Ann. of Math.* **71** (60), 536-551.
[3] P. A. Griffiths and J. Harris, "Principles of Algebraic Geometry", Wiley, 1978.
[4] H. I. Levine, Complex projective space, *Ann. of Math.* **71** (60), 529-535.
[5] W. Stoll, A general first main theorem on value distribution I-II, *Acta Math.* **118** (67), 111-191.

ON THE STRUCTURE OF RATIONAL TOEPLITZ
OPERATORS*

By Douglas N. Clark

---

1. Introduction. Let $L^2 = L^2(0, 2\pi)$, equipped with its usual orthonormal basis e^{int}, $n = 0, \pm 1, \pm 2, \ldots$, and let H^2 denote the subspace spanned by e^{int}, $n \geqslant 0$. Equivalently, H^2 may be thought of as the space of analytic functions $x(z) = \sum a_n z^n$, with $\sum |a_n|^2 < \infty$. For $F(e^{it}) \in L^\infty(0, 2\pi)$, we define the Toeplitz operator T_F, acting on $x \in H^2$, as follows

$$T_F x = PFx$$

where P is the orthogonal projection of L^2 on H^2.

In two papers in the early fifties [7], [8], Hartman and Wintner brought Toeplitz operators to the attention of modern analysts. Hartman and Wintner proposed the study of selfadjoint T_F (F real valued) and they proved in [8] that such a Toeplitz operator has spectrum equal to the interval [ess inf F, ess sup F] and has no point spectrum. The study of (bounded) selfadjoint Toeplitz operators was completed by Rosenblum [11], who proved that (i) the spectrum of T_F is absolutely continuous and (ii) the multiplicity of a real λ in the spectrum of T_F is n if the set where $F \leqq \lambda$ is the union of n intervals (modulo zero sets and modulo 2π) and is ∞ otherwise.

Hartman and Wintner focused on the selfadjoint case, but several of their results extend at once to general bounded T_F. For example, the proof of [7, (II)] proves the "spectral inclusion theorem" (to the effect that the spectrum of T_F contains that of the operator of multiplication by F on L^2) and the proof given by Hartman and Wintner for [8, (ii)] also proves that $T_F - \lambda I$ and $T_F^* - \bar{\lambda} I$ cannot both have a nonzero kernel. In 1958, Krein [9] began a more systematic study of non-selfadjoint Toeplitz operators. Stated slightly more generally than he proved it, Krein's result is this: if F is continuous (and periodic), then the essential spectrum of T_F is the curve $\gamma_F : t \to F(e^{it})$ and the index of

*Partially supported by NSF grants.

$T_F - \lambda I$, for $\lambda \notin \gamma_F$, is minus the winding number of γ_F with respect to λ.

More recent investigations of Toeplitz operators have involved generalizations of Krein's index theorem, see [5], and certain problems of a non-spectral nature. The latter include commutant and invariant subspace problems for analytic ($F \in H^\infty$) Toeplitz operators.

The subject of the present paper is a study of the structure, up to similarity, for *rational* Toeplitz operators, that is for T_F, where $F(e^{it})$ is the boundary value of a rational function $F(z)$. In many cases that will be delineated in the next section, such a T_F can be shown to be similar to an orthogonal direct sum of analytic Toeplitz operators, coanalytic Toeplitz operators and normal operators. Thus results about invariant subspaces and commutant for analytic Toeplitz operators can be generalized to non-analytic ones. Furthermore, Rosenblum's spectral theory for selfadjoint (rational) Toeplitz operators is a special case.

Many of the cases have been treated in earlier papers [4, 2, 3]; the principal new contribution here is the consideration of intersecting loops of the curve γ_F defined above (case IV). Proof will be given here only for the case that has not appeared before.

2. Similarity theory. We list in this section a number of (non-disjoint) cases depending, for the most part, on the nature of the curve $\gamma_F : t \to F(e^{it})$, and we describe the state of the similarity problem in each case. Throughout, $F(z)$ is a rational function, with no poles on $|z| = 1$ (since $F(e^{it}) \in L^\infty$).

Case I. γ_F is a simple closed analytic curve (so that the Riemann mapping function $\tau(z)$ from $|z| < 1$ to the interior of γ_F is analytic and $\tau'(z) \neq 0$ in $|z| \leqslant 1$), the winding number ω of γ_F about each interior point is positive and F never backs up.

The last condition means that the argument of $\tau^{-1} \circ F$ is monotone increasing on $|z| = 1$.

We have in Case I

THEOREM 1. T_F *is similar to*

$$T_\tau \oplus \cdots \oplus T_\tau,$$

the orthogonal direct sum of ω copies of the analytic Toeplitz operator T_τ acting on H^2.

Theorem 1 is proved in [2]. In [6], Duren considered the case where $F(z) = \alpha z + \beta/z$, $|\alpha| > |\beta|$. In this case, γ_F is an ellipse and $\omega = 1$. Duren did not obtain similarity, but proved $T_F = L^{-1}T_{\tau_0}L$, where L is not linear, but preserves invariant subspaces and τ_0 is the mapping function from $|z| < 1$ to the interior of $\bar{\gamma}_F$, the conjugate of γ_F. In [4], Clark and Morrel obtained the conclusion of Theorem 1 under the hypothesis that $\omega > 0$ and F is ω-to-one in an annulus $s \leqslant |z| \leqslant 1$, for some $s < 1$.

Case II. γ_F is a subset of a simple closed curve γ, the winding number of γ_F about any point off γ_F is 0 and the mapping function τ from $|z| < 1$ to the interior of γ is analytically continuable across γ in a neighborhood of γ_F.

In this case, we say F *backs up* at $e^{i\theta}$ if $\arg \tau^{-1} \circ F(e^{it})$ is monotone decreasing at $t = \theta$.

THEOREM 2. *In Case II, T_F is similar to a normal operator V whose spectrum is absolutely continuous (with respect to arc length on γ_F). The multiplicity of a point $\lambda \in \gamma_F$ in the spectrum of V is equal to the number of points $e^{i\theta}$ where F backs up and $F(e^{i\theta}) = \lambda$.*

For the proof, see [2]. The special case where $F(z)$ is the quotient of two "Blaschke factors" of the form $(z - \alpha)(1 - \bar{\alpha}z)^{-1}$, was proved in [1], where similarity (with a unitary operator) was obtained using Sz.-Nagy-Foiaş theory.

Another special case is that in which γ_F is an interval on the real line. In this case (for appropriate choice of γ) F backs up if $F(e^{it})$ is decreasing at $t = \theta$, and the theorem reduces to the result of Rosenblum [11] cited above (by noting that two similar selfadjoint operators are unitarily equivalent).

Case I $\cup$ II. γ_F is a simple closed analytic curve, $\omega > 0$, and $F(\Sigma) \neq \gamma_F$, where Σ is the set where F backs up.

In this case Theorems 1 and 2 can be combined as

THEOREM 3. *T_F is similar to*

$$T_\tau \oplus \cdots \oplus T_\tau \oplus V$$

where there are ω copies of T_τ and where V is a normal operator with (absolutely continuous) spectrum equal to $F(\Sigma)$, and with the spectral multiplicity of V given in Theorem 2.

The proof is given in [2].

Case III. Label the bounded components of the complement of γ_F as ℓ_i, if the winding number W_i of γ_F with respect to an interior point of ℓ_i is negative and l_i if the winding number $\omega_i > 0$. Assume that two of the closures $\mathrm{cl}\,\ell_i, \mathrm{cl}\,l_j$ intersect only at a finite number of points (called the multiple points of F), that the boundaries, $\partial\ell_i$ and ∂l_i, are analytic curves except at the multiple points and that no two arcs of $\partial\ell_i$, ∂l_i intersect tangentially at the multiple points. Assume also that no multiple point has the form $F(e^{i\theta})$ where F backs up near $e^{i\theta}$ or where $F'(e^{i\theta}) = 0$.

Backing up at $e^{i\theta}$, where $F(e^{i\theta})$ is not a multiple point, is defined just as before: if $F(e^{i\theta}) \in \partial l_j$ and if τ_j is the Riemann mapping function from $|z| < 1$ to l_j, then F backs up at $e^{i\theta}$ if $\arg \tau_j^{-1} \circ F(e^{it})$ is decreasing at $t = \theta$; if $F(e^{i\theta}) \in \partial\ell_i$ and if $\mathfrak{I}_i$ is the Riemann mapping function for ℓ_i, then F backs up at $e^{i\theta}$ if $\arg \mathfrak{I}_i^{-1} \circ F(e^{it})$ is increasing at $t = \theta$.

THEOREM 4. *T_F is similar to*

$$\Sigma T_{\tau_i}^{(\omega_i)} \oplus \Sigma T_{\mathfrak{I}_i}^{(W_i)} \oplus V$$

where $T_{\tau_i}^{(\omega_i)}$ denotes the sum of ω_i copies of T_{τ_i}, $T_{\mathfrak{I}_i}^{(W_i)}$ denotes the sum of $-W_i$ copies of $T_{\mathfrak{I}_i}^*$ and where V is a normal operator with spectrum and spectral multiplicity as described in Case II.

The theorem is proved in [3]. Special cases include the classical N-leaved roses of analytic geometry and figure eights. A consequence concerning the structure of T_F is

COROLLARY (Loop Independence). *If M_i [resp. m_i] denotes the closed span of the eigenvectors of T_F [resp. T_F^*] corresponding to $\lambda \in \ell_i$ [resp. $\lambda \in l_i$], then any two subspaces from $\{M_i, m_i\}$ have 0 intersection.*

Case IV. γ_F is the image of $|z| = 1$ under a function $\tau(z)$, analytic in $|z| \leq 1$ with non zero derivative $\tau'(z)$ on $|z| = 1$, $F'(z) \neq 0$, for $|z| = 1$, F is one-to-one on $|z| = 1$, except at finitely many points, and the winding number of γ_F about any point $\lambda \notin \gamma_F$ is non negative.

Conditions on the curve γ_F guaranteeing the existence of such a $\tau(z)$ are known (Marx [10]). They are combinatorial in nature, involving the structure of a "word" associated with γ_F.

THEOREM 5. *In Case IV, T_F is similar to T_τ.*

The proof, which we shall give in §3, consists of lifting the curve γ_F to a simple closed curve on the Riemann surface for τ^{-1} and applying the arguments of [4]. Corresponding to the annulus condition of [4] mentioned above, a slightly more general condition than $F'(z) \neq 0$ for $|z| = 1$ can be substituted in Theorem 5. This will be mentioned in §3.

In contrast with the corollary on Loop Independence which holds in Case III, we have in Case IV,

COROLLARY (Loop Dependence). *If Ξ is any open set lying inside γ_F, then the closed linear span of the eigenvectors of T_F^* corresponding to eigenvalues λ with $\bar{\lambda} \in \Xi$ is all of H^2.*

We close this section by mentioning two cases in which the similarity properties of rational T_F are unknown.

Case I'. γ_F is a simple closed curve having a cusp, F never backs up and the winding number of γ_F about an interior point is > 0.

In certain instances, the results of [4] apply. For example, if $F(z) = z^2 - 2(1 + \epsilon)z - 2\epsilon z^{-1}$, with $0 < \epsilon < \frac{1}{2}$, the annulus hypothesis of [4] holds and T_F is similar to T_τ. Whether T_F is always similar to a direct sum of analytic T_τ in case I' is not known.

Case IV'. The union of the interiors of the loops of γ_F is a connected set, but γ_F is not the image of $\mathbf{T} = \{|z| = 1\}$ under an analytic function.

An example of such a γ_F is shown in figure 1, where the fact that γ_F is not the image of $\mathbf{T}$ under an analytic function follows from Marx' Theorem [10]. Curves of the type pictured may be realized as γ_F for rational F by translations and squarings applied to a rational figure eight. (I am grateful to J. R. Quine for this observation.)

I do not have a conjecture about the structure of the corresponding T_F. Evidently it is not similar to a direct sum of normal operators and analytic and coanalytic Toeplitz operators.

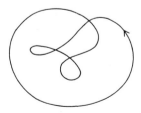

Figure 1.

3. Proof of Theorem 5. We are assuming γ_F is the image of $\mathbf{T} = \{|z| = 1\}$ under a function $\tau(z)$, analytic in the closure of $D = \{|z| < 1\}$. By a theorem of Stoilow [12, p. 121], there is a Riemann surface $\mathfrak{R}$, forming a branched covering of γ_F and its interior, and a lifting τ^* of τ with $\tau^* : D \to \mathfrak{R}$ such that $\tau = r\tau^*$, where r is the projection of $\mathfrak{R}$ into $\mathbb{C}$. Specifically, $\mathfrak{R} = \{(z, \tau(z)) \mid z \in \operatorname{cl} D\}$, equipped with a suitable topological and manifold structure, and $r(w_1, w_2) = w_2$, for $(w_1, w_2) \in \mathfrak{R}$. τ^* induces a homeomorphism of D and its image $\tau^*(D)$ in $\mathfrak{R}$.

We show how to lift F to a map F^* from

$$\Omega = \{z \mid z \in D, F(z) \in \operatorname{int} \gamma_F\}$$

to $\mathfrak{R}$. The map τ is one-to-one on $\mathbf{T}$ except at finitely many points $z_1, \ldots, z_k$. If $z \in \partial\Omega$ and $F(z) \neq \tau(z_1), \ldots, \tau(z_k)$, then there is a unique $z^* \in \tau^*(\mathbf{T})$ such that $z^* = \tau^* \tau^{-1} F(z)$ [unique, since τ^* is one-to-one and τ is one-to-one on $\mathbf{T} \setminus \{z_1, \ldots, z_k\}$]. We define

$$(3.1) \qquad F^*(z) = z^* = \tau^* \tau^{-1} F(z).$$

If z is interior to Ω, connect z to a point $z_0 \in \partial\Omega$ such that

$$(3.2) \qquad F(z_0) \neq \tau(z_1), \ldots, \tau(z_k)$$

by an arc γ such that

$$(3.3) \qquad \begin{aligned} &\gamma \setminus \{z_0\} \subset \operatorname{int} \Omega \\ &\tau' \neq 0 \quad \text{on} \quad \gamma \setminus \{z\}. \end{aligned}$$

The curve $F(\gamma) \setminus \{F(z_0)\}$ lies in $\operatorname{int} \gamma_F$ and there is just one curve $\tau^{-1} F(\gamma)$ in D with endpoint on $\mathbf{T}$ [there is just one possibility for the endpoint $(\tau^{-1} F(z_0))$ and hence just one such curve since $\tau' \neq 0$ on $\gamma \setminus \{z\}$]. Now let w be the endpoint of the curve $\tau^{-1} F(\gamma)$ lying inside D and define $F^*(z) = \tau^*(w)$ [so that (3.1) is again satisfied]. It follows immediately from (3.1) that

$$rF^*(z) = r\tau^* \tau^{-1} F(z) = \tau \tau^{-1} F(z) = F(z).$$

$F^*(z)$ is analytic in Ω and is defined at every z of modulus 1 except z such that (3.2) holds. We can define $F^*(z_j), j = 1, \ldots, k$ by continuity.

Since $F'(z) \neq 0$ for $|z| = 1$, and $F^*(z)$ is one-to-one on $\mathbf{T}$, we have that

(3.4) $F^*(z)$ is defined and 1 - to - 1 in an annulus $r \leqslant |z| \leqslant 1$.

Condition (3.4) is the more general hypothesis on $F(z)$ alluded to after the statement of Theorem 5.

We claim that there is a positive integer s, such that $F^*(z)$ is s-to-1 from Ω to $\mathscr{R}$, except at those points where $F' = 0$. In fact, we shall prove

LEMMA 1.   *If w_1, w_2 are two points in $\tau^*(D)$ such that $F'(z) \neq 0$ for every z such that $F^*(z) = w_1$ or w_2, and if, in a neighborhood of w_1, there are s analytic functions $d_1^*(w), \ldots, d_s^*(w)$ which, for $i = 1, \ldots, s$, satisfy*

(3.5) $|d_i^*(w)| < 1, F^*(d_i^*(w)) = w$

then analytic continuation of the d_i^* along a suitable curve γ^* from w_1 to w_2 yields s distinct functions (again denoted $d_1^*, \ldots, d_s^*$) which satisfy (3.5) in a neighborhood of w_2.

First of all, the lemma implies the claim. Secondly, it shows that the image of Ω under F is equal to all of $\tau^*(D)$.

Proof.   Let γ^* lie interior to $\tau^*(\mathbf{T})$ and let $F'(z) \neq 0$ whenever $F^*(z) \in \gamma^*$. We claim that analytic continuation of any two of the d_i^* (say d_1^* and d_2^*) along γ^* from w_1 to w_2 yields distinct functions which satisfy (3.5) in a neighborhood of w_2. Indeed, $d_1^*(w)$ and $d_2^*(w)$ must be distinct for every $w \in \gamma^*$; they surely must continue to satisfy $F^*(d_i^*(w)) = w$, $i = 1, 2$; and $|d_i^*(w)| < 1$ holds for all $w \in \gamma^*$: for if $|d_i^*(w_0)| = 1$ at some $w_0 \in \gamma^*$, then $F^*(d_i^*(w_0)) = w_0 \in \tau^*(\mathbf{T})$ contradicting the construction of γ^*.

The functions $d_1^*, \ldots, d_s^*$, which, in a neighborhood of any point of $\tau^*(D)$, satisfy (3.5), can be continued analytically to each point of $F^*(\Omega)$ except, possibly, those points

(3.6) $w = F^*(z)$ where $F'(z) = 0$,

but they cannot, in general, be defined as single-valued functions, since continuation around points w satisfying (3.6) may permute the d_i^*. However, if $\jmath(x_1, \ldots, x_s)$ is a symmetric function of $x_1, \ldots, x_s$, then $\jmath(d_1^*(w), \ldots, d_s^*(w))$ is single-valued and analytic in $\tau^*(D)$. The $d_i^*(w)$

have the further property that

$$rw = rF^*(d_i^*(w)) = F(d_i^*(w)).$$

Thus, locally, the map $rw \to d_i^*(w)$ is the same as $\lambda \to d_j(\lambda)$, where $d_j(\lambda)$ is a function satisying

(3.7) $\lambda = F(d_j(\lambda)), \lambda \in \tau(D)$;

that is, $d_i^*(w) = d_j(rw)$. In particular, if $\,{}_{s}(x_1, \ldots, x_s)$ is a symmetric function, $\,{}_{s}(d_1^*(w), \ldots, d_s^*(w)) = \,{}_{s}(d_1(rw), \ldots, d_s(rw))$, for a suitable set of $d_1, \ldots, d_s$ satisfying (3.7).

We have proved

LEMMA 2.   *There is a set $d_1, \ldots, d_s$ of functions satisfying (3.7) such that, if $\,{}_{s}(x_1, \ldots, x_s)$ is symmetric in $x_1, \ldots, x_s$, then $\,{}_{s}(d_1(rw), \ldots, d_s(rw))$ is analytic for $w \in \tau^*(D)$. Thus*

$$\,{}_{s}(d_1(r\tau^*(z)), \ldots, d_s(r\tau^*(z))) = \,{}_{s}(d_1(\tau(z)), \ldots, d_s(\tau(z)))$$

is analytic in $|z| < 1$.

Now we proceed to the construction of the similarity operator L satisfying

(3.8) $LT_F = T_\tau L.$

Let $f(z) = p(z)/q(z)$ be the rational function which generates the adjoint of $T_F : f(z) = \bar{F}(\bar{z}^{-1})$. Write

$$q(z) = \prod_1^m (z - \gamma_i) \prod_1^n (z - \delta_1) \qquad |\gamma_i| < 1, |\delta_i| > 1$$

and define, for $x \in H^2$ and $\zeta \in D$,

$$(Lx)(\zeta) = \left(x(z), \prod(1 - \delta_i^{-1}z) / \prod_1^s [1 - \bar{d}_i(\tau(\zeta))z] \right),$$

where $d_1, \ldots, d_s$ are defined in Lemma 2.

The rest of the proof closely follows the one given in [4] and will only be sketched. (Note that the notation for the functions d_i satisfying (3.7) differs here from that of [4]. The functions $d_i(\lambda)$ in [4] are given by $\bar{d}_i(\bar{\lambda})$ in our present construction.)

Let $k_\lambda(z)$ denote the function

$$k_\lambda(z) = \prod(1 - \delta_i^{-1}z) / \prod_1^s \left[1 - \bar{d}_i(\bar{\lambda})z\right].$$

From among the $d_i^*(w)$, we can select one, which we label d_1^*, such that $|d_1^*(w)| = 1$ on $\tau^*(\mathbf{T})$. Renumber the d_i, if necessary, so that $d_1(rw) = d_1^*(w)$, at least for w in the image, under F^*, of the annulus (3.4), where $d_1^*(w) = F^{*-1}(w)$ is single-valued. By partial fractions, we can write

$$(3.9) \qquad k_\lambda(z) = c(\lambda)\left(1 - \bar{d}_1(\bar{\lambda})z\right)^{-1} + B(\lambda, z).$$

As in [4], using (3.4), we see that $B(\lambda, z)$ is continuous in $f(A) \times \operatorname{cl} D$, where $A = \{1 \leqslant |z| < r^{-1}\}$. Thus if $p(z)$ is a polynomial, we have that

$$(Lp)(\zeta) = \left(p(z), k_{\bar{\tau}(\zeta)}(z)\right)$$

$$= \bar{c}(\bar{\tau}(\zeta))p(d_1(\tau(\zeta))) + \left(p(z), B(\bar{\tau}(\zeta), z)\right),$$

so that the image of p under L is analytic in $|\zeta| < 1$ (by Lemma 2) and continuous in $|\zeta| \leqslant 1$. Thus L maps polynomials into H^2.

Again following [4], we have that

$$c_1\|p\| \leqslant \|\bar{c}(\bar{\tau}(\zeta))p(d_1(\tau(\zeta)))\| \leqslant c_2\|p\|,$$

and that $p \to (p, B(\bar{\tau}(\zeta), z))$ is Hilbert-Schmidt class. Therefore, L has a unique extension as a bounded operator from H^2 into H^2 and L is a Fredholm operator.

The relation (3.8) follows from the fact that k_λ is an eigenvector of $T_f = T_F^*$, with eigenvalue λ. Indeed,

$$LT_F(x) = \left(T_F x, k_{\bar{\tau}(\zeta)}\right) = \left(x, T_f k_{\bar{\tau}(\zeta)}\right) = \tau(\zeta)Lx.$$

Thus, all that remains to be proved is that L is one-to-one and has index 0.

To prove 1-to-1ness, note that the kernel K of L is the orthogonal complement of the $\{k_\lambda\}$, and hence it is invariant under T_F, and, since L is Fredholm, K is finite dimensional. Therefore K must contain an eigenvector of T_F. By [4, Lemma 2.2], T_F has no eigenvalues in the boundary of its spectrum. By an application of [4, Lemma 2.1], we can show that T_F has no eigenvalues in the set $F(\mathbf{T})$ other than, possibly, some coming from the span of the $\{k_\lambda\}$. Since all the eigenvectors for eigenvalues off $F(\mathbf{T})$ lie in the span of the $\{k_\lambda\}$, we have that $K = \{0\}$ and L is one-to-one.

The proof that L has index 0 is identical to the corresponding proof in [4, §6], with appropriate change of notation.

This proves Theorem 5.

UNIVERSITY OF VIRGINIA
UNIVERSITY OF GEORGIA

REFERENCES

[1] D. N. Clark, Sz.-Nagy-Foiaş theory and similarity for a class of Toeplitz operators, to appear in proceedings of the 1977 Spectral Theory Semester, Banach Center Publications, Warsaw.

[2] ———, On a similarity theory for rational Toeplitz operators, *J. Reine Angew. Math.* **320** (1980), 6–31.

[3] ———, On Toeplitz operators with loops, *J. Operator Theory* **4** (1980), 37–54.

[4] D. N. Clark and J. H. Morrel, On Toeplitz operators and similarity, *Amer. J. Math.* **100** (1978), 973–986.

[5] R. G. Douglas, *Banach Algebra Techniques in Operator Theory*, Academic Press, New York, 1972.

[6] P. L. Duren, Extension of a result of Beurling on invariant subspaces, *Trans. Amer. Math. Soc.* **99** (1961), 320–324.

[7] P. Hartman and A. Wintner, On the spectra of Toeplitz's matrices, *Amer. J. Math.* **72** (1950), 359–366.

[8] ———, The spectra of Toeplitz's matrices, *ibid.* **76** (1954), 867–882.

[9] M. G. Krein, Integral equations on a half line with difference kernel, *Uspehi Mat. Nauk* (N.S.) **13** (1958) No. 5 (83), 3–120 (Russian) = *Amer. Math. Soc. Transl.* (2) **22** (1962), 163–288.

[10] M. L. Marx, Extensions of normal immersions of S^1 into R^2, *Trans. Amer. Math. Soc.* **187** (1974), 309–326.

[11] M. Rosenblum, A concrete spectral theory for self-adjoint Toeplitz operators, *Amer. J. Math.* **87** (1965), 709–718.

[12] S. Stoilow, *Lecons sur les principes topologiques de la theorie des fonctions analytiques*, Gauthier-Villars, Paris, 1938.

THE REPRODUCING KERNEL SPACE OF THE CLAMPED PLATE*

By C. V. COFFMAN

---

1. Introduction. The Green's function for a sufficiently regular positive definite elliptic boundary value problem of order exceeding the dimension of the space can generally be represented as a reproducing kernel. In particular, this is the case for the problem of the "clamped plate" i.e. for the biharmonic problem

$$(1.1) \qquad \Delta^2 u = f \quad \text{in } \Omega, \qquad u = \frac{\partial u}{\partial n} = 0 \quad \text{on } \partial\Omega,$$

on a plane region Ω.

This natural approach to the study of (1.1) leads to unification, simplification and generalization of many of the "classical" results concerning (1.1) and/or its Green's function. With the exception of a few brief discussions, see, e.g., [4], [9], there seems to be no such analysis in the literature. It is the purpose of this note to provide such a treatment.

For the general theory of reproducing kernel spaces the reader is referred to [1] or [13]. We recall however, that a real *reproducing kernel space* or *proper functional Hilbert space* is a Hilbert space X of real valued functions u on a basic set S such that for each $x \in S$

$$u \to u(x)$$

is a continuous linear functional on X. The *reproducing kernel* $k(x, y)$ is characterized as follows: for each $x \in S, k(x, \cdot) \in X$ and for any $u \in X$,

$$u(x) = (u, k(x, \cdot)),$$

where $(\cdot, \cdot)$ is the inner product on X.

---

*Supported by National Science Foundation Grant MCS 77-03643

The space of the clamped plate, i.e. the space associated with (1.1), is the proper functional completion, when it exists, of $C_0^\infty(\Omega)$ with respect to the norm

$$(1.2) \qquad \|u\| = \left(\int \int_\Omega (\Delta u)^2 \, dx \, dy \right)^{1/2}.$$

Naturally some restriction on Ω is necessary in order that this proper functional completion should exist, however the necessary restriction is surprisingly mild. The construction could be based on the general theory developed in [2], but special features of this problem make possible the somewhat simpler treatment which follows. When the plane region Ω is such that $C_0^\infty(\Omega)$ has a proper functional completion with respect to the norm defined by (1.2) we shall denote the resulting space by $X(\Omega)$; for brevity we shall say that $X(\Omega)$ exists in this case and $X(\Omega)$ does not exist otherwise.

2. An existence criterion for X(Ω). We shall use the following notations and conventions: z and ζ denote complex variables with $z = x + iy$, $\zeta = \xi + i\eta$; $C_0^\infty(\Omega)$ is the space of real infinitely differentiable functions with compact support in Ω and these functions are considered to be defined on the entire plane $\mathcal{C}$; $M(\Omega)$ denotes the $L^2(\Omega)$-closure of

$$\{ f : f = (\Delta u) | \Omega, u \in C_0^\infty(\Omega) \}.$$

As with $C_0^\infty(\Omega)$, $X(\Omega)$, when it exists, will be regarded as consisting of functions which are defined on all of $\mathcal{C}$ and vanish outside of Ω.

Suppose that Ω is bounded and that $G(z, \zeta)$ is the harmonic Green's function of Ω or some bounded region that contains it. Then $X(\Omega)$ can be constructed directly, [4], as follows: $X(\Omega)$ consists of those functions representable in the form

$$u(z) = \int \int_\Omega G(z, \zeta) f(\zeta) \, d\xi \, d\eta, \qquad z \in \Omega,$$

with $f \in M(\Omega)$ and is normed by

$$\|u\| = \left(\int \int_\Omega |f|^2 \, dx \, dy \right)^{1/2}.$$

By the results of section 5 we can reduce any case in which $\overline{\Omega} \neq \mathcal{C}$ to the case of a bounded region. Thus the only non-trivial case is that of a region Ω whose closure is the entire plane. We shall now develop an existence criterion which will aid in the discussion of that case.

For $z \in \mathcal{C}$ we put

$$m(z) = \sup\{u(z) : u \in C_0^\infty(\Omega), \|u\| \leqslant 1\}$$

where $\|u\|$ is defined by (1.2). Of course $m(z) = 0$ if $z \notin \Omega$; note also that $m(z)$ is lower semi-continuous.

LEMMA 2.1.   *If $m(z)$ is finite on a set of positive measure in $\mathcal{C}$ then it is finite everywhere and uniformly bounded on compact sets.*

Proof.   Put

$$S_n = \{z : m(z) \leqslant n\}, \qquad n = 1, 2, \ldots,$$

then S_n is closed, hence measurable,

$$\bigcup_{n=1}^\infty S_n$$

has positive measure, hence S_n has positive measure when n is sufficiently large. If

$$D_R = \{z : |z| < R\}$$

then it follows by a standard argument (cf. e.g. the proof of Theorem 3.6.4, p. 82, [14]) that for any set $\sigma \subset D_R$ of positive measure

$$\left(\int\int_\sigma |u|^2 \, dx \, dy + \int\int_{D_R} \left(u_{xx}^2 + 2u_{xy}^2 + u_{yy}^2\right) dx \, dy\right)^{1/2}$$

is a norm on $H^{2,2}(D_R)$ equivalent to the standard norm on that space

$$\|u\|_{2,2}^{D_R} = \left(\int\int_{D_R} \left(u^2 + u_x^2 + u_y^2 + u_{xx}^2 + u_{xy}^2 + u_{yy}^2\right) dx \, dy\right)^{1/2}.$$

When R is sufficiently large, we can choose σ of positive measure such that $m(z)$ is bounded on σ, thus it follows that there exists a constant K_R

such that for any $u \in C_0^\infty(\Omega)$,

$$\|u \mid D_R\|_{2,2}^{D_R} \leqslant K_R \left(\int \int_\Omega (\Delta u)^2 \, dx \, dy \right)^{1/2}.$$

Since $H^{2,2}(D_R)$ embeds continuously in $C(\overline{D}_R)$, and since $R > 0$ is essentially arbitrary, the result follows.

PROPOSITION 2.2. *If for each $z \in \mathcal{C}$ (or merely in some set of positive measure) the linear functional*

$$u \rightarrow u(z)$$

is bounded over $C_0^\infty(\Omega)$ with respect to the norm $\| \cdot \|$ then $C_0^\infty(\Omega)$ has a proper functional completion with respect to that norm.

Proof. Let $\{u_n\}$ be a $\| \cdot \|$-Cauchy sequence in $C_0^\infty(\Omega)$ and let f be the $L^2(\Omega)$-limit of the sequence $\{\Delta u_n \mid \Omega\}$. By Lemma 2.1 there exists a continuous function u on $\mathcal{C}$ such that

$$u(z) = \lim_{n \to \infty} u_n(z)$$

uniformly on compact subsets of $\mathcal{C}$. In view of this uniformity it follows that f is the distribution Laplacian of u on Ω, i.e. for every $\varphi \in C_0^\infty(\Omega)$

$$\int \int_\Omega u \Delta \varphi \, dx \, dy = \int \int_\Omega \varphi f \, dx \, dy.$$

If we take $X(\Omega)$ to be the set of functions u which are the limits of such sequences and norm $X(\Omega)$ by (1.2), where Δu denotes the distribution Laplacian, then $X(\Omega)$ is complete and hence is a proper functional Hilbert space.

The sufficient condition of the above proposition is obviously necessary, thus, from the proof of the proposition we have the following.

COROLLARY 2.3. *If $X(\Omega)$ exists then its elements are continuous real valued functions on $\mathcal{C}$ which vanish outside of Ω and have distribution Laplacians over Ω which belong to $L^2(\Omega)$. The norm of an element in $X(\Omega)$ is the $L^2(\Omega)$-norm of its distribution Laplacian. Norm convergence in $X(\Omega)$ implies uniform convergence on compact sets. Finally, if Ω' is a subregion of Ω then $X(\Omega')$ exists and is a subspace of $X(\Omega)$.*

3. Regions for which X(Ω) exists. It is immediate from Proposition 2.2 that $X(\Omega)$ exists whenever $\overline{\Omega} \neq \mathcal{C}$ since in such a case there is a set of positive measure on which $m(z) = 0$. Regarding the general case, the hypothesis of Proposition 2.2 can be put in more concrete form as follows. If $z \in \Omega$ is fixed and

$$(3.1) \qquad \gamma(z,\zeta) = -\frac{1}{2\pi}\log|z - \zeta| + v(\zeta)$$

where v is harmonic on Ω then for $u \in C_0^\infty(\Omega)$

$$u(z) = \int\int_\Omega \gamma(z,\zeta)\Delta u(\zeta)\,d\xi\,d\eta.$$

Thus, in order for the hypothesis of Proposition 2.2 to hold it suffices that for each $z \in \Omega$ there exists a $\gamma(z, \circ) \in L^2(\Omega)$ having the form (3.1) with v harmonic on Ω. As we shall see, this is also necessary.

If $\mathcal{C}\backslash\Omega$ has a connected component C_0 which is not a singleton then $\gamma(z,\zeta)$ can be constructed as follows. Let $a,b \in C_0, a \neq b$ and for $z \in \Omega$ take

$$\gamma(z,\zeta) = \frac{1}{2\pi}\,\mathrm{Re}\left[\log\frac{\zeta - a}{\zeta - z} - \frac{z - a}{\sqrt{(\zeta - a)(\zeta - b)}}\right],$$

where $1/\sqrt{(\zeta - a)(\zeta - b)}$ denotes the branch of that function whose Laurent expansion at ∞ begins with $1/\zeta$: this is single-valued on $\Omega\backslash C_0$. On a neighborhood of infinity we take that branch of $\log(\zeta - a/\zeta - z)$ whose development begins with $(z - a)/\zeta$, the real part of this function has a single valued harmonic extension to Ω. With the implied interpretation of the expression (3.2) we see easily that $\gamma(z, \circ) \in L^2(\Omega)$. Thus $X(\Omega)$ exists in this case.

Suppose now that $\mathcal{C}\backslash\Omega$ is finite, say

$$\mathcal{C}\backslash\Omega = \{z_1, \ldots, z_n\}$$

where $z_1, \ldots, z_n$ are distinct. If $\gamma(z, \circ) \in L^2(\Omega)$ and has the form (3.1) with v harmonic on $L^2(\Omega)$ then it follows readily from Liouville's theorem that v must be of the form

$$(3.3) \qquad v(\zeta) = \sum_{n=1}^{N} a_n \log|\zeta - z_n|.$$

It is an elementary exercise to show that the a_n can be chosen so that $\gamma(z, \cdot) \in L^2(\Omega)$ when (3.1), (3.3) hold provided $\{z_1, \ldots, z_n\}$ does not lie on a line; when this set does lie on a line the a_n cannot be so chosen unless z is on the same line. In summary we have the following.

PROPOSITION 3.1.   *The space $X(\Omega)$ exists when $\mathcal{C}\backslash\Omega$ has a connected component which is not a singleton or when $\mathcal{C}\backslash\Omega$ does not lie on a line.*

We have used implicitly the fact that if $X(\Omega)$ exists then $X(\Omega')$ exists for any region Ω' such that $\Omega' \subset \Omega$.

Remark.   In connection with the first condition of Proposition 3.1 we note that Hadamard, [10], had exhibited the existence of a biharmonic Green's function on the complement of a half-line. He did this by exhausting the region by an increasing sequence of cardioids and showing that the Green's functions of these cardioids, with the singularity fixed, were uniformly bounded on any compact set. The biharmonic Green's function of a cardioid can be computed explicitly since the Riemann mapping function is rational, [11].

4. Properties of the reproducing kernel.   When $X(\Omega)$ exists we shall denote its reproducing kernel by $\Gamma_\Omega(z, \zeta)$. The fundamental formula that characterizes Γ_Ω is

$$(4.1) \qquad u(z) = (u, \Gamma_\Omega(z, \cdot)),$$

$$= \int\int_\Omega \Delta u(\zeta) \Delta_\zeta \Gamma_\Omega(z, \zeta) \, d\xi \, d\eta.$$

A number of fundamental relations for Γ_Ω can be deduced immediately from the defining property (4.1), these are tabulated below.

PROPOSITION 4.1.   (i) *The kernel Γ_Ω is symmetric i.e.*

$$\Gamma_\Omega(z, \zeta) = \Gamma_\Omega(\zeta, z),$$

(ii) *I_Ω satisfies*

$$(4.2) \qquad \Gamma_\Omega(z, z) = \|\Gamma_\Omega(z, \cdot)\|^2,$$

(iii) *If $\Omega' \subset \Omega$ then for $z, \zeta \in \mathcal{C}$,*

$$(4.3) \qquad \Gamma_{\Omega'}(z,z) \leqslant \Gamma_\Omega(z,z),$$

and

$$(4.4) \qquad |\Gamma_\Omega(z,\zeta) - \Gamma_{\Omega'}(z,\zeta)|^2 \leqslant (\Gamma_\Omega(z,z) - \Gamma_{\Omega'}(z,z))$$
$$\times (\Gamma_\Omega(\zeta,\zeta) - \Gamma_{\Omega'}(\zeta,\zeta)),$$

(iv) *If $f \in C_0^\infty(\Omega)$ and $u \in X(\Omega)$ is a weak solution of (1.1) then*

$$(4.5) \qquad u(z) = \int\int_\Omega \Gamma_\Omega(z,\zeta) f(\zeta)\, d\xi\, d\eta.$$

Proof. The symmetry and (4.2) follow by taking $u = \Gamma_\Omega(\zeta, \cdot)$ in (4.1), using the symmetry of the inner product, and then taking $\zeta = z$. When $\Omega' \subset \Omega$ we take $u = \Gamma_{\Omega'}(z, \cdot)$ in (4.1) and apply the Schwarz inequality to get (4.3). We use the relation

$$\Gamma_\Omega(z,\zeta) - \Gamma_{\Omega'}(z,\zeta) = (\Gamma_\Omega(z,\cdot) - \Gamma_{\Omega'}(z,\cdot), \Gamma_\Omega(\zeta,\cdot) - \Gamma_{\Omega'}(\zeta,\cdot))$$

to obtain (4.4).

By definition a weak solution u of (1.1) belongs to $X(\Omega)$ and satisfies

$$(u,v) = \int\int_\Omega \Delta u\, \Delta v\, dx\, dy = \int\int_\Omega v\, f\, dx\, dy$$

for all $v \in C_0^\infty(\Omega)$ and hence for all $v \in X(\Omega)$, taking $v = \Gamma_\Omega(z, \cdot)$ yields (4.5).

COROLLARY 4.2. *If $X(\Omega)$ exists and $\Omega_1 \subset \Omega_2 \subset \cdots$ is an increasing sequence of subregions of Ω with*

$$\Omega = \bigcup_{n=1}^\infty \Omega_n$$

then

$$\Gamma_\Omega(z,\zeta) = \lim_{n\to\infty} \Gamma_{\Omega_n}(z,\zeta)$$

uniformly on compact subsets of $\mathcal{C} \times \mathcal{C}$.

Proof. It follows from (4.1) and (4.2) that

$$(\Gamma_\Omega(z,z))^{1/2} = \max\{u(z) : u \in X(\Omega), \|u\| \leqslant 1\}$$

$$= \sup\{u(z) : u \in C_0^\infty(\Omega), \|u\| \leqslant 1\}$$

$$= \lim_{n\to\infty} \sup\{u(z) : u \in C_0^\infty(\Omega_n), \|u\| \leqslant 1\}$$

$$= \lim_{n\to\infty} (\Gamma_{\Omega_n}(z,z))^{1/2}.$$

The result then follows immediately from (4.3) and (4.4).

Remarks. 1. The assertions (ii) and (iii) of Proposition 4.1 were proved for the biharmonic Green's function by Hadamard, [10], the proofs above however are simpler and are valid for more general regions.

2. For results relating to Corollary 4.2 see Babuška, [5].

5. The effect of Möbius transformation.

Let Ω be a region in $\mathcal{C}$, $f(z)$ a Möbius transformation with pole in $\mathcal{C}\backslash\Omega$ and $\Omega' = f(\Omega)$. Then if $u(\zeta)$ is biharmonic on Ω',

$$(5.1) \qquad v(z) = |f'(z)|^{-1}u(f(z))$$

will be biharmonic on Ω, (to see this it suffices to consider $f(z) = 1/z$ and use the representation

$$u(\zeta) = |\zeta|^2 v(\zeta) + w(\zeta),$$

v, w harmonic, for biharmonic functions). This fact was used by Loewner, [11], to relate the biharmonic Green's functions of Ω and Ω'.

To obtain the same result using the reproducing kernel characterization of the Green's function we observe that if $u \in C_0^\infty(\Omega')$ and if v is defined by (5.1) for $z \in \Omega$ and set equal to zero outside Ω then $v \in C_0^\infty(\Omega)$ and

$$(5.2) \qquad \Delta_z^2 v(z) = |f'(z)|^3 \Delta_\zeta^2 u(f(z)).$$

Using (5. 2) and integrating by parts we see that the one-to-one correspondence $u \rightleftarrows v$ given by (5.1) is an isometry with respect to $\|\cdot\|$. It follows immediately that $X(\Omega)$ and $X(\Omega')$ exist or fail to exist together and when they exist (5.1) extends to an isometry between them. Loewner's formula,

$$(5.3) \qquad \Gamma_\Omega(z, z') = |f'(z)|^{-1}|f'(z')|^{-1}\Gamma_{\Omega'}(f(z), f(z'))$$

follows immediately from (4.1) and (5.1).

The formula (5.3) yields a simple computation of Γ_Ω when Ω is the unit disk, $\Omega = \{z : |z| < 1\}$. In view of the uniqueness of the Green's function and Proposition 4.1, (iv) we can use the "classical" characterization [8], [9]. One can then write immediately

$$(5.4) \qquad \Gamma_\Omega(0, \zeta) = \frac{1}{8\pi}(|\zeta|^2\log|\zeta| + \frac{1}{2}(1 - |\zeta|^2), \qquad |\zeta| < 1,$$

and then from (5.3)

$$(5.5) \qquad \Gamma_\Omega(z, \zeta) = |1 - \bar{z}\zeta|^2\Gamma_\Omega\left(0, \frac{\zeta - z}{1 - \bar{z}\zeta}\right).$$

It follows immediately from (5.4) and (5.5) that $\Gamma_\Omega(z, \zeta) > 0$ for $|z|$, $|\zeta| < 1$.

Remark. These considerations enable one to define $X(\Omega)$ and/or the biharmonic Green's function Γ_Ω when Ω is a subset of the extended plane. Some advantage accrues from throwing the point of singularity of Γ_Ω to ∞ since the fundamental singularity at ∞ is harmonic, cf. Loewner, [11].

6. Zaremba's formula. S. Zaremba, [15], showed that the computation of the biharmonic Green's function can be reduced to the computation of the reproducing kernel $h_\Omega(z, \zeta)$ of the space $H(\Omega)$ of functions which are harmonic over Ω and belong to $L^2(\Omega)$. His formula is in fact a direct consequence of the orthogonal decomposition

$$(6.1) \qquad L^2(\Omega) = M(\Omega) \oplus H(\Omega)$$

which follows in turn from Weyl's lemma, [Th. 2.3.1, p. 42, 14].

For a point $z \in \Omega$ we denote

$$\Omega_z = \Omega\backslash\{z\}.$$

It is clear from (4.1) that $\Delta_\zeta \Gamma_\Omega(z, \zeta)$ is orthogonal to $M(\Omega_z)$ and thus in view of (6.1), with Ω replaced by $\Omega_z, \Delta_\zeta \Gamma_\Omega(z, \cdot)$ must be harmonic over Ω_z. By a second appeal to (4.1) we easily see that $\Delta_\zeta \Gamma_\Omega(z, \circ)$ must have the form (3.1) with v harmonic on Ω. On the other hand, $\Delta_\zeta I_\Omega(z, \cdot) \in M(\Omega)$ and since

$$f \to \int \int_\Omega h_\Omega(z, \zeta) f(\zeta) \, d\xi \, d\eta$$

is the orthogonal projection of $L^2(\Omega)$ onto $H(\Omega)$ it follows that if $\gamma(z, \cdot) \in L^2(\Omega)$ and has the form (3.1) with v harmonic on Ω then

(6.2) $\qquad \Delta_\zeta \Gamma_\Omega(z, \zeta) = \gamma(z, \zeta) - \int \int_\Omega h_\Omega(\zeta, \zeta') \gamma(z, \zeta') \, d\xi' \, d\eta'.$

If $\bar{\gamma}(z, \cdot)$ is a function of the same form as γ then

(6.3) $\qquad \Gamma_\Omega(z, \zeta) = \int \int_\Omega \bar{\gamma}(\zeta, \zeta') \Delta_\zeta \Gamma_\Omega(z, \zeta') \, d\xi' \, d\eta'.$

If Ω is bounded and has the harmonic Green's function $G(z, \zeta)$ we combine (6.2) and (6.3) and take $\bar{\gamma} = \gamma = G$ to obtain the formula of Zaremba. It is worth noting however that, contrary to the impression given by the latter usual formulation, explicit knowledge of G is not necessary for the computation of Γ_Ω when h_Ω is known.

7. Rectangular regions.

The intractability of the biharmonic problem on a rectangle has long been recognized and was given special notice by the French Academy in setting the Prix Vaillant, [6].

The most satisfactory theoretical resolution of this problem to date is that of Aronszajn, Brown and Butcher, [3]. Their solution consists in computing the kernel h_Ω for the region Ω and then using Zaremba's formula to obtain Γ_Ω. Suppose that

$$\Omega = \{ z = x + iy : 0 < x < a, 0 < y < b \},$$

the fundamental lemma of [3] is as follows.

PROPOSITION 7.1. *If $u \in H(\Omega)$ then*

(7.1) $\qquad u = v + w$

where $v, w \in H(\Omega)$ have the respective forms

$$(7.2) \qquad v(z) = \sum_{n=1}^{\infty} \left(A_n \sinh \frac{n\pi y}{b} + B_n \cosh \frac{n\pi y}{b} \right) \sin \frac{n\pi x}{a}$$

$$(7.3) \qquad w(z) = \sum_{n=1}^{\infty} \left(C_n \sinh \frac{n\pi x}{a} + D_n \cosh \frac{n\pi x}{a} \right) \sin \frac{n\pi y}{b}.$$

(Compare the attempt of Lauricella, [12], submitted to the French Academy in response to [6], see also [7].)

The functions of the forms (7.2) and (7.3) respectively span closed subspaces of $H(\Omega)$, the kernels of which are easily computed. Given the general validity of the representation (7.1), the kernel of $H(\Omega)$ can be expressed as an infinite series in terms of the former two kernels.

Let $H_1(\Omega)$ and $H_2(\Omega)$ denote the subspaces of $H(\Omega)$ that consist of functions of the forms (7.2) and (7.3) respectively. It is easily seen, [3], that

$$(7.4) \qquad H_1(\Omega) \cap H_2(\Omega) = \{0\}$$

and that $H_1(\Omega) + H_2(\Omega)$ is dense in $H(\Omega)$, however a substantial part of [3] is devoted to a proof of the equality

$$(7.5) \qquad H(\Omega) = H_1(\Omega) + H_2(\Omega),$$

which is absolutely essential to the solution outlined above. We shall show here that this equality is implied by the existence theorem for

$$(7.6) \qquad \Delta^2 \varphi = \rho \qquad \text{on} \quad \Omega$$

$$(7.7) \qquad \varphi = f, \qquad \frac{\partial \varphi}{\partial n} = g \qquad \text{on} \quad \partial\Omega$$

on a rectangle Ω.

Let $B(\Omega)$ denote the Hilbert space of functions which are biharmonic with square integrable Laplacian on Ω and vanish on $\partial\Omega$; $B(\Omega)$ is furnished with the inner product

$$\langle \varphi, \psi \rangle = \int\int_{\Omega} \Delta\varphi \Delta\psi \, dx \, dy.$$

(The reproducing kernels of $B(\Omega)$ and $X(\Omega)$ sum to

$$\int\int_\Omega G(z,\zeta')G(\zeta',\zeta)\,d\xi'\,d\eta',$$

where G is the harmonic Green's function of Ω.)

The existence theorem for (7.6), (7.7), [3], specialized to the case $\rho = 0$, $f = 0$, gives a necessary and sufficient condition for a function g on $\partial\Omega$ to be the normal derivative of a function $\varphi \in B(\Omega)$. Without quoting this condition explicitly (another formulation will be given below) we note that it involves no compatibility requirement between the restrictions of g to the four individual segments of $\partial\Omega$. Thus, in particular, if $\varphi \in B(\Omega)$ then there exists a $\psi \in B(\Omega)$ whose normal derivative, $\partial\psi/\partial n$, vanishes on the vertical segments of $\partial\Omega$ and coincides with $\partial\varphi/\partial n$ on the horizontal segments of $\partial\Omega$. It follows from the closed graph theorem that

$$P : \varphi \to \psi,$$

is a bounded (but not orthogonal) projection on $B(\Omega)$.

Let $u \in H(\Omega)$, $\varphi \in B(\Omega)$, then by Green's theorem,

$$(7.8)\qquad \int\int_\Omega u\Delta\varphi\,dx\,dy = \int_{\partial\Omega} u\,\frac{\partial\varphi}{\partial n}\,ds.$$

If u has the representation (7.1), (7.2), (7.3) and $\psi = P\varphi$ then it follows from (7.8) that

$$(7.9)\qquad \int\int_\Omega u\Delta\psi\,dx\,dy = \int\int_\Omega v\Delta\varphi\,dx\,dy$$

If we put

$$\varphi(z) = \int\int_\Omega G(z,\zeta)v(\zeta)\,dx\,dy,$$

then (7.9) implies

$$(7.10)\qquad \int\int_\Omega |v|^2\,dx\,dy \leqslant \|P\|^2 \int\int_\Omega |u|^2\,dx\,dy$$

where $\|P\|$ is the norm of the projection P. Because of (7.4) and the density of $H_1(\Omega) + H_2(\Omega)$ in $H(\Omega)$, the equality (7.5) follows from the inequality (7.10).

Let g be a function on $\partial\Omega$ and suppose that on the segment of $\partial\Omega$ where $y = 0$

$$g(x,0) = \sum_{n=1}^{\infty} a_n \sin \frac{n\pi x}{a} , \qquad \sum_{n=1}^{\infty} na_n^2 < \infty$$

and g has similar representations on the remaining segments. It is easily seen (by expanding $\varphi \in B(\Omega)$ in a double sine series) that this is necessary in order that $g = \partial\varphi/\partial n$ on $\partial\Omega$, $\varphi \in B(\Omega)$. The above analysis shows that this condition is also sufficient. Equivalently, the boundary values of the functions $u \in H(\Omega)$ are precisely those generalized functions with the representation

$$\sum_{n=1}^{\infty} b_n \sin \frac{n\pi x}{a} , \qquad \sum_{n=1}^{\infty} n^{-1}b_n^2 < \infty$$

on the segment where $y = 0$ and similar representations on the remaining segments. This latter assertion is equivalent to Proposition 7.1.

CARNEGIE-MELLON UNIVERSITY

REFERENCES

[1] Aronszajn, N., Theory of reproducing kernels. *Trans. Amer. Math. Soc.* **68** (1950), 337–404.

[2] Aronszajn, N., Smith, K. T., Functional spaces and functional completion, *Ann. Inst. Fourier,* **6** (1956), 125–185.

[3] Aronszajn, N., Brown, R. D., Butcher, R. S., Construction of the solution of boundary value problems for the biharmonic operator in a rectangle. *Ann. Inst. Fourier,* **23**, 3 (1973), 49–89.

[4] Aronszajn, N., Green's functions and reproducing kernels, Proceedings of the Symposium on Spectral Theory and Differential Problems, Oklahoma, A. and M. College, Stillwater, Oklahoma, (1951), 355–411.

[5] Babuška, L., The theory of small changes in the domain of existence in the theory of partial differential equations and its applications. Differential Equations and Their Applications. Proceedings of a Conference held in Prague, September 1962, Academic Press, New York, 1963.

[6] Comptes Rendus Hebdomadaires des Seances de l'Academie des Sciences, **141** (1905), 1145.

[7] Comptes Rendus Hebdomadaires des Seances de l'Academie des Sciences, **145** (1907),
     983–991.
[8] Garabedian, P. R., Partial Differential Equations, Wiley, New York, 1964.
[9] Gould, S. H., Variational Methods for Eigenvalue Problems; an Introduction to the
     Weinstein Method of Intermediate Problems. University of Toronto Press,
     Toronto, 1966.
[10] Hadamard, H., Mémoire sur le problème d'analyse relatif a l'équilibre des plaques
     élastiques encastrées. Mémoirés presentes par divers savants à l'Academe des
     Sciences, **33** (1908), 1–128. Oeuvres de Jacques Hadamard, Editions du
     Centre National de la Recherche Scientifique, Paris, 1968, II, 515–641.
[11] Loewner, C., On generation of solutions of the biharmonic equation, *Pacific J. Math.*,
     3 (1953), 417–436.
[12] Lauricella, G., Sur l'integration de l'équation relative à l'équilibre des plaques élasti-
     ques encastrées, *Acta Math.* **32**, 201–256.
[13] Meschkowski, H., Hilbertsche Raume mit Kernfunction, Die Grundlehren der
     Mathematischen Wissenschaften in Einzeldarstellungen, Band 113, Springer,
     Berlin, 1962.
[14] Morrey, C. B., Multiple Integrals in the Calculus of Variations, Die Grundlehren der
     Mathematischen Wissenschaften in Einzeldarstellungen, Vol. 130, Springer,
     New York, 1966.
[15] Zaremba, S., L'équation biharmonique et une classe remarquable de functions
     fondamentales harmoniques, *Bull. Acad. Cracovie*, (1907), 147–196.

A NONLINEAR HYPERBOLIC VOLTERRA EQUATION IN VISCOELASTICITY

By C. M. Dafermos* and J. A. Nohel**

Abstract. A general model for the nonlinear motion of a one dimensional, finite, homogeneous, viscoelastic body is developed and analysed by an energy method. It is shown that under physically reasonable conditions the nonlinear boundary initial-value problem has a unique, smooth solution (global in time), provided the given data are sufficiently "small" and smooth; moreover, the solution and its derivatives of first and second order decay to zero as $t \to \infty$. Various modifications and generalizations, including two and three dimensional problems, are also discussed.

1. Introduction. In nonlinear systems of "hyperbolic" type, characteristic speeds are not constant so that weak waves are amplified and smooth solutions may blow up in finite time due to the formation of shock waves. It would be interesting to consider situations where this destabilizing mechanism coexists (and thus competes) with dissipation.

In certain cases (e.g., viscosity of the rate type) dissipation is so powerful that waves cannot break and solutions remain globally smooth. A more interesting situation arises when the amplification and decay mechanisms have comparable power so that the outcome of their confrontation cannot be predicted at the outset. Elementary dimensional considerations indicate that breaking of waves develops on a time scale inversely proportional to wave amplitude while dissipation proceeds at a roughly constant time scale. It should thus be expected that dissipation prevails and waves do not break when the initial data are "small." Results of this type for quasilinear wave equations with frictional damp-

*Brown University, research sponsored by the National Science Foundation under Grant No. MCS79-05774, and by the United States Army under Grant No. DAAG27-76-G-0294 and under Contract No. DAAG29-80-C-0041.

**Mathematics Research Center, research sponsored by the United States Army under Contract No. DAAG29-80-C-0041.

ing were first obtained by Nishida [1] and, subsequently, by Matsumura [2], who uses methodology that goes back to Schauder [3]. The more delicate situation of thermal damping (one dimensional thermoelasticity) is discussed in Slemrod [4].

A different, subtler type of dissipation mechanism is induced by memory effects and arises in nonlinear viscoelasticity. A simple, one dimensional, model corresponds to the constitutive relation

$$(1.1) \qquad \sigma(t,x) = \varphi(e(t,x)) + \int_{-\infty}^{t} a'(t - \tau)\psi(e(\tau,x))\,d\tau,$$

where σ is the stress, e the strain, a the relaxation function with $' = d/dt$, and φ, ψ assigned constitutive functions. We normalize the relaxation function so that $a(\infty) = 0$. When the reference configuration is a natural state, $\varphi(0) = \psi(0) = 0$. Experience indicates that $\varphi(e), \psi(e)$, as well as the equilibrium stress

$$(1.2) \qquad \chi(e) \overset{\text{def}}{=} \varphi(e) - a(0)\psi(e)$$

are increasing functions of e, at least near equilibrium (e small). Moreover, the effect of viscosity is dissipative. To express mathematically the above physical requirements, we impose upon $a(t)$, $\varphi(e)$, $\psi(e)$ and $\chi(e)$ the following assumptions:

$$(1.3) \qquad a(t) \in W^{2,1}(0, \infty), a(t) \text{ is strongly positive definite on } [0, \infty);$$

$$(1.4) \qquad \varphi(e) \in C^{3}(-\infty, \infty), \qquad \varphi(0) = 0, \qquad \varphi'(0) > 0;$$

$$(1.5) \qquad \psi(e) \in C^{3}(-\infty, \infty), \qquad \psi(0) = 0, \qquad \psi'(0) > 0;$$

$$(1.6) \qquad \chi'(0) = \varphi'(0) - a(0)\psi'(0) > 0.$$

Assumption (1.3), which requires that $a(t) - \alpha \exp(-t)$ be a positive definite kernel on $[0, \infty)$ for some $\alpha > 0$, expresses the dissipative character of viscosity. Smooth, integrable, nonincreasing, convex relaxation functions, e.g.,

$$(1.7) \qquad a(t) = \sum_{k=1}^{K} \nu_k \exp(-\mu_k t), \qquad \nu_k > 0, \quad \mu_k > 0,$$

which are commonly employed in the application of the theory of viscoelasticity, satisfy (1.3).

It is often convenient to express σ, given by (1.1), in terms of equilibrium stress, namely (integrate (1.1) by parts and use (1.2)),

$$(1.8) \qquad \sigma(t,x) = \chi(e(t,x)) + \int_{-\infty}^{t} a(t - \tau)\psi(e(\tau,x))_{\tau} \, d\tau.$$

We now consider a homogeneous, one dimensional body (string or bar) with reference configuration $[0, 1]$ of density $\rho = 1$ (for simplicity) and constitutive relation (1.1) which is moving under the action of an assigned body force $g(t,x)$, $-\infty < t < \infty$, $0 \leqslant x \leqslant 1$. We let $u(t,x)$ denote the displacement of particle x at time t in which case the strain is $e(t,x) = u_x(t,x)$. Thus the equation of motion $\rho u_{tt} = \sigma_x + \rho g$ here takes the form

$$(1.9) \qquad u_{tt} = \varphi(u_x)_x + \int_{-\infty}^{t} a'(t - \tau)\psi(u_x)_x \, d\tau + g,$$

$$-\infty < t < \infty, \quad 0 \leqslant x \leqslant 1,$$

or, if one uses representation (1.8) for the stress,

$$(1.10) \qquad u_{tt} = \chi(u_x)_x + \int_{-\infty}^{t} a(t - \tau)\psi(u_x)_{\tau x} \, d\tau + g,$$

$$-\infty < t < \infty, \quad 0 \leqslant x \leqslant 1.$$

The history of the motion of the body up to time $t = 0$ is assumed known, i.e.,

$$(1.11) \qquad u(t,x) = v(t,x), \qquad -\infty < t \leqslant 0, \quad 0 \leqslant x \leqslant 1,$$

where $v(t,x)$ is a given function which satisfies equation (1.9) together with appropriate boundary conditions, for $t \leqslant 0$. Our task is to determine a smooth extension $u(t,x)$ of $v(t,x)$ on $(-\infty, \infty) \times [0, 1]$ which satisfies (1.9) together with assigned boundary conditions, for $-\infty < t < \infty$.

Upon setting

$$(1.12) \qquad h = \int_{-\infty}^{0} a'(t - \tau)\psi(v_x)_x \, d\tau + g, \qquad t \geqslant 0, \quad 0 \leqslant x \leqslant 1,$$

$$(1.13) \qquad u_0(x) = v(0,x), \qquad u_1(x) = v_t(0,x), \qquad 0 \leqslant x \leqslant 1,$$

the history-value problem (1.9), (1.11) reduces to the initial-value problem

$$(1.14) \quad u_{tt} = \varphi(u_x)_x + \int_0^t a'(t - \tau)\psi(u_x)_x \, d\tau + h,$$

$$0 \leqslant t < \infty, \quad 0 \leqslant x \leqslant 1,$$

$$(1.15) \quad u(0, x) = u_0(x), \qquad u_t(0, x) = u_1(x), \qquad 0 \leqslant x \leqslant 1.$$

Conversely, (1.14), (1.15) can be reduced to (1.9), (1.11) by constructing a function $v(t, x)$ on $(-\infty, 0] \times [0, 1]$ which satisfies $v(0, x) = u_0(x)$, $v_t(0, x) = u_1(x)$,

$$(1.16) \quad \begin{cases} v_{tt}(0, x) = \varphi(u_{0x})_x + h(0, x), & 0 \leqslant x \leqslant 1, \\ v_{ttt}(0, x) = \varphi''(u_{0x})u_{0xx}u_{1x} + \varphi'(u_{0x})u_{1xx} \\ \qquad\qquad + a'(0)\psi(u_{0x})_x + h_t(0, x), & 0 \leqslant x \leqslant 1, \end{cases}$$

together with appropriate boundary conditions, for $t \leqslant 0$, and then defining $g(t, x)$ on $(-\infty, \infty) \times [0, 1]$ by

$$(1.17) \quad g(t, x) = \begin{cases} v_{tt} - \varphi(v_x)_x - \int_{-\infty}^t a'(t - \tau)\psi(v_x)_x \, d\tau, \\ \qquad\qquad\qquad t \leqslant 0, \quad 0 \leqslant x \leqslant 1, \\ h - \int_{-\infty}^0 a'(t - \tau)\psi(v_x)_x \, d\tau, \\ \qquad\qquad\qquad t \geqslant 0, \quad 0 \leqslant x \leqslant 1. \end{cases}$$

The purpose of (1.16) is to ensure that $g(t, x)$, as defined by (1.17), has the smoothness properties, across $t = 0$, which will be required below in the existence theorem.

For the special case $\psi(e) \equiv \varphi(e)$ variants of existence theorems for (1.14), (1.15) were established by MacCamy [5], Dafermos and Nohel [6] and Staffans [7]. The assumption $\psi \equiv \varphi$ allows one to invert the linear Volterra integral operator on the right-hand side of (1.14) and thus express $\varphi(u_x)_x$ in terms of $u_{tt} - h$ through an inverse Volterra integral operator using the resolvent kernel associated with a'. One may then transfer time derivatives from u_{tt} to the resolvent kernel via integration by parts. This procedure reveals the instantaneous character of dissipation and, at the same time, renders the memory term linear and milder, thus simplifying the analysis considerably. On the other hand, the above approach is somewhat artificial: By inverting the right-hand side of

(1.14), one loses sight of the original equation and of the physical interpretation of the derived a priori estimates. More importantly, the physical appropriateness of the restriction $\psi = \varphi$ is by no means clear.

The present normalization of the kernel a with $a(\infty) = 0$ is different from that in the existing literature (see [5], [6], [7]). The reader should note a', not a, enters the constitutive relation (1.1) as well as the equation of motion (1.9). The present normalization is more convenient for technical reasons; for, the equivalent form (1.8) of the constitutive equation in which a (rather than a') enters, and the corresponding equation of motion (1.10), are extremely convenient for obtaining the crucial a priori estimates in our analysis, when a satisfies assumption (1.3). In the earlier literature in which only the special case $\psi \equiv \varphi$ was studied, the normalization

$$a(t) = a_\infty + A(t), \qquad 0 \leqslant t < \infty,$$

$a(0) = 1, a_\infty > 0, A \in W^{2,1}(0, \infty)$, A strongly positive was used.

In another noteworthy special case, when $a(t) = \exp(-\mu t)$, (1.9) is equivalent to the third order partial differential equation

$$u_{ttt} + \mu u_{tt} = \varphi(u_x)_{tx} + \mu\chi(u_x)_x + g_t + \mu g$$

studied by Greenberg [8].

In this paper we show how one may deal with Equation (1.9) directly and establish existence of solutions without the assumption $\varphi \equiv \psi$. We will consider in detail the case where the boundary of the body is free of traction which leads to boundary conditions

$$(1.18) \qquad \sigma(t,0) = \sigma(t,1) = 0, \qquad -\infty < t < \infty.$$

Other types of boundary conditions will be discussed in Section 4. The change of variable (superposition of a rigid motion)

$$(1.19) \qquad u(t,x) = \bar{u}(t,x) + m_0 + m_1 t + \int_0^t \int_0^\tau \int_0^1 g(s, y)\, dy\, ds\, d\tau$$

shows that without loss of generality we may assume

$$(1.20) \qquad \int_0^1 g(t,x)\, dx = 0, \qquad -\infty < t < \infty,$$

$$(1.21) \qquad \int_0^1 u(t,x)\, dx = 0, \qquad -\infty < t < \infty.$$

Of the body force g we require

$$(1.22) \quad \begin{cases} g(t,\cdot),\, g_t(t,\cdot),\, g_x(t,\cdot) \\ \quad \text{in } C\big((-\infty,\infty);\, L^2(0,1)\big) \cap L^2\big((-\infty,\infty);\, L^2(0,1)\big) \\ g(t,x) = g_1(t,x) + g_2(t,x) \\ \quad \text{with } g_{1tt}(t,\cdot),\, g_{2tx}(t,\cdot) \text{ in } L^2\big((-\infty,\infty);\, L^2(0,1)\big). \end{cases}$$

As noted above, despite the presence of viscous dissipation, it is not to be expected that a global smooth solution to (1.9), (1.11), (1.18) will exist unless the amplitude of waves remains small. Consequently, one may only hope to obtain global existence results under the restriction that $g(t,x)$ be appropriately "small". We "measure" $g(t,x)$ by

$$(1.23) \quad G \overset{\text{def}}{=} \sup_{(-\infty,\infty)} \int_0^1 \{ g^2 + g_t^2 + g_x^2 \}(t,x)\, dx$$

$$+ \int_{-\infty}^{\infty} \int_0^1 \{ g^2 + g_t^2 + g_x^2 + g_{1tt}^2 + g_{2tx}^2 \}\, dx\, dt.$$

Our main result is

THEOREM 1.1. *Under assumptions* (1.3), (1.4), (1.5), (1.6), *there exists a constant* $\mu > 0$ *with the following property: For every* $g(t,x)$ *on* $(-\infty,\infty) \times [0,1]$ *which satisfies* (1.20) *and* (1.22) *with*

$$(1.24) \quad G \leqslant \mu^2,$$

and for any $v(t,x)$ *on* $(-\infty,0] \times [0,1]$, *with* $v(t,\cdot), v_t(t,\cdot), v_x(t,\cdot),$ $v_{tt}(t,\cdot), v_{tx}(t,\cdot), v_{xx}(t,\cdot), v_{ttt}(t,\cdot), v_{ttx}(t,\cdot), v_{txx}(t,\cdot), v_{xxx}(t,\cdot)$ *in* $C((-\infty, 0]; L^2(0,1)) \cap L^2((-\infty,0]; L^2(0,1))$, *which satisfies Equation* (1.9) *together with the boundary conditions* (1.18) *for* $t \leqslant 0$, *there exists a unique* $u(t,x)$ *on* $(-\infty,\infty) \times [0,1]$, *with* $u(t,\cdot), u_t(t,\cdot), u_x(t,\cdot), u_{tt}(t,\cdot), u_{tx}(t,\cdot),$ $u_{xx}(t,\cdot), u_{ttt}(t,\cdot), u_{ttx}(t,\cdot), u_{txx}(t,\cdot), u_{xxx}(t,\cdot)$ *in* $C((-\infty,\infty); L^2(0,1)) \cap$ $L^2((-\infty,\infty); L^2(0,1))$, *which satisfies* (1.9), (1.11), (1.18), *as well as* (1.21). *Furthermore,*

$$(1.25) \quad u(t,\cdot), u_t(t,\cdot), u_x(t,\cdot), u_{tt}(t,\cdot), u_{tx}(t,\cdot), u_{xx}(t,\cdot) \xrightarrow[[0,1]]{\text{unif.}} 0,$$

$$t \to \infty.$$

The proof of the above theorem employs the general strategy developed in [2, 6, 7]. We first establish, in Section 2, the existence of a local solution, defined on a maximal interval $(-\infty, T_0)$, with the property that when $T_0 < \infty$ a certain norm blows up as $t \uparrow T_0$. Then, in Section 3, we show that, due to viscous dissipation, the aforementioned norm remains uniformly bounded on the maximal interval, provided that (1.24) holds with μ sufficiently small. In particular, $T_0 = \infty$ and the smooth solution exists globally.

In the final Section 4, we have collected information on various extensions of the above results. We show how one can handle boundary conditions other than (1.18). We indicate how alternative sets of assumptions on $v(t, x)$ and $g(t, x)$ lead to variants of Theorem 1.1 rendering information on the smoothness of solutions. Finally, we explain how the present techniques may be used to establish existence theorems for the equations of multidimensional viscoelasticity as well as abstract integrodifferential equations in Hilbert space.

2. Local Solutions. In this section we establish a local existence theorem on a maximal interval. It is more convenient to work with Equation (1.14) to which, as we have seen, (1.9) may be reduced. Also we shall impose here boundary conditions

(2.1) $u_x(t, 0) = u_x(t, 1) = 0, \quad t \geqslant 0,$

which, though apparently stronger than (1.18), are actually equivalent to (1.18), as will be shown in Section 3. Finally, we temporarily strengthen assumption (1.4) into

(2.2) $\varphi(e) \in C^3(-\infty, \infty), \quad \varphi(0) = 0, \quad \varphi'(e) \geqslant \kappa > 0,$

$$-\infty < e < \infty.$$

On the other hand, assumptions $\psi'(0) > 0, \chi'(0) > 0$ and the positivity of the kernel $a(t)$ will not play any role in this section.

THEOREM 2.1.   Let $u_0(x), u_{0x}(x), u_{0xx}(x), u_{0xxx}(x), u_1(x), u_{1x}(x), u_{1xx}(x)$ be in $L^2(0, 1)$ and assume

(2.3) $u_{0x}(0) = u_{0x}(1) = 0, \quad u_{1x}(0) = u_{1x}(1) = 0.$

Moreover, let $h(t, x)$ be defined on $[0, \infty) \times [0, 1]$ with $h(t, \cdot), h_t(t, \cdot), h_x(t, \cdot)$ in $C([0, \infty); L^2(0, 1))$ and $h(t, x) = h_1(t, x) + h_2(t, x), h_{1tt}, h_{2tx}$ in

$L^2([0, \infty); L^2(0, 1))$. *Then there is* $T_0, 0 < T_0 \leqslant \infty$, *and a unique function* $u(t, x) \in C^2([0, T_0) \times [0, 1])$, *with* $u_{ttt}(t, \cdot), u_{ttx}(t, \cdot), u_{txx}(t, \cdot), u_{xxx}(t, \cdot)$ *in* $C([0, T]; L^2(0, 1))$, *for every* $0 < T < T_0$, *such that* u *satisfies* (1.14) *on* $[0, T_0) \times [0, 1]$ *together with initial conditions* (1.15) *and boundary conditions* (2.1) *on* $[0, T_0)$. *Furthermore, if* $T_0 < \infty$,

$$
(2.4) \qquad \int_0^1 \{ u^2(t, x) + u_t^2(t, x) + u_x^2(t, x) + u_{tt}^2(t, x)
$$

$$
+ u_{tx}^2(t, x) + u_{xx}^2(t, x) + u_{ttt}^2(t, x)
$$

$$
+ u_{ttx}^2(t, x) + u_{txx}^2(t, x) + u_{xxx}^2(t, x) \} \, dx \to \infty, \quad as \quad t \uparrow T_0.
$$

We note that $h(t, x)$ and $u_0(x), u_1(x)$, defined by (1.12), (1.13) with $v(t, x)$ and $g(t, x)$ as in Theorem 1.1, do satisfy the assumptions of Theorem 2.1.

The proof of Theorem 2.1 which is a variant of the local result in [6] will be based upon the Banach fixed point theorem. We begin with some preparation. For $M, T > 0$, we let $X(M, T)$ denote the set of functions $w(t, x)$ on $[0, T] \times [0, 1]$, with $w(t, \cdot), w_t(t, \cdot), w_x(t, \cdot), w_{tt}(t, \cdot), w_{tx}(t, \cdot),$ $w_{xx}(t, \cdot), w_{ttt}(t, \cdot), w_{ttx}(t, \cdot), w_{txx}(t, \cdot), w_{xxx}(t, \cdot)$ in $L^\infty([0, T]; L^2(0, 1))$ which assume initial data $w(0, x) = u_0(x), w_t(0, x) = u_1(x)$ and boundary conditions $w_x(t, 0) = w_x(t, 1) = 0, t \in [0, T]$, and satisfy

$$
(2.5) \qquad \text{ess-sup}_{[0, T]} \int_0^1 \{ w_{ttt}^2(t, x)
$$

$$
+ w_{ttx}^2(t, x) + w_{txx}^2(t, x) + w_{xxx}^2(t, x) \} \, dx \leqslant M^2.
$$

For $w(t, x) \in X(M, T)$, (2.5) and the Poincaré inequality yield

$$
(2.6) \qquad w_x^2(t, x) + w_{tx}^2(t, x) + w_{xx}^2(t, x) \leqslant M^2,
$$

$$
0 \leqslant t \leqslant T, \quad 0 \leqslant x \leqslant 1.
$$

We now consider the map $S : X(M, T) \to C^2([0, T] \times [0, 1])$ which carries $w(t, x) \in X(M, T)$ into the solution $u(t, x)$ of the linear equation

$$
(2.7) \qquad u_{tt} - \varphi'(w_x) u_{xx} = \int_0^t a'(t - \tau) \psi(w_x)_x \, d\tau + h
$$

satisfying initial conditions (1.15) and boundary conditions (2.1). We note that $\varphi'(w_x(t,x))$ is in $W^{1,\infty}$ and $\varphi'(w_x)_{tt}, \varphi'(w_x)_{tx}$ are in $L^\infty([0,T];$ $L^2(0,1))$. Furthermore, if $f(t,x)$ denotes the right-hand side of (2.7), then $f(t,\cdot), f_t(t,\cdot), f_x(t,\cdot)$ are in $C([0,T]; L^2(0,1))$ and $f(t,x) = f_1(t,x) +$ $f_2(t,x)$ with f_{1tt}, f_{2tx} in $L^2([0,T]; L^2(0,1))$. It then follows by standard theory that $u_{ttt}(t,\cdot), u_{ttx}(t,\cdot), u_{txx}(t,\cdot)$ and $u_{xxx}(t,\cdot)$ are in $C([0,T]; L^2(0,1))$. Our strategy is to show that, under proper conditions, S has a unique fixed point in $X(M,T)$ which will obviously be the solution to (1.14), (1.15), (2.1) with the desired properties.

LEMMA 2.1.   *When M is sufficiently large and T is sufficiently small, S maps $X(M,T)$ into itself.*

Proof.   We fix $\eta > 0$ and apply to (2.7) the forward difference operator Δ, $(\Delta\omega)(t) \stackrel{\text{def}}{=} \omega(t+\eta) - \omega(t)$, thus obtaining

(2.8) $\Delta u_{tt} - \varphi'(w_x)\Delta u_{xx} = \Delta\varphi'(w_x)u_{xx} + \Delta\varphi'(w_x)\Delta u_{xx}$

$$+ \Delta \int_0^t a'(t-\tau)\psi(w_x)_x \, d\tau + \Delta h.$$

We multiply (2.8) by Δu_{txx} and integrate over $[0,s] \times [0,1], 0 \leqslant s \leqslant T$. After appropriate integrations by parts, we divide through by η^2 and we let $\eta \downarrow 0$. We give the details of the computation of one term:

(2.9) $\displaystyle\int_0^s\int_0^1 \Delta u_{tt}\Delta u_{txx}\,dx\,dt = -\int_0^s\int_0^1 \Delta u_{ttx}\Delta u_{tx}\,dx\,dt$

$$= -\frac{1}{2}\int_0^1 (\Delta u_{tx})^2(s,x)\,dx$$

$$+ \frac{1}{2}\int_0^1 (\Delta u_{tx})^2(0,x)\,dx,$$

whence

(2.10) $\displaystyle\lim_{\eta\downarrow 0}\frac{1}{\eta^2}\int_0^s\int_0^1 \Delta u_{tt}\Delta u_{txx}\,dx\,dt = -\frac{1}{2}\int_0^1 u_{ttx}^2(s,x)\,dx$

$$+ \frac{1}{2}\int_0^1 u_{ttx}^2(0,x)\,dx$$

where

(2.11) $u_{ttx}(0, x) = \varphi(u_{0x}(x))_{xx} + h_x(0, x).$

We apply the same procedure to the remaining terms of (2.8) thus obtaining

(2.12) $\dfrac{1}{2} \displaystyle\int_0^1 u_{ttx}^2(s, x)\, dx + \dfrac{1}{2} \int_0^1 \varphi'(w_x(s, x)) u_{txx}^2(s, x)\, dx$

$$= \dfrac{1}{2} \int_0^1 u_{ttx}^2(0, x)\, dx + \dfrac{1}{2} \int_0^1 \varphi'(u_{0x}(x)) u_{1xx}^2(x)\, dx$$

$$+ \dfrac{1}{2} \int_0^s \int_0^1 \varphi''(w_x) w_{tx} u_{txx}^2 \, dx\, dt$$

$$+ \int_0^s \int_0^1 \varphi'''(w_x) w_{tx} w_{xx} u_{xx} u_{ttx} \, dx\, dt$$

$$+ \int_0^s \int_0^1 \varphi''(w_x) w_{txx} u_{xx} u_{ttx} \, dx\, dt$$

$$+ \int_0^s \int_0^1 \varphi''(w_x) w_{tx} u_{xxx} u_{ttx} \, dx\, dt$$

$$- \int_0^1 a'(s) \psi(u_{0x}(x))_x u_{txx}(s, x)\, dx$$

$$+ \int_0^1 a'(0) \psi(u_{0x}(x))_x u_{1xx}(x)\, dx$$

$$+ \int_0^s \int_0^1 a''(t) \psi(u_{0x})_x u_{txx} \, dx\, dt$$

$$- \int_0^1 u_{txx}(s, x) \int_0^s a'(s - t) \psi(w_x(t, x))_x \, dt\, dx$$

$$+ \int_0^s \int_0^1 a'(0) \psi(w_x)_x u_{txx} \, dx\, dt$$

$$+ \int_0^s \int_0^1 u_{txx} \int_0^t a''(t - \tau) \psi(w_x)_x \, d\tau\, dx\, dt$$

$$+ \int_0^s \int_0^1 h_{2tx} u_{ttx} \, dx\, dt$$

$$- \int_0^1 h_{1t}(s, x) u_{txx}(s, x)\, dx + \int_0^1 h_{1t}(0, x) u_{1xx}(x)\, dx$$

$$+ \int_0^s \int_0^1 h_{1tt} u_{txx} \, dx\, dt.$$

We now differentiate (2.7) with respect to t and x to obtain

$$(2.13) \quad u_{ttt} - \varphi'(w_x)u_{txx} - \varphi''(u_{0x})u_{0xx}u_{1x} - \int_0^t \{\varphi''(w_x)w_{\tau x}u_{xx}\}_\tau \, d\tau$$

$$= a'(t)\psi(u_{0x})_x + \int_0^t a'(t - \tau)\psi(w_x)_{\tau x} \, d\tau + h_t,$$

$$(2.14) \quad u_{ttx} - \varphi'(w_x)u_{xxx} - \varphi''(u_{0x})u_{0xx}^2 - \int_0^t \{\varphi''(w_x)w_{xx}u_{xx}\}_\tau \, d\tau$$

$$= \int_0^t a'(t - \tau)\psi(w_x)_{xx} \, d\tau + h_x,$$

from which we easily get the estimates

$$(2.15) \quad \int_0^1 u_{ttt}^2(s, x) \, dx - 6 \int_0^1 \varphi'(w_x(s, x))^2 u_{txx}^2(s, x) \, dx$$

$$\leqslant 6 \int_0^1 \varphi''(u_{0x}(x))^2 u_{0xx}^2(x)u_{1x}^2(x) \, dx$$

$$+ 6 \int_0^1 \left\{ \int_0^s [\varphi''(w_x)w_{tx}u_{xx}]_t \, dt \right\}^2 dx$$

$$+ 6 \int_0^1 a'(t)^2 \psi(u_{0x}(x))_x^2 \, dx$$

$$+ 6 \int_0^1 \left\{ \int_0^s a'(s - t)\psi(w_x)_{tx} \, dt \right\}^2 dx + 6 \int_0^1 h_t^2(s, x) \, dx,$$

$$(2.16) \quad \int_0^1 \varphi'(w_x(s, x))^2 u_{xxx}^2(s, x) \, dx - 5 \int_0^1 u_{ttx}^2(s, x) \, dx$$

$$\leqslant 5 \int_0^1 \varphi''(u_{0x}(x))^2 u_{0xx}^4 \, dx + 5 \int_0^1 \left\{ \int_0^s [\varphi''(w_x)w_{xx}u_{xx}]_t \, dt \right\}^2 dx$$

$$+ 5 \int_0^1 \left\{ \int_0^s a'(s - t)\psi(w_x)_{xx} \, dt \right\}^2 dx + 5 \int_0^1 h_x^2(s, x) \, dx.$$

Let us set

$$(2.17) \quad N^2 = \int_0^1 \{u_0^2 + u_{0x}^2 + u_{0xx}^2 + u_{0xxx}^2 + u_1^2 + u_{1x}^2 + u_{1xx}^2\} \, dx$$

$$+ \sup_{[0,\infty)} \int_0^1 \{h^2(t, x) + h_t^2(t, x) + h_x^2(t, x)\} \, dx$$

$$+ \int_0^\infty \int_0^1 \{h_{1tt}^2 + h_{2tx}^2\} \, dx \, dt,$$

$$(2.18) \quad V^2 = \sup_{[0,T]} \int_0^1 \{u_{ttt}^2(t, x) + u_{ttx}^2(t, x) + u_{txx}^2(t, x) + u_{xxx}^2(t, x)\} \, dx.$$

Then, by virtue of (2.5), (2.6), the Poincaré inequality and Schwarz's inequality, every term on the right-hand side of (2.12), (2.15), and (2.16) can be majorized by one of $p(N), Tq(M)V^2, p(N)V, Tq(M)V, T^2q(M),$ $T^{1/2}p(N)V, T^2q(M)V^2, Tp(N)$, where $p(\cdot)$ and $q(\cdot)$ are locally bounded functions on $[0, \infty)$. Thus, combining (2.12), (2.15) and (2.16) and using (2.2), we arrive at an estimate of the form

$$(2.19) \quad V^2 \leqslant c\{ p(N) + Tq(M)V^2 + p(N)V$$

$$+ Tq(M)V + T^2q(M)V^2$$

$$+ T^{1/2}p(N)V + T^2q(M) + Tp(N)\}.$$

Applying the Cauchy-Schwarz inequality,

$$(2.20) \quad \{1 - cTq(M) - 2cT^2q(M) - cTp(N) - \tfrac{1}{2}\}V^2$$

$$\leqslant c\{2p(N) + cp^2(N) + q(M) + T^2q(M) + Tp(N)\}.$$

Thus, if one fixes $M^2 \geqslant 8c\{2p(N) + cp^2(N) + q(M)\}$ and then selects T so small that, at the same time, $cTq(M) + 2cT^2q(M) + cTp(N) < \tfrac{1}{4}$ and $cT^2q(M) + cTp(N) \leqslant M^2/8$, (2.20) yields $V^2 \leqslant M^2$ and $u(t,x) \in X(M,T)$. The proof of the lemma is complete.

We now equip $X(M,T)$ with the metric

$$(2.21) \quad \rho(u, \bar{u}) = \max_{[0,T]} \left\{ \int_0^1 \left[(u_{tt} - \bar{u}_{tt})^2 + (u_{tx} - \bar{u}_{tx})^2 \right. \right.$$

$$\left. \left. + (u_{xx} - \bar{u}_{xx})^2 \right](t, x)\, dx \right\}^{1/2},$$

where $u, \bar{u} \in X(M,T)$. On account of the lower semicontinuity property of norms in Banach space, $X(M,T)$ is complete under ρ.

LEMMA 2.2. *For M sufficiently large and T sufficiently small, the map $S: X(M,T) \to X(M,T)$ is a strict contraction with respect to the metric ρ.*

Proof. Let $w(t, x), \overline{w}(t, x) \in X(M, T)$. We set $u = Sw, \overline{u} = S\overline{w}$, $W = w - \overline{w}, U = u - \overline{u}$. Then $U(t, x)$ is the solution of the problem:

$$(2.22) \quad U_{tt} - \varphi'(w_x)U_{xx} = A(t, x)\overline{u}_{xx} \int_0^t W_{\tau x} \, d\tau$$

$$+ \int_0^t a'(t - \tau) \left[\psi'(w_x) W_{xx} + B(\tau, x)\overline{w}_{xx} W_x \right] d\tau,$$

$$(2.23) \quad U(0, x) = 0, \qquad U_t(0, x) = 0, \qquad 0 \leqslant x \leqslant 1,$$

$$(2.24) \quad U_x(t, 0) = U_x(t, 1) = 0, \qquad 0 \leqslant t \leqslant T,$$

where

$$(2.25) \quad A(t, x) = \begin{cases} \dfrac{\varphi'(w_x(t, x)) - \varphi'(\overline{w}_x(t, x))}{w_x(t, x) - \overline{w}_x(t, x)}, & w_x(t, x) \neq \overline{w}_x(t, x) \\ \varphi''(w_x(t, x)), & w_x(t, x) = \overline{w}_x(t, x), \end{cases}$$

$$(2.26) \quad B(t, x) = \begin{cases} \dfrac{\psi'(w_x(t, x)) - \psi'(\overline{w}_x(t, x))}{w_x(t, x) - \overline{w}_x(t, x)}, & w_x(t, x) \neq \overline{w}_x(t, x) \\ \psi''(w_x(t, x)), & w_x(t, x) = \overline{w}_x(t, x). \end{cases}$$

Furthermore,

$$(2.27) \quad U_{ttt} - \varphi'(w_x)U_{txx} = A(t, x)\overline{u}_{txx} W_x + \varphi''(w_x)w_{tx} U_{xx}$$

$$+ \varphi''(w_x)\overline{u}_{xx} W_{tx} - C(t, x)\overline{w}_{tx} W_x$$

$$+ a'(0)\psi'(w_x)W_{xx} + a'(0)B(t, x)\overline{w}_{xx} W_x$$

$$+ \int_0^t a''(t - \tau)\psi'(w_x) W_{xx} \, d\tau$$

$$+ \int_0^t a''(t - \tau)B(\tau, x)\overline{w}_{xx} W_x \, d\tau$$

where

$$(2.28) \quad C(t,x) = \begin{cases} \dfrac{\varphi''(w_x(t,x)) - \varphi''(\overline{w}_x(t,x))}{w_x(t,x) - \overline{w}_x(t,x)}, & w_x(t,x) \neq \overline{w}_x(t,x), \\ \varphi'''(w_x(t,x)), & w_x(t,x) = \overline{w}_x(t,x). \end{cases}$$

Multiplying Equation (2.27) by U_{tt} and integrating over $[0,1] \times [0,s]$, $0 < s \leqslant T$, we obtain, after certain integrations by parts,

$$(2.29) \quad \frac{1}{2} \int_0^1 U_{tt}^2(s,x)\,dx + \frac{1}{2} \int_0^1 \varphi'(w_x(s,x)) U_{tx}^2(s,x)\,dx$$

$$= \frac{1}{2} \int_0^s \int_0^1 \varphi''(w_x) w_{tx} U_{tx}^2\,dx\,dt - \int_0^s \int_0^1 \varphi''(w_x) w_{xx} U_{tt} U_{tx}\,dx\,dt$$

$$+ \int_0^s \int_0^1 A\overline{u}_{txx} W_x U_{tt}\,dx\,dt + \int_0^s \int_0^1 \varphi''(w_x) w_{tx} U_{tt} U_{xx}\,dx\,dt$$

$$+ \int_0^s \int_0^1 \varphi''(w_x) \overline{u}_{xx} W_{tx} U_{tt}\,dx\,dt - \int_0^s \int_0^1 C\overline{w}_{tx} W_x U_{tt}\,dx\,dt$$

$$+ a'(0) \int_0^s \int_0^1 \psi'(w_x) W_{xx} U_{tt}\,dx\,dt$$

$$+ a'(0) \int_0^s \int_0^1 B\overline{w}_{xx} W_x U_{tt}\,dx\,dt$$

$$+ \int_0^s \int_0^1 U_{tt} \int_0^t a''(t-\tau)\psi'(w_x) W_{xx}\,d\tau\,dx\,dt$$

$$+ \int_0^s \int_0^1 U_{tt} \int_0^t a''(t-\tau) B\overline{w}_{xx} W_x\,d\tau\,dx\,dt.$$

Moreover, from (2.22) we get

$$(2.30) \quad \int_0^1 \varphi'(w_x(s,x))^2 U_{xx}^2(s,x)\,dx$$

$$\leqslant 3 \int_0^1 U_{tt}^2(s,x)\,dx + 3 \int_0^1 A^2 \overline{u}_{xx}^2 \left(\int_0^s W_{\tau x}\,d\tau \right)^2 dx$$

$$+ 3 \int_0^1 \left\{ \int_0^s a'(s-t)\left[\psi'(w_x) W_{xx} + B\overline{w}_{xx} W_x \right]dt \right\}^2 dx.$$

Combining (2.29) with (2.30) and using (2.2), (2.5), (2.6), the Poincaré inequality and the Cauchy-Schwarz inequality, we arrive, after a long computation, at an estimate of the form

(2.31) $\int_0^1 \{ U_{tt}^2(s,x) + U_{tx}^2(s,x) + U_{xx}^2(s,x) \} \, dx$

$$\leqslant (T + T^2) \max_{[0,T]} \int_0^1 \{ W_{tt}^2(t,x) + W_{tx}^2(t,x) + W_{xx}^2(t,x) \} \, dx$$

$$+ m \int_0^s \int_0^1 \{ U_{tt}^2(t,x) + U_{tx}^2(t,x) + U_{xx}^2(t,x) \} \, dx \, dt.$$

where m depends solely upon $a(t)$, M, and bounds of φ, ψ and their derivatives on the interval $[-M, M]$. In order to assist the reader to see how (2.31) is derived from (2.29), (2.30), we give the details of the estimation of one of the most complicated terms on the right-hand side of (2.29):

(2.32) $\int_0^s \int_0^1 U_{tt} \int_0^t a''(t - \tau)\psi'(w_x) W_{xx} \, d\tau \, dx \, dt$

$$\leqslant \frac{2}{\epsilon} \int_0^s \int_0^1 U_{tt}^2 \, dx \, dt$$

$$+ 2\epsilon \int_0^s \int_0^1 \left\{ \int_0^t a''(t - \tau)\psi'(w_x) W_{xx} \, d\tau \right\}^2 dx \, dt$$

$$\leqslant \frac{2}{\epsilon} \int_0^s \int_0^1 U_{tt}^2 \, dx \, dt + 2\epsilon \int_0^s \left\{ \int_0^t |a''(t - \tau)| \, d\tau \right\}$$

$$\times \left\{ \int_0^t |a''(t - \tau)| \int_0^1 \psi'(w_x)^2 W_{xx}^2 \, d\tau \right\} dx \, dt$$

$$\leqslant \frac{2}{\epsilon} \int_0^s \int_0^1 U_{tt}^2 \, dx \, dt + 2\epsilon s \left\{ \int_0^\infty |a''(\tau)| \, d\tau \right\}^2$$

$$\times \left\{ \max_{[-M, M]} \psi'(e)^2 \right\} \left\{ \sup_{[0, T]} \int_0^1 W_{xx}^2(t,x) \, dx \right\}.$$

From (2.31) and Gronwall's inequality we deduce

$$(2.33) \quad \max_{[0,T]} \int_0^1 \left\{ U_{tt}^2(t,x) + U_{tx}^2(t,x) + U_{xx}^2(t,x) \right\} dx$$

$$\leqslant (T + T^2) e^{mT} \max_{[0,T]} \int_0^1 \left\{ W_{tt}^2(t,x) + W_{tx}^2(t,x) \right.$$

$$\left. + W_{xx}^2(t,x) \right\} dx.$$

Thus, when T is so small that $(T + T^2)\exp(mT) < \frac{1}{4}$, (2.33) yields

$$(2.34) \quad \rho(Sw, S\overline{w}) \leqslant \tfrac{1}{2}\rho(w, \overline{w}), \qquad \text{for} \quad w, \overline{w} \in X(M, T)$$

and the proof of the lemma is complete.

Proof of Theorem 2.1. From Lemma 2.2 and the Banach fixed point theorem we deduce the existence of a unique fixed point of S in $X(M, T)$, for conveniently large M and appropriately small T, which will be the unique solution of (1.14), (1.15), (2.1) on $[0, T] \times [0, 1]$. Let $T_0 \leqslant \infty$ be the maximal interval of existence of a solution $u(t, x)$ to (1.14), (1.15), (2.1) with $u_t(t, \cdot), u_x(t, \cdot), u_{tt}(t, \cdot), u_{tx}(t, \cdot), u_{xx}(t, \cdot), u_{ttt}(t, \cdot),$ $u_{ttx}(t, \cdot), u_{txx}(t, \cdot), u_{xxx}(t, \cdot)$ in $L^\infty([0, T]; L^2(0, 1))$ for every $0 < T < T_0$. If $T_0 < \infty$ and (2.4) is not satisfied, we can extend $u(t, x)$ up to $t = T_0$ so that $u(t, x) \in C^1([0, T_0] \times [0, 1])$. Moreover, by weak convergence in $L^2(0, 1), u(T_0, x), u_x(T_0, x), u_{xx}(T_0, x), u_{xxx}(T_0, x), u_t(T_0, x), u_{tx}(T_0, x),$ $u_{txx}(T_0, x)$ are all in $L^2(0, 1)$. But then, using $u(T_0, x), u_t(T_0, x)$ as new initial data, we may extend $u(t, x)$ to some interval $[T_0, T_0 + \epsilon]$, beyond T_0, and this is a contradiction since $[0, T_0)$ is assumed maximal. The function $u(t, x)$ will be a solution of (2.7), with $w(t, x) \equiv u(t, x)$, and thus, as noted above, $u_{ttt}(t, \cdot), u_{ttx}(t, \cdot), u_{txx}(t, \cdot), u_{xxx}(t, \cdot)$ are all in $C([0, T]; L^2(0, 1))$, for every T in $(0, T_0)$. The proof is complete.

3. Global Solutions. Our objective in this section is to show that when the body force is "small" the maximal interval of existence of solution to (1.9), (1.11), (1.18) is $(-\infty, \infty)$ and solutions decay as $t \to \infty$. For that purpose, the dissipative character of viscosity, embodied in assumptions (1.3) on the relaxation function a, plays the crucial role. Assumption (1.3) will be exploited here through its consequences recorded in the following

LEMMA 3.1.   *There exist positive constants* β, γ *such that*

(3.1) $$\int_{-\infty}^{s}\left[\int_{-\infty}^{t} a(t-\tau)w(\tau)\,d\tau\right]^{2}dt$$

$$\leqslant \beta\int_{-\infty}^{s} w(t)\int_{-\infty}^{t} a(t-\tau)w(\tau)\,d\tau\,dt,$$

(3.2) $$\int_{-\infty}^{s}\left[\int_{-\infty}^{t} a'(t-\tau)w(\tau)\,d\tau\right]^{2}dt$$

$$\leqslant \gamma\int_{-\infty}^{s} w(t)\int_{-\infty}^{t} a(t-\tau)w(\tau)\,d\tau\,dt,$$

for any $s \in (-\infty, \infty)$ *and every* $w(t) \in L^{2}(-\infty, s)$.

The proof can be read off, for example, from Lemma 4.2 of [7], recalling that $a(t), a'(t), a''(t)$ are in $L^{1}(0, \infty)$, and that (by assumption (1.3)) $a(t) - \alpha\exp(-t)$ is a positive definite kernel on $[0, \infty)$ for some $\alpha > 0$. As a matter of fact, we may use

(3.3) $$\beta = \frac{1}{\alpha}\left\{\int_{0}^{\infty}|a(t)|\,dt\right\}^{2} + \frac{4}{\alpha}\left\{\int_{0}^{\infty}|a'(t)|\,dt\right\}^{2},$$

(3.4) $$\gamma = \frac{1}{\alpha}\left\{\int_{0}^{\infty}|a'(t)|\,dt\right\}^{2} + \frac{4}{\alpha}\left\{\int_{0}^{\infty}|a''(t)|\,dt\right\}^{2}.$$

Another important implication of a combination of (1.3), (1.5) and (1.6) is the property:

LEMMA 3.2.   *Let* $k(t)$ *be the resolvent kernel of the operator*

(3.5) $$\varphi'(0)\omega(t) + \int_{-\infty}^{t} a'(t-\tau)\psi'(0)\omega(\tau)\,d\tau;$$

that is, k is the unique solution of the linear Volterra equation

(3.6) $$\varphi'(0)k(t) + \int_{0}^{t} a'(t-\tau)\psi'(0)k(\tau)\,d\tau = -\psi'(0)a'(t).$$

Then $k(t) \in L^{1}(0, \infty)$.

The proof of Lemma 3.2 follows by a standard argument: Since $a'(t) \in L^{1}(0, \infty)$, the Paley-Wiener theorem states that $k(t) \in L^{1}(0, \infty)$

if and only if

$$(3.7) \qquad P(z) \stackrel{\text{def}}{=} \varphi'(0) + \psi'(0)\hat{a}'(z) = \chi'(0) + \psi'(0)z\hat{a}(z)$$

does not vanish on the half plane $\operatorname{Re} z \geqslant 0$. (In (3.7) $z = \xi + i\zeta$ and $\hat{\ }$ denotes the Laplace transform).

A simple calculation yields

$$(3.8) \qquad \operatorname{Re} P(z) = \chi'(0) + \psi'(0)\xi \operatorname{Re} \hat{a}(z) - \psi'(0)\zeta \operatorname{Im} \hat{a}(z),$$

$$(3.9) \qquad \operatorname{Im} P(z) = \psi'(0)\zeta \operatorname{Re} \hat{a}(z) + \psi'(0)\xi \operatorname{Im} \hat{a}(z).$$

On account of (1.3), (1.5) and (1.6),

$$(3.10) \qquad \operatorname{Re} P(\xi + i0) = \chi'(0) + \psi'(0)\xi\hat{a}(\xi) > 0, \qquad 0 \leqslant \xi < \infty.$$

As regards $\operatorname{Im} P(z)$, since by the strong positivity of $a(t)$, $\operatorname{Re} \hat{a}(i\zeta) > 0$, we have $\operatorname{Im} P(0 + i\zeta) = \psi'(0)\zeta \operatorname{Re} \hat{a}(i\zeta)$ is positive for $\zeta > 0$ and negative for $\zeta < 0$. On the other hand, $\operatorname{Im} P(\xi + i0) = 0, 0 \leqslant \xi < \infty$. Furthermore, since $a'(t) \in L^1(0, \infty)$, we deduce by the Riemann-Lebesgue lemma that $\lim_{|z| \to \infty} \operatorname{Im} P(z) = \psi'(0) \lim_{|z| \to \infty} \operatorname{Im} \hat{a}'(z) = 0$, uniformly on $\operatorname{Re} z \geqslant 0$. But $\operatorname{Im} P(z)$ is harmonic on $\operatorname{Re} z \geqslant 0$ so that, by the maximum principle, we conclude that $\operatorname{Im} P(z) > 0$ on $\{z = \xi + i\zeta \,|\, \xi \geqslant 0, \zeta > 0\}$ and $\operatorname{Im} P(z) < 0$ on $\{z = \xi + i\zeta \,|\, \xi \geqslant 0, \zeta < 0\}$. In conjunction with (3.10) this yields $P(z) \neq 0$ on $\operatorname{Re} z \geqslant 0$ and the proof of the lemma is complete.

Before proceeding to the proof of Theorem 1.1, let us show that the boundary conditions (1.18) are equivalent to

$$(3.11) \qquad u_x(t, 0) = u_x(t, 1) = 0, \qquad -\infty < t < \infty.$$

We multiply (1.8) by $\psi(e(t, x))_t$, integrate over $(-\infty, s), -\infty < s < \infty$, and use the positivity of $a(t)$ to get

$$(3.12) \qquad \int_{-\infty}^{s} \chi(e(t, x))\psi(e(t, x))_t \, dt \leqslant 0, \qquad -\infty < s < \infty, \quad x = 0, 1,$$

or

$$(3.13) \qquad \Psi(e(s, x)) \leqslant 0, \qquad -\infty < s < \infty, \quad x = 0, 1,$$

where

$$(3.14) \qquad \Psi(e) \stackrel{\text{def}}{=} \int_0^e \chi(\eta)\psi'(\eta)\,d\eta.$$

On account of (1.5), (1.6), $\Psi(e) > 0$ on $(-\delta, \delta) \setminus \{0\}$, δ positive small. Thus (3.13) yields $e(s, x) = 0$, $-\infty < s < \infty$, $x = 0, 1$, and (3.11) has been established.

Proof of Theorem 1.1. By virtue of (1.4), (1.5), (1.6), there are positive δ and κ such that

$$(3.15) \qquad \varphi'(e) > \kappa, \quad \psi'(e) > \kappa, \quad \chi'(e) > \kappa, \quad |e| \le \delta.$$

We modify $\varphi(e)$ outside the interval $[-\delta, \delta]$ so that (2.2) be satisfied and we let $u(t, x)$ be the solution to (1.9), (1.11), (3.11) on a maximal time interval $(-\infty, T_0)$.

For $T \in (-\infty, T_0)$, we set

$$(3.16) \qquad U(T) = \sup_{(-\infty, T]} \int_0^1 \left\{ u^2(t, x) + u_t^2(t, x) + u_x^2(t, x) \right.$$

$$+ u_{tt}^2(t, x) + u_{tx}^2(t, x) + u_{xx}^2(t, x) + u_{ttt}^2(t, x)$$

$$+ u_{ttx}^2(t, x) + u_{txx}^2(t, x) + u_{xxx}^2(t, x) \right\} dx$$

$$+ \int_{-\infty}^T \int_0^1 \left\{ u^2 + u_t^2 + u_x^2 + u_{tt}^2 + u_{tx}^2 \right.$$

$$+ u_{xx}^2 + u_{ttt}^2 + u_{ttx}^2 + u_{txx}^2 + u_{xxx}^2 \right\} dx\,dt.$$

Our strategy is to show that there are positive constants v, K, $v \le \delta$, such that, if

$$(3.17) \qquad |u_x(t, x)|^2 + |u_{tx}(t, x)|^2 + |u_{xx}(t, x)|^2 \le v^2,$$

$$-\infty < t \le T, \quad 0 \le x \le 1,$$

then

(3.18) $U(T) \leqslant KG$

where G is defined by (1.23). Once this claim has been established, we may complete the proof of the theorem by the following line of argument similar to that previously used in [6]: First we note that, by virtue of our assumptions on $v(t, x)$, (3.17) is automatically satisfied, as a strict inequality, when t is sufficiently small. Next we observe that, in view of the Poincaré inequality

(3.19) $|u_x(t,x)|^2 + |u_{tx}(t,x)|^2 + |u_{xx}(t,x)|^2$

$$\leqslant \int_0^1 \left\{ |u_{xx}(t,y)|^2 + |u_{txx}(t,y)|^2 + |u_{xxx}(t,y)|^2 \right\} dy,$$

and when $G \leqslant \mu^2$ with $\mu^2 < \nu^2/K$, (3.18) implies (3.17) (as a strict inequality). Thus, for $G \leqslant \mu^2 < \nu^2/K$, (3.17) and (3.18) will hold for every T on the maximal interval of existence in which case Theorem 2.1 (in particular (2.4)) implies $T_0 = \infty$. From (3.16), (3.18) we have $u(t, \cdot)$, $u_t(t, \cdot), u_x(t, \cdot), u_{tt}(t, \cdot), u_{tx}(t, \cdot), u_{xx}(t, \cdot), u_{ttt}(t, \cdot), u_{ttx}(t, \cdot), u_{txx}(t, \cdot), u_{xxx}$ $\cdot(t, \cdot)$ in $L^2((-\infty, \infty); L^2(0,1)) \cap L^\infty((-\infty, \infty); L^2(0,1))$. Now $u(t, \cdot)$, $u_t(t, \cdot), u_x(t, \cdot), u_{tt}(t, \cdot), u_{tx}(t, \cdot), u_{xx}(t, \cdot), u_{ttt}(t, \cdot), u_{ttx}(t, \cdot), u_{txx}(t, \cdot)$ in $L^2((-\infty, \infty); L^2(0,1))$ implies

(3.20) $u(t, \cdot), u_t(t, \cdot), u_x(t, \cdot), u_{tt}(t, \cdot), u_{tx}(t, \cdot), u_{xx}(t, \cdot) \xrightarrow{L^2(0,1)} 0, t \to \infty,$

which, in conjunction with $u_x(t, \cdot), u_{tx}(t, \cdot), u_{xx}(t, \cdot), u_{ttx}(t, \cdot), u_{txx}(t, \cdot)$, $u_{xxx}(t, \cdot)$ in $L^\infty((-\infty, \infty); L^2(0,1))$, yields (1.25).

It thus remains to verify (3.18) under the assumption (3.17). We fix s in $(-\infty, T]$. The first estimate is obtained by multiplying (1.10) by $\psi(u_x)_{tx}$, integrating over $(-\infty, s] \times [0, 1]$ and integrating by parts. The reader should be cautioned that in these and the many integrations by parts which follow, there are several possible ways to carry out such integrations. The ones selected in this section are chosen for the purpose of using the same estimates when considering the boundary conditions

(4.1), (4.2) below (see Theorem 4.1). The result of this calculation is

$$(3.21) \quad \frac{1}{2} \int_0^1 \psi'(u_x(s,x)) u_{tx}^2(s,x)\, dx$$

$$+ \frac{1}{2} \int_0^1 \chi'(u_x(s,x)) \psi'(u_x(s,x)) u_{xx}^2(s,x)\, dx$$

$$+ \int_{-\infty}^s \int_0^1 \psi(u_x)_{tx} \int_{-\infty}^t a(t-\tau) \psi(u_x)_{\tau x}\, d\tau\, dx\, dt$$

$$= \frac{1}{2} \int_{-\infty}^s \int_0^1 \psi''(u_x) u_{tx}^3\, dx\, dt$$

$$- \frac{1}{2} \int_{-\infty}^s \int_0^1 \{ \chi'(u_x)\psi''(u_x) - \chi''(u_x)\psi'(u_x) \} u_{tx} u_{xx}^2\, dx\, dt$$

$$- \int_0^1 g(s,x)\psi(u_x(s,x))_x\, dx + \int_{-\infty}^s \int_0^1 g_t \psi(u_x)_x\, dx\, dt.$$

To motivate our next estimate, we differentiate (1.10) with respect to t and then integrate formally by parts to get

$$(3.22) \quad u_{ttt} = \chi(u_x)_{tx} + \int_{-\infty}^t a(t-\tau)\psi(u_x)_{\tau\tau x}\, d\tau + g_t.$$

We would like to multiply (3.22) by $\psi(u_x)_{ttx}$ and then integrate over $(-\infty, s] \times [0,1]$ in order to arrive at an estimate analogous to (3.21). Unfortunately, this operation is not legitimate since $\psi(u_x)_{ttx}$ does not necessarily exist as a function. Consequently, same as with the derivation of (2.12) in Section 2, we shall have to work first with a discrete analog of (3.22) and then pass to the limit. To this end we apply to (1.10) the forward difference operator Δ, of step $\eta > 0$, thus arriving at

$$(3.23) \quad \Delta u_{tt} = \Delta \chi(u_x)_x + \int_{-\infty}^t a(t-\tau) \Delta \psi(u_x)_{\tau x}\, d\tau + \Delta g.$$

We now multiply (3.23) by $\Delta \psi(u_x)_{tx}$, we integrate over $(-\infty, s] \times [0,1]$, we perform a number of integrations by parts, we divide through by η^2, and we pass to the limit as $\eta \to 0$. The outcome of this tedious but straightforward calculation is

$$(3.24) \quad \frac{1}{2} \int_0^1 \psi'(u_x(s,x)) u_{ttx}^2(s,x)\, dx$$

$$+ \frac{1}{2} \int_0^1 \chi'(u_x(s,x)) \psi'(u_x(s,x)) u_{txx}^2(s,x)\, dx$$

$$+ \lim_{\eta \downarrow 0} \frac{1}{\eta^2} \int_{-\infty}^s \int_0^1 \Delta\psi(u_x)_{tx} \int_{-\infty}^t a(t-\tau) \Delta\psi(u_x)_{\tau x}\, d\tau\, dx\, dt$$

$$= \frac{5}{2} \int_{-\infty}^s \int_0^1 \psi''(u_x) u_{tx} u_{ttx}^2\, dx\, dt$$

$$- \int_0^1 \psi''(u_x(s,x)) u_{tx}^2(s,x) u_{ttx}(s,x)\, dx$$

$$+ \int_{-\infty}^s \int_0^1 \psi'''(u_x) u_{tx}^3 u_{ttx}\, dx\, dt$$

$$+ \frac{3}{2} \int_{-\infty}^s \int_0^1 \{\chi''(u_x)\psi'(u_x) - \chi'(u_x)\psi''(u_x)\} u_{tx} u_{txx}^2\, dx\, dt$$

$$+ \int_{-\infty}^s \int_0^1 \{\chi''(u_x)\psi'(u_x) - \chi'(u_x)\psi''(u_x)\} u_{xx} u_{ttx} u_{txx}\, dx\, dt$$

$$+ \int_{-\infty}^s \int_0^1 \{\chi'''(u_x)\psi'(u_x) - \chi''(u_x)\psi''(u_x)$$

$$- \chi'(u_x)\psi'''(u_x)\} u_{tx}^2 u_{xx} u_{txx}\, dx\, dt$$

$$- \int_{-\infty}^s \int_0^1 \chi''(u_x)\psi''(u_x) u_{tx} u_{xx}^2 u_{ttx}\, dx\, dt$$

$$- \int_{-\infty}^s \int_0^1 \chi''(u_x)\psi'''(u_x) u_{tx}^3 u_{xx}^2\, dx\, dt$$

$$- \int_0^1 \chi''(u_x(s,x))\psi'(u_x(s,x)) u_{tx}(s,x) u_{xx}(s,x) u_{txx}(s,x)\, dx$$

$$- \int_0^1 g_{1t}(s,x)\psi(u_x(s,x))_{tx}\, dt + \int_{-\infty}^s \int_0^1 g_{1tt}\psi(u_x)_{tx}\, dx\, dt$$

$$+ \int_{-\infty}^s \int_0^1 g_{2tx}\psi(u_x)_{tt}\, dx\, dt.$$

The reader should note that because $\psi(u_x)_{ttx}$ does not exist as a function, one cannot, after dividing (3.23) by η, pass to the limit as $\eta \to 0$ under the integral on the right hand side of (3.23). However, the limit of every other term in (3.23) exists as $\eta \to 0$, and therefore,

$$\lim_{\eta \to 0} \frac{1}{7} \int_{-\infty}^{t} a(t - \tau) \Delta \psi(u_x)_{\tau x}(\tau, x) \, d\tau$$

exists for $t \in [0, T], 0 \leqslant x \leqslant 1$. The same comment (arrived at by the same reasoning) applies to the limit as $\eta \to 0$ of the multiple (quadratic form) integral on the left hand side of (3.24). It is necessary for the subsequent estimates (in particular (3.27), (3.28)) to know that this limit exists and is finite (in fact positive). This is important for the concluding part of the proof of Theorem 1.1 (see argument preceding (3.37) below).

To get our next estimate we multiply by u_{txx} the identity

$$(3.25) \quad a(0)\Delta\psi(u_x)_x = -\int_{-\infty}^{t} a'(t - \tau)\Delta\psi(u_x)_x \, d\tau$$

$$+ \int_{-\infty}^{t} a(t - \tau)\Delta\psi(u_x)_{\tau x} \, d\tau$$

and we integrate over $(-\infty, s] \times [0, 1]$. We majorize the right-hand side of the resulting equation by first applying Schwarz's inequality and then using (3.1) and (3.2). The result is

$$(3.26) \quad a(0)\int_{-\infty}^{s}\int_{0}^{1} u_{txx}\Delta\psi(u_x)_x \, dx \, dt$$

$$\leqslant \left\{ \int_{-\infty}^{s}\int_{0}^{1} u_{txx}^2 \, dx \, dt \right\}^{1/2}$$

$$\times \left\{ \gamma \int_{-\infty}^{s}\int_{0}^{1} \Delta\psi(u_x)_x \int_{-\infty}^{t} a(t - \tau)\Delta\psi(u_x)_x \, d\tau \, dx \, dt \right\}^{1/2}$$

$$+ \left\{ \int_{-\infty}^{s}\int_{0}^{1} u_{txx}^2 \, dx \, dt \right\}^{1/2}$$

$$\times \left\{ \beta \int_{-\infty}^{s}\int_{0}^{1} \Delta\psi(u_x)_{tx} \int_{-\infty}^{t} a(t - \tau)\Delta\psi(u_x)_{\tau x} \, d\tau \, dx \, dt \right\}^{1/2}.$$

Dividing through by η (the step of the forward difference operator Δ), letting $\eta \downarrow 0$, and using (3.15) and the Cauchy-Schwarz inequality, we end up with the estimate

$$(3.27) \quad \frac{1}{2} \kappa a(0) \int_{-\infty}^{s} \int_{0}^{1} u_{txx}^2 \, dx \, dt$$

$$- \frac{\gamma}{\kappa a(0)} \int_{-\infty}^{s} \int_{0}^{1} \psi(u_x)_{tx} \int_{-\infty}^{t} a(t - \tau)\psi(u_x)_{\tau x} \, d\tau \, dx \, dt$$

$$- \frac{\beta}{\kappa a(0)}$$

$$\times \lim_{\eta \downarrow 0} \frac{1}{\eta^2} \int_{-\infty}^{s} \int_{0}^{1} \Delta \psi(u_x)_{tx} \int_{-\infty}^{t} a(t - \tau)\Delta \psi(u_x)_{\tau x} \, d\tau \, dx \, dt$$

$$\leqslant - a(0) \int_{-\infty}^{s} \int_{0}^{1} \psi''(u_x) u_{tx} u_{xx} u_{txx} \, dx \, dt.$$

Next we integrate over $(-\infty, s] \times [0, 1]$ the square of (3.23), we use (3.1), then we divide through by η^2 and we let $\eta \downarrow 0$ to get

$$(3.28) \quad \int_{-\infty}^{s} \int_{0}^{1} u_{ttt}^2 \, dx \, dt - 4 \int_{-\infty}^{s} \int_{0}^{1} \chi'(u_x)^2 u_{txx}^2 \, dx \, dt$$

$$- 4\beta \lim_{\eta \downarrow 0} \frac{1}{\eta^2} \int_{-\infty}^{s} \int_{0}^{1} \Delta \psi(u_x)_{tx} \int_{-\infty}^{t} a(t - \tau)\Delta \psi(u_x)_{\tau x} \, d\tau \, dx \, dt$$

$$\leqslant 4 \int_{-\infty}^{s} \int_{0}^{1} \chi''(u_x)^2 u_{tx}^2 u_{xx}^2 \, dx \, dt + 4 \int_{-\infty}^{s} \int_{0}^{1} g_t^2 \, dx \, dt.$$

To the above estimate we append

$$(3.29) \quad \int_{-\infty}^{s} \int_{0}^{1} u_{ttx}^2 \, dx \, dt - \int_{-\infty}^{s} \int_{0}^{1} u_{ttt} u_{txx} \, dx \, dt$$

$$+ \int_{0}^{1} u_{tt}(s, x) u_{txx}(s, x) \, dx = 0$$

which can be derived from

$$(3.30) \quad \int_{-\infty}^{s} \int_{0}^{1} \Delta u_{tx}^2 \, dx \, dt = \int_{-\infty}^{s} \int_{0}^{1} \Delta u_{tt} \Delta u_{xx} \, dx \, dt$$

$$- \int_{0}^{1} \Delta u_t(s, x) \Delta u_{xx}(s, x) \, dx$$

by passing to the limit. In turn, (3.30) can be easily verified via integrations by parts.

We now differentiate (1.9) with respect to t,

$$(3.31) \quad u_{ttt} = \varphi(u_x)_{tx} + \int_{-\infty}^{t} a'(t - \tau)\psi(u_x)_{\tau x} \, d\tau + g_t,$$

and we easily get the estimate

$$(3.32) \quad \int_0^1 u_{ttt}^2(s, x) \, dx - 5\int_0^1 \varphi'(u_x(s, x))u_{txx}^2(s, x) \, dx$$

$$-5\left\{ \int_0^\infty |a'(t)| \, dt \right\}^2 \sup_{(-\infty, s]} \int_0^1 \psi'(u_x)^2 u_{txx}^2 \, dx$$

$$\leqslant 5\int_0^1 \varphi''(u_x(s, x))u_{tx}^2(s, x)u_{xx}^2(s, x) \, dx$$

$$+5\left\{ \int_0^\infty |a'(t)| \, dt \right\}^2 \sup_{(-\infty, s]} \int_0^1 \psi''(u_x)^2 u_{tx}^2 u_{xx}^2 \, dx$$

$$+5\int_0^1 g_t^2(s, x) \, dx.$$

The final set of estimates is derived by the following procedure: We differentiate Equation (1.9) with respect to x and then add and subtract appropriate terms to arrive at

$$(3.33) \quad \varphi'(0)u_{xxx} + \int_{-\infty}^{t} a'(t - \tau)\psi'(0)u_{xxx} \, d\tau$$

$$= u_{ttx} - [\varphi'(u_x) - \varphi'(0)]u_{xxx}$$

$$- \varphi''(u_x)u_{xx}^2 - \int_{-\infty}^{t} a'(t - \tau)[\psi'(u_x) - \psi'(0)]u_{xxx} \, d\tau$$

$$- \int_{-\infty}^{t} a'(t - \tau)\psi''(u_x)u_{xx}^2 \, d\tau - g_x \overset{\text{def}}{=} X(t, x).$$

Thus, if $k(t)$ is the resolvent kernel of the operator (3.5),

$$(3.34) \quad \varphi'(0)u_{xxx}(t, x) = X(t, x) + \int_{-\infty}^{t} k(t - \tau)X(\tau, x) \, d\tau.$$

By Lemma 3.2, $k(t) \in L^1(0, \infty)$ so that we have estimates

(3.35) $\varphi'(0)^2 \int_0^1 u_{xxx}^2(s, x)\, dx \leqslant 2 \int_0^1 X^2(s, x)\, dx$

$$+ 2\left\{ \int_0^\infty |k(t)|\, dt \right\}^2 \sup_{(-\infty, s]} \int_0^1 X^2(t, x)\, dx,$$

(3.36) $\varphi'(0)^2 \int_{-\infty}^s \int_0^1 u_{xxx}^2\, dx\, dt$

$$\leqslant 2\left\{ 1 + \left[\int_0^\infty |k(t)|\, dt \right]^2 \right\} \int_{-\infty}^s \int_0^1 X^2(t, x)\, dx\, dt.$$

We are now ready to prove (3.18) under assumption (3.17). First we note that, on account of (3.15), $U(T)$ can be majorized, with the help of the Poincaré inequality, by the supremum over $(-\infty, T]$ of an appropriate linear combination of the left-hand sides of the estimates (3.21), (3.24), (3.27), (3.28), (3.29), (3.32), (3.35) and (3.36). On the other hand, each term on the right-hand sides of these estimates can be majorized, by means of the Cauchy-Schwarz inequality and (3.17), by either cG, or $O(\nu)U(T)$, or $\epsilon U(T) + c(\epsilon)G$ for any $\epsilon > 0$. We thus arrive at an estimate of the form

(3.37) $U(T) \leqslant \{O(\nu) + O(\epsilon)\}U(T) + c(\epsilon)G$

from which one can get (3.18) by fixing ν and ϵ sufficiently small. The proof of Theorem 1.1 is complete.

4. Remarks and Extensions. When the endpoints of the body are fixed, in the place of (1.18) we have boundary conditions

(4.1) $u(t, 0) = u(t, 1) = 0, \qquad -\infty < t < \infty.$

Similarly, when one endpoint (say $x = 0$) is fixed and the other is free,

(4.2) $u(t, 0) = 0, \qquad \sigma(t, 1) = 0, \qquad -\infty < t < \infty.$

In these cases no rigid motions are possible so we don't have to assume (1.20) nor should we expect that (1.21) will generally hold.

When the body force satisfies (1.22) with $g_2(t, x) \equiv 0$, all estimates derived in Section 3 for the case (3.11) are also valid under (4.1) or (4.2). We thus have

THEOREM 4.1.   *There is* $\mu > 0$ *with the property that for every* $g(t, x)$ *on* $(-\infty, \infty) \times [0, 1]$, *which satisfies* (1.22) *with* $g_2(t, x) \equiv 0$ *and* (1.24), *and any* $v(t, x)$ *on* $(-\infty, 0] \times [0, 1]$, *with* $v(t, \cdot), v_t(t, \cdot), v_x(t, \cdot), v_{tt}(t, \cdot),$ $v_{tx}(t, \cdot), v_{xx}(t, \cdot), v_{ttt}(t, \cdot), v_{ttx}(t, \cdot), v_{txx}(t, \cdot), v_{xxx}(t, \cdot)$ *in* $C((-\infty, 0]; L^2(0, 1)) \cap L^2((-\infty, 0]; L^2(0, 1))$, *which satisfies* (1.9) *together with* (4.1) (*or* (4.2)) *for* $t \leq 0$, *there exists a unique* $u(t, x)$ *on* $(-\infty, \infty) \times [0, 1]$, *with* $u(t, \cdot), u_t(t, \cdot), u_x(t, \cdot), u_{tt}(t, \cdot), u_{tx}(t, \cdot), u_{xx}(t, \cdot), u_{ttt}(t, \cdot), u_{ttx}(t, \cdot),$ $u_{txx}(t, \cdot), u_{xxx}(t, \cdot)$ *in* $C((-\infty, \infty); L^2(0, 1)) \cap L^2((-\infty, \infty); L^2(0, 1))$; *which satisfies* (1.9), (1.11) *and* (4.1) (*or* (4.2)). *Moreover,* (1.25) *holds.*

As history and body force get smoother, solutions become smoother. Regularity results can be obtained by establishing a priori estimates for derivatives of u of order 4, 5, etc. Such estimates fall into three categories: those derived by differentiating (1.10) a number of times with respect to t and/or x and then multiplying by the appropriate multiplier (recall the derivation of (3.24)); those derived by expressing certain derivatives in terms of other derivatives through the equation itself (compare with (3.32) or (3.35)); those obtained through interpolation (such as (3.29)). The program is feasible because, since the problem is autonomous (kernel of convolution type), differentiations with respect to x or t essentially preserve the form of the equation (compare, for example, (3.22) with (1.10)). Time derivatives of u satisfy the same homogeneous boundary conditions as u, at $x = 0, 1$, so that differentiating the equation with respect to t is generally a better prospect than differentiating with respect to x. In any event there are so many possible combinations of differentiations, integrations by parts, etc., that one may establish several variants of regularity theorems. Here is a typical one:

THEOREM 4.2.   *Suppose the assumptions of Theorem* 1.1 *hold and, in addition,* φ *and* ψ *are* C^4 *smooth,* $v_{ttt}(t, \cdot), v_{ttx}(t, \cdot), v_{txx}(t, \cdot), v_{txxx}(t, \cdot),$ $v_{xxxx}(t, \cdot)$ *are in* $C((-\infty, 0]; L^2(0, 1)) \cap L^2((-\infty, 0]; L^2(0, 1))$ *and*

(4.3)
$$
\begin{cases}
g_{tt}(t, \cdot), g_{tx}(t, \cdot), g_{xx}(t, \cdot) \\[4pt]
in\ C\big((-\infty, \infty); L^2(0, 1)\big) \cap L^2\big((-\infty, \infty); L^2(0, 1)\big) \\[4pt]
g(t, x) = g_1(t, x) + g_2(t, x) \\[4pt]
with\ g_{1ttt}(t, \cdot), g_{2ttx}(t, \cdot) \\[4pt]
in\ L^2\big((-\infty, \infty); L^2(0, 1)\big).
\end{cases}
$$

Then, when (1.24) *is satisfied with* μ *sufficiently small, the solution* $u(t, x)$
of (1.9), (1.11), (1.18) *possesses* $u_{ttt}(t, \cdot), u_{ttx}(t, \cdot), u_{txx}(t, \cdot), u_{txxx}(t, \cdot),$
$u_{xxxx}(t, \cdot)$ *in* $C((-\infty, \infty); L^2(0, 1)) \cap L^2((-\infty, \infty); L^2(0, 1))$ *and*

$$(4.4) \qquad u_{ttt}(t, \cdot), u_{ttx}(t, \cdot), u_{txx}(t, \cdot), u_{xxx}(t, \cdot) \xrightarrow[\;[0, 1]\;]{\text{unif.}} 0, \qquad t \to \infty.$$

It is noteworthy that the extra derivatives (4.3) of g that are required
in order to guarantee smoothness of the solution need not be "small".
This is due to the fact that all energy integrals are quadratic forms in the
higher order derivatives of u, with coefficients that are solely controlled
by ν of (3.17). As we have seen in Section 3, ν is controlled by U which,
in turn, is controlled by G.

We close with remarks on the multidimensional situation. The
configuration of the body is now a bounded set $\Omega \subset R^n$ with smooth
boundary $\partial\Omega$ and the displacement is an n-dimensional vector field $\mathbf{u}$. A
typical problem is to determine $\mathbf{u}(t, \mathbf{x}), -\infty < t < \infty, \mathbf{x} \in \Omega$, such that

$$(4.5) \qquad \frac{\partial^2 u_i}{\partial t^2} = \sum_{j=1}^{n} \left\{ \frac{\partial \Phi_{ij}}{\partial x_j} + \int_{-\infty}^{t} a'(t - \tau) \frac{\partial \Psi_{ij}}{\partial x_j} \, d\tau \right\} + g_i,$$

$$i = 1, \ldots, n, \; -\infty < t < \infty, \mathbf{x} \in \Omega,$$

$$(4.6) \qquad \mathbf{u}(t, \mathbf{x}) = \mathbf{v}(t, \mathbf{x}), \qquad -\infty < t \leqslant 0, \quad \mathbf{x} \in \Omega,$$

$$(4.7) \qquad \mathbf{u}(t, \mathbf{x}) = \mathbf{0}, \qquad -\infty < t < \infty, \quad \mathbf{x} \in \partial\Omega,$$

where Φ and Ψ are known, smooth, matrix valued functions of the
matrix $\mathbf{e} = \nabla\mathbf{u}$ (strain), $\mathbf{g}(t, \mathbf{x})$ is an assigned body force and $\mathbf{v}(t, \mathbf{x})$ is the
given history.

We assume $\Phi(0) = \Psi(0) = 0$, set

$$(4.8) \qquad C_{ijkl}(\mathbf{e}) = \frac{\partial \Phi_{ij}(\mathbf{e})}{\partial e_{kl}}, \qquad D_{ijkl}(\mathbf{e}) = \frac{\partial \Psi_{ij}(\mathbf{e})}{\partial e_{kl}},$$

$$(4.9) \qquad E_{ijkl}(\mathbf{e}) = C_{ijkl}(\mathbf{e}) - a(0)D_{ijkl}(\mathbf{e}),$$

and impose the symmetry restrictions

$$(4.10) \qquad C_{ijkl} = C_{klij}, \qquad D_{ijkl} = D_{klij}.$$

Assumptions (1.4), (1.5), (1.6) will here turn into coercivity conditions for the partial differential operators associated with $C_{ijkl}(\mathbf{0}), D_{ijkl}(\mathbf{0})$ and $E_{ijkl}(\mathbf{0})$. Under boundary conditions (4.7), coercivity is equivalent to strong ellipticity

$$(4.11) \qquad \sum_{i,j,k,l} C_{ijkl}(\mathbf{0})\xi_i\xi_k\zeta_j\zeta_l > 0, \qquad |\xi| = |\zeta| = 1,$$

$$(4.12) \qquad \sum_{i,j,k,l} D_{ijkl}(\mathbf{0})\xi_i\xi_k\zeta_j\zeta_l > 0, \qquad |\xi| = |\zeta| = 1,$$

$$(4.13) \qquad \sum_{i,j,k,l} E_{ijkl}(\mathbf{0})\xi_i\xi_k\zeta_j\zeta_l > 0, \qquad |\xi| = |\zeta| = 1.$$

Assumptions (4.10), (4.11), (4.12) and (4.13) can be motivated by Mechanics. However, in order to carry through the analysis, we require an additional condition whose physical interpretation is less clear. We define

$$(4.14) \qquad F_{jklpqr}(\mathbf{e}) = \sum_i D_{ijkl}(\mathbf{e})E_{ipqr}(\mathbf{e})$$

and assume that $\mathbf{F}$ is symmetric,

$$(4.15) \qquad F_{jklpqr} = F_{pqrjkl},$$

and that its value at $\mathbf{e} = \mathbf{0}$ corresponds to a coercive operator. We note that in the special case $\mathbf{\Phi} \equiv \mathbf{\Psi}$ the resulting $\mathbf{F}$ automatically satisfies the above conditions.

Under the above assumptions it is possible to establish the existence of globally defined smooth solutions to (4.5), (4.6), (4.7) by the procedure followed here in the one-dimensional case. The strategy is to establish a priori energy estimates for the $L^2(\Omega)$ norms of derivatives of $\mathbf{u}$ of sufficiently high order (depending upon n) that would guarantee, via Sobolev's lemma, pointwise bounds analogous to (3.17). The calculations, however, are very long.

It is easy to discern the essential ingredients in the proofs and it thus seems feasible to develop an existence theory for the history-value problem in a class of abstract nonlinear integrodifferential equations

$$(4.16) \qquad \frac{d^2u}{dt^2} = A(u(t)) + \int_{-\infty}^t a'(t - \tau)B(u(\tau))\,d\tau + g(t)$$

on a Hilbert space H, where A and B are nonlinear operators defined on a scale of Hilbert spaces (abstracting the scale of Sobolev spaces $[W^{k,2}(\Omega)]^n$) and satisfying appropriate symmetry and coercivity conditions. We remark also that the general and physically interesting case in which the stress-strain relaxation function is a $n \times n$ matrix $\mathbf{a}$ (in (4.5) $\mathbf{a} = aI$) is considerably more complicated than the situation considered here.

BROWN UNIVERSITY

UNIVERSITY OF WISCONSIN, MADISON

REFERENCES

[1] Nishida, T., Global smooth solutions for the second-order quasilinear wave equation with the first order dissipation (unpublished).

[2] Matsumura, A., Global existence and asymptotics of the solutions of the second order quasilinear hyperbolic equations with first order dissipation, *Publ. Res. Inst. Math. Sci. Kyoto Univ.*, Ser. A **13** (1977), 349–379.

[3] Schauder, J., Das Anfangswertproblem einer quasilinearen hyperbolischen Differentialgleichung zweiter Ordnung in beliebiger Anzahl von unabhängigen Veränderlichen, *Fund. Math.* **24** (1935), 213–246.

[4] Slemrod, M., Global existence, uniqueness and asymptotic stability of classical smooth solutions in one-dimensional non-linear thermoelasticity, *Arch. Rat. Mech. Anal.* **76** (1981), 97–133.

[5] MacCamy, R. C., A model for one-dimensional, nonlinear viscoelasticity, *Q. Appl. Math.* **35** (1977), 21–33.

[6] Dafermos, C. M. and J. A. Nohel, Energy methods for nonlinear hyperbolic Volterra integrodifferential equations, *Comm. P. D. E.* **4** (1979), 219–278.

[7] Staffans, O., On a nonlinear hyperbolic Volterra equation, *SIAM J. Math. Anal.* **11** (1980), 793–812.

[8] Greenberg, J. M., A priori estimates for flows in dissipative materials, *J. Math. Anal. Appl.* **60** (1977), 617–630.

R-CONVEXITY OF THE INTEGRAL OF SET-VALUED FUNCTIONS

By Halina Frankowska and Czesław Olech

---

1. Introduction. The reachable set of a linear control system is, as is well known, the image by a linear map of the integral of a set-valued map. As such it is convex and under some general assumptions also compact.

For nonlinear control system this is no more the case in general. Studying the problem of convexity of reachable sets in the nonlinear case, A. Pliś [5] introduced a notion of R-convexity. A set is R-convex if it is equal to the intersection of balls of fixed radius R. He proved that if the right hand side of a control system (differential inclusion) is R-convex then under some regularity assumptions the reachable set is R-convex locally.

A natural question to ask is when the reachable set of a linear control system is R-convex, which is equivalent to the question of whether the integral of a set-valued-function is R-convex.

Łojasiewicz, Jr. [1] gave an answer to this question for 2-dimensional case when values of the function are polygons with vertices changing smoothly. He also proved that if the dimension of the system is at least 3 the integral of smoothly turning line segment cannot be R-convex.

The purpose of the present paper is to consider this question for arbitrary set-valued functions. In section 2 we present a lemma which is basic for our consideration. This lemma is a more precise version of a lemma, which the second of the authors proved and used for establishing convexity and closedness properties of the integral of a set-valued map (cf. [2]).

In section 3 we give some useful characterizations of R-convexity and we prove some necessary and sufficient conditions for the integral of set-valued functions to be R-convex.

Section 4 contains some special cases and corollaries. In particular, we show how the results of Łojasiewicz we mentioned above can be obtained from ours.

117

2. A Lemma. Below we denote by $U(t)$ a set-valued function from $[0, 1]$ into closed subsets of R^k and by K the set of integrable selections of U; that is $K = \{u : [0, 1] \to R^k \,|\, u \text{ integrable and } u(t) \in U(t) \text{ a.e. in } [0, 1]\}$.

The integral of U, denoted by $I(U)$, is defined as the set $\{\int_0^1 u(t) \,|\, u \in K\}$. It is well known that $I(U)$ is convex.

By $I^{-1}(x)$, $x \in R^n$ we denote the set $\{u \in K \,|\, \int_0^1 u(\tau) \,d\tau = x\}$. $I^{-1}(x)$ is not empty iff $x \in I(U)$. In general $I^{-1}(x)$ is a set containing more than one point. In [2] a lemma is proved which states that for any $x, y \in I(U)$ being near to an extreme point of $\operatorname{cl} I(U)$ the L_1-norm $\|u - v\|$ is small for any $u \in I^{-1}(x)$ and $v \in I^{-1}(y)$. This implies, in particular that if z is an extreme point of $\operatorname{cl} I(U)$ then $z \in I(U)$ and $I^{-1}(z)$ reduces to a one point set. Thus extreme points of $\operatorname{cl} I(U)$ belong to $I(U)$, hence if $I(U)$ is bounded then $I(U)$ is compact and convex.

The lemma which follows is a more precise version of the mentioned result (cf. [2], Lemma 1) in the sense that it gives an estimate of the norm $\|u - v\|$ by a certain function which is determined by the shape of $I(U)$ around an extreme point. We shall define now the function in question.

Let z be a fixed boundary point of $I(U)$ and let $n \in R^k$, $|n| = 1$ be any normal vector to $I(U)$ at z; that is $\langle n, z \rangle = \max_{x \in I(U)} \langle n, x \rangle$; here $\langle , \rangle$ stands for the scalar product in R^k.

Let

(1) $\varphi(\rho) = \sup\{|x - z| \,|\, x \in I(U), \langle z - x, n \rangle \leqslant \rho\}$

The function φ is non negative, nondecreasing and concave, since $I(U)$ is convex. It depends on the set $I(U)$, $z \in \partial I(U)$ and n; but to avoid unnecessary complication we omit parameters in the notation since they are fixed. Notice also that for bounded $I(U)$ $\varphi(\rho) \to 0$ as $\rho \to 0$ if and only if z is an exposed point of $I(U)$. Clearly by (1) we have the inequality

(2) $|x - z| \leqq \varphi(\langle z - x, n \rangle)$ for each $x \in I(U)$.

We have the following

LEMMA 2.1. *Let z be a boundary point of $I(U)$ and n be a normal vector to $I(U)$ at z. Let φ be given by (1). Then for each $x \in I(U)$ we* *have the inequality*

(3) $$\|v - u\| \leqslant ((k - 1)\gamma + 1)\varphi(\langle z - x, n\rangle))$$

for any $v \in I^{-1}(z)$ and $u \in I^{-1}(x)$, where $\gamma = \sup_{\rho > 0} 2\varphi(\rho/2)/\varphi(\rho)$ $\leqslant 2$.

Proof. Let $A \subset [0, 1]$ be measurable, $u \in I^{-1}(z)$ fixed. Put $u_A = \chi_A u + \chi_{A'} v$, where $A' = [0, 1]\backslash A$ and χ_A stands for the characteristic function of A. Let $x_A = \int_0^1 u_A(\tau)\,d\tau$. Since $u_A \in K$, we have $x_A \in I(u)$ and we have the equality $z - x_A + z - x_{A'} = z - x$, therefore

(4) $$\langle z - x_A, n\rangle + \langle z - x_{A'}, n\rangle = \langle z - x, n\rangle.$$

Notice that all terms in (4) are nonnegative.

Let us fix $1 \leqslant i \leqslant k$, and take $A = \{t \mid v_i(t) - u_i(t) > 0\}$, where v_i and u_i denote the i-th coordinate of v and u, respectively.

We have the inequality

$$\int_A |v_i(\tau) - u_i(\tau)|\,d\tau \leqslant \left|\int_0^1 (v(\tau) - u_A(\tau)\,d\tau)\right| = |z - x_A|$$

$$\leqslant \varphi(\langle z - x_A, n\rangle)$$

Similarly $\int_{A'} |v_i(\tau) - u_i(\tau)|\,d\tau \leqslant \varphi(\langle z - x_{A'}, n\rangle)$. The above inequalities imply that

$$\int_0^1 |v_i(\tau) - u_i(\tau)|\,d\tau \leqslant \varphi(\langle z - x_A, n\rangle) + \varphi(\langle z - x_{A'}, n\rangle)$$

and because of (4) and concavity of φ

(5) $$\int_0^1 |v_i(\tau) - u_i(\tau)|\,d\tau \leqslant 2\varphi\left(\frac{\langle z - x, n\rangle}{2}\right) \leqslant \gamma\varphi(\langle z - x, n\rangle)$$

Since $\|v - u\| \leqslant \sum \int_0^1 |u_i(\tau) - v_i(\tau)|\,d\tau$, inequality (5) implies (3). Notice that if $v_i(\tau) - u_i(\tau) \geqslant 0$ almost everywhere in $[0, 1]$ for some i then (5) holds also with γ replaced by 1 and that without any loss of generality we may assume that this is the case for one i. Thus the constant $(k - 1)\gamma + 1$ in (3).

Notice that the above lemma holds also for any φ which is nonnegative and concave provided the inequality (2) is assumed. On the other hand inequality (3) implies that $|x - z| \leqslant ((k - 1)\gamma + 1)\varphi(\langle z - x, n\rangle)$. Hence if (3) holds true for some function φ nonnegative concave then $((k - 1)\gamma + 1)\varphi(\langle z - x, u\rangle)$ estimates from above the right-hand side of (1).

3. R-convexity of the integral $I(U)$. In this section we shall apply the lemma to get a characterization of R-convexity of the integral $I(U)$ of a set-valued function $U(t)$; that is we will formulate necessary and sufficient conditions for $I(U)$ to be an intersection of balls of a fixed radius R.

Before we state the results we shall prove a proposition which gives a few useful characterizations of R-convexity itself.

PROPOSITION 3.1. *For $A \subset R^k$ convex and compact the following conditions are equivalent*

(i) *A is an intersection of balls of radius R*
(ii) *For any two points $x, y \in A$, such that $|x - y| \leqslant 2R$ each arc of a circle of radius R which joins x and y and whose length is not greater than πR is contained in A.*
(ii′) *For each $x, y, \in A$ such that $|x - y| \leqslant 2R$ we have the inequality*

$$(6) \qquad \sup\left\{ r \mid \frac{x + y}{2} + rn \in A \right\} \geqslant \frac{|x - y|^2}{4R + 2\sqrt{4R^2 - |x - y|^2}}$$

for any n such that $|n| = 1$ and $\langle x - y, n\rangle = 0$,

(iii) *For each z on the boundary of A and any vector n, $|n| = 1$ normal to A at z we have the inequality*

$$(7) \qquad |z - x| \leqslant \sqrt{2R} \langle z - x, n\rangle^{1/2} \qquad \text{for each } x \in A.$$

(iv) *For each z, y from the boundary of A any vectors, $n, m, |n| = |m| = 1$ which are normal to A at z and y, respectively, the inequality holds*

$$(8) \qquad |z - y| \leqslant R|n - m|.$$

(v) *For each $z \in \partial A$ and any n, $|n| = 1$ normal to A at z the ball of center at $z - Rn$ and radius R contains A; that is $|z - Rn - x| \leqslant R$ for each $x \in A$.*

Proof. Condition (ii') is an equivalent analytical version of condition (ii). Indeed, the inequality (6) says that the middle point of each arc of a circle of radius R passing through x, y and of length smaller or equal πR belongs to A; but that implies easily that each such arc is contained in A.

Every ball of radius R contains any arc of a circle of radius R whose length is not greater than πR and whose end points are in the ball. This and (i) implies (ii). Hence (ii) follows from (i).

Assume (ii). It implies together with boundness of A that $|x - z| \leqslant 2R$ for any $x, y \in A$. To prove (iii) we take z, n and $x \in A$ fixed. Since n is normal to A at z we have the inequality $\langle x, n \rangle \leqslant \langle z, n \rangle$. There is $m \in R^k$, $|m| = 1$, $\langle m, n \rangle = 0$ such that $x - z = \alpha n + \beta m$ and $\beta \leqslant 0$. Clearly, the above inequality implies that $\alpha \leqslant 0$, too.

From (ii) it follows that there is $y - z = \gamma n + \delta m$ such that $\gamma < 0$ and $|z - y| = |x - y| = R$. Denote by C the arc of the circle of radius R centered at y with length less or equal πR and with end points at x and z. Notice that

(9) $\langle y + Rn, n \rangle \geqslant \langle z, n \rangle$

Indeed, $\langle y + Rn - z, n \rangle = R + \langle y - z, n \rangle \geqslant 0$, since $|\langle y - z, n \rangle| \leqslant R$. Equality in (9) holds only if $\langle y - z, m \rangle = 0$. Now if $\langle x - z, m \rangle = \beta < 0$ and $\langle y - z, m \rangle = \delta < 0$, then $y + Rn \in C$ does not belong to A since $\langle y + Rn, n \rangle > \langle z, n \rangle$. Therefore (ii) implies that $\beta\delta \leqslant 0$ and in consequence that

(10) $\langle x - z, y + Rn - z \rangle \leqslant 0$

Notice that $|x - y|^2 = |x - z|^2 + |y - z|^2 + 2\langle x - z, z - y \rangle$. But $|x - y|^2 = |y - z|^2 = R^2$, thus

$$\langle x - z, y - z \rangle = \tfrac{1}{2}|z - x|^2$$

The latter together with (10) implies (7). Thus (ii) implies (iii).

Assume now (iii) and let y, z and n, m be as in (iv). Then (iii) can be applied to both z, n and y, m and we get $|z - y|^2 \leqslant 2R\langle z - y, n \rangle$ and

$|z - y|^2 \leqslant 2R\langle y - z, m \rangle$. Hence

$$|z - y|^2 \leqslant R\langle z - y, n - m \rangle$$

which implies (8), and we proved that (iv) follows from (iii).

To prove the implication (iv) $\Rightarrow$ (v) consider the ball $B(x, R)$ of radius R centered at $x = z - Rn$ where n, $|n| = 1$ is normal to A at z. If A were not contained in $B(x, R)$ then there would exists $\overline{R} > R$ and $y \in A$ such that $A \subset B(x, \overline{R})$ and $|x - y| = \overline{R}$. Clearly $y \neq z$ and $m = (y - x)/|y - x|$ is normal to A at y. Applying (iv) we obtain

$$R|m - n| \geqslant |y - z| = |R(m - n) + (\overline{R} - R)m| > R|m - n|$$

Hence a contradiction, therefore $A \subset B(x, R)$ which proves (v).

Implication (v) $\Rightarrow$ (i) is obvious. Therefore the proof is completed.

Now we are in the position to state the announced characterization of R-convexity of the integral $I(U)$. To state the theorem let us introduce the following notation: for each n, $|n| = 1$ we denote by z_n any point of $I(U)$ such that n is normal to $I(U)$ at z_n.

THEOREM 3.1.   *Let U be a set-valued function from $[0, 1]$ into closed subsets of R^k and assume that $I(U)$ is not empty. Then the following conditions are equivalent.*

(i) *There is R such that $I(U)$ is R-convex.*

(ii) *There is L such that for each n and any $u \in I^{-1}(z_n)$ the inequality holds*

$$\|u - v\| \leqslant \sqrt{2L} \left\langle z_n - \int_0^1 v(t)\, dt, n \right\rangle^{1/2} \qquad \textit{for each } v \in K.$$

(iii) *There is M such that for each $n, m, |m| = 1$ and any $u \in I^{-1}(z_n)$, $v \in I^{-1}(z_m)$*

$$\|u - v\| \leqslant M|n - m|.$$

Proof.   The implication (i) $\Rightarrow$ (ii) follows from Lemma 2.1 and condition (iii) of Proposition 3.1.

If $u \in I^{-1}(z_n)$ and $v \in I^{-1}(z_m)$ then we obtain from (ii) the two inequalities $\|u - v\|^2 \leqslant 2L\langle z_n - z_m, n \rangle$ and $\|u - v\|^2 \leqslant 2L\langle z_m - z_n,$

$m\rangle$. Thus

$$\|u - v\|^2 \leqslant L < z_n - z_m, n - m\rangle \leqslant L|z_n - z_m||n - m|$$

Since $|z_n - z_m| \leqslant \|u - v\|$, therefore the above inequality implies $\|u - v\| \leqslant L|n - m|$. Thus (ii) implies (iii). Assuming (iii) we have inequality (8) in condition (iv) of Proposition 3.1. Hence (iii) implies that $I(U)$ is M-convex. Hence (i) follows and the proof is completed.

Remark. Condition (ii) implies that z is an exposed point of $I(u)$ and that $I^{-1}(z)$ reduces to a one point set; that is if $u_1, u_2 \in I^{-1}(z)$ then $u_1(t) = u_2(t)$ a.e. in $[0, 1]$. For the same reasons condition (iii) implies that the mapping $n \to u_n \in K$ where n belongs to unit sphere $\{n \mid |n| = 1\}$ and $\langle u_n(t), n \rangle \geqslant \langle u(t), n \rangle$ a.e. in $[0, 1]$ for each $u \in K$, is single valued and satisfies the Lipschitz condition $\|u_n - u_m\| \leqslant M|n - m|$.

Both characterizations (ii) and (iii) are expressed in terms of normal vectors to $I(U)$ and boundary points of the integral. However condition (ii) may be reformulated without explicit reference to normal vectors. Indeed (ii) implies that $I^{-1}(x)$ as a set valued function from $I(U)$ into subsets of K satisfies a Hölder condition with constant $M = \sqrt{2L}$ at boundary points of $I(U)$, that is,

$$(11) \qquad h\big(I^{-1}(x), I^{-1}(z)\big) \leqslant M|x - z|^{1/2},$$

for $z \in \partial I(U)$ and any $x \in I(U)$, where h stands for the Hausdorff distance of subsets of the space L_1. Indeed, since $I^{-1}(z)$ is a one-point set $h(I^{-1}(x), I^{-1}(z)) = \sup_{v \in I^{-1}(x)} \|v - u\|$, where $\{u\} = I^{-1}(z)$, and by (ii) this is estimated by $M\langle z - x, n\rangle^{1/2} \leqslant M|z - x|^{1/2}$. However (11) itself does not imply that $I^{-1}(z)$ reduces to a one point set. If the latter property is additionally assumed then (11) implies (ii) also. Indeed, we have the following

THEOREM 3.2. *Under the assumption of Theorem 3.1, $I(U)$ is R-convex for some $R < +\infty$ if and only if $I(U)$ is strictly convex and there is $M > 0$ such that $I^{-1}(x)$ satisfies a Hölder condition with exponent $1/2$ and constant M; that is (11) holds for every $x, z \in I(U)$.*

Proof. Let us fix $x, z \in I(U)$ and take any $u \in I^{-1}(x)$. Since $I_t(U) = \int_0^t U(\tau) d\tau$ depends continuously on t with respect to the Hausdorff metric there is $v \in K$ and $t_0 \leqslant 1$ such that $y = \int_0^{t_0} v(t) dt \in \partial I(U)$

and $\int_0^{t_0}(v(t) - u(t))\,dt = z - x$. Denote $A = [0, t_0]$. Then $u_A = v\chi_A + \chi_{A'}u \in I^{-1}(z)$, where $A' = [t_0, 1]$. Indeed $\int_0^1 u_A(t)\,dt = \int_0^{t_0}(v(t) - u(t))\,dt + \int_0^1 u(t)\,dt = z - x + x = z$. But $u_A - u = v - (\chi_A u + \chi_{A'}v)$, therefore if $I(U)$ is R-convex then by (ii) of Theorem 3.1 we get

$$\|u_A - u\| = \|v - (\chi_A u + \chi_{A'}v)\|$$

$$\leqslant \sqrt{2L}\left\langle \int_0^{t_0}(v(t) - u(t))\,dt, n \right\rangle^{1/2}$$

where n is any normal vector to $I(U)$ at z. Hence

$$\|u_A - u\| \leqslant M|z - x|^{1/2}, \qquad \text{where } M = \sqrt{2L}.$$

Therefore we proved that for each $u \in I^{-1}(x)$ there is $\tilde{u} = u_A \in I^{-1}(z)$ such that $\|u - \tilde{u}\| \leqslant M|z - x|^{1/2}$. Since the same arguments show that for each $\tilde{u} \in I^{-1}(z)$ there is $u \in I^{-1}(x)$ such that $\|u - \tilde{u}\| \leqslant M|z - x|^{1/2}$ and thus

$$h\big(I^{-1}(x), I^{-1}(z)\big) \leqslant M|z - x|^{1/2},$$

which was to be proved.

To prove the opposite implication let us take any $x, y \in I(U)$ and $n, |n| = 1$ such that $\langle n, x - y \rangle = 0$. Let $r > 0$ be such that $z = (x + y)/2 + rn$ belongs to the boundary of $I(U)$. Take any $u \in I^{-1}(x)$ and $v \in I^{-1}(y)$. There is $A \subset [0, 1]$ such that

$$(x + y)/2 = \int_0^1 u_A(t)\,dt = \int_0^1 u_{A'}(t)\,dt,$$

where $u_A = \chi_A u + \chi_{A'}v$ and $u_{A'} = \chi_{A'}u + \chi_A v$. Since $z \in \partial I(U)$ and $I(U)$ is assumed to be strictly convex therefore z is an extreme point of $I(U)$. Thus $I^{-1}(z)$ reduces to a one-point set, which we denote by w. Therefore from the assumed Hölder condition for I^{-1} we have the inequality

$$\|w - u_A\| + \|w - u_{A'}\| \leqslant 2M|z - (x + y)/2|^{1/2} = 2Mr^{1/2}.$$

But the left hand side of the above inequality is greater or equal to $\|u_A - u_{A'}\| = \|u - v\| \geqslant |x - y|$. Therefore we conclude that $|x - y|$

$\leqslant 2Mr^{1/2}$. If $|x - y| \leqslant 2M^2$ then the above inequality gives

$$r \geqslant \frac{|x - y|^2}{4M^2 + 2\sqrt{2M^4 - |x - y|^2}}.$$

Hence condition (ii′) of Proposition 3.1 implies that $I(U)$ is R-convex with $R = M^2$ and completes the proof.

The assumption that $I(U)$ is strictly convex in Theorem 3.2 is essential. Indeed suppose $I(U)$ is R-convex and put $\tilde{U}(t) = U(t) \times \{0\}$ $\subset R^{k+1}$ then $I(\tilde{U}) = I(U) \times \{0\}$ and $I^{-1}(x)$ satisfies a Hölder condition on $I(\tilde{U})$ but is not R-convex since it is not strictly convex subset of R^{k+1}.

4. Some corollaries and examples. In this section we shall discuss some examples and also described some special cases, when $I(U)$ is not R-convex.

We start with a trivial application of Theorem 3.1.

COROLLARY 4.1. *Assume that $U(t)$ is measurable and $R(t)$-convex. Then $I(U)$ is R-convex for some $R < +\infty$ if $\int_0^1 R(t)\,dt < +\infty$.*

Proof. Take n, m such that $|n| = |m| = 1$. Since $U(t)$ is $R(t)$-convex, hence strictly convex, we have $z_n = \int_0^1 u_n(t)\,dt$, where n is normal to $I(U)$ at z_n and normal to $U(t)$ at $u_n(t)$ for each t. Both z_n and $u_n(t)$ for each t are uniquely defined. Similarly $z_m = \int_0^1 u_m(t)\,dt$. But by (iv) of Proposition 3.1 $|u_n(t) - u_m(t)| \leqslant R(t)|n - m|$, hence $\|u_n - u_m\|$ $= \int_0^1 |u_n(t) - u_m(t)|\,dt \leqslant |n - m|\int_0^1 R(t)\,dt$, and condition (iii) of Theorem 3.1 holds with $L = \int_0^1 R(t)\,dt$, which completes the proof.

Since above $R \leqslant \int_0^1 R(t)\,dt$, thus $\int_0^1 R(\tau)\,d\tau = +\infty$ does not necessarily imply that $I(U)$ is not R-convex for any $R > 0$. In particular, $I(U)$ may be R-convex for $U(t)$ being of dimension less than k.

We shall discuss now the case when $U(t) = \{g(t), f(t)\}$. Since R-convexity is invariant with respect to translation, we can consider equivalently the case $U(t) = \{0, h(t)\}$ where $h(t) = f(t) - g(t)$. Assume that h is integrable. We denote by m the Lebesgue measure in the unit sphere $S_{k-1} = \{n \mid n \in R^k, |n| = 1\}$. Without any loss of generality we may assume that $h(t) \neq 0$ everywhere. For $A \subset S_k$ put $h^{-1}A$

$$= \{t \in [0,1] | h(t)/|h(t)| \in A\} \text{ and let}$$

$$\mu(A) = \int_{h^{-1}A} |h(\tau)| d\tau.$$

μ is a measure on S.

COROLLARY 4.2.   *The integral of* $U(t) = \{0, h(t)\}$ *is R-convex for some R if the inequality*

(12) $\mu(A) \leqslant Km(A)$,   *K constant*

holds for each measurable $A \subset S_{k-1}$.

Proof. Fix $m, n \in S_{k-1}$. If $U(t) = \{0, h(t)\}$ then $u_m(t) = h(t)$ if $\langle m, h(t) \rangle > 0$ and 0 otherwise. Similarly $u_n(t) = h(t)$ if $\langle n, h(t) \rangle > 0$ and 0 if $\langle n, h(t) \rangle \leqslant 0$. Therefore $\| u_m - u_n \| = \int_{h^{-1}A(m,n)} |h(t)| dt$, where $A(m,n) = \{x \in S_k \, | \, \langle x, m \rangle \langle x, n \rangle < 0 \}$. Hence by (12) $\| u_m - u_n \| \leqslant Km(A(m,n))$. But $m(A(m,n)) \leqslant L|m - n|$ for some constant L. Therefore Corollary 4.2 follows from Theorem 3.1.

If $k = 2$ and h is of class C^1 then (12) holds if and only if $h(t)$ and $h'(t)$ are linearly independent for each $t \in [0,1]$. Indeed, let $h(t) = (\rho(t) \cos \omega(t), \rho(t) \sin \omega(t))$. The linear independence of $h(t)$ and $h'(t)$ is equivalent to $\dot\omega(t) \neq 0$ for each $t \in [0,1]$. But $h^{-1}(A) = \{t \, | \, \omega(t) \in A\}$ if S_1 is identified with $[0, 2\pi]$, and if $\dot\omega(t) \neq 0$ everywhere and continuous then (12) follows. On the other hand if $\dot\omega(t_0) = 0$ for some t_0 then $|\omega(t) - \omega(t_0)| \leqslant \delta(a)a$ if $|t - t_0| \leqslant a$ where $\delta(a) \to 0$ if $a \to 0$. Therefore for any $K > 0$ there is small enough a such that

$$\mu(A) \geqslant \int_{t_0}^{t_0 + a} |h(t)| dt \geqslant \alpha a > K2\delta(a)a = Km(A)$$

where $A = \{\omega \, \| \, \omega - \omega(t_0)| \leqslant \delta(a) \}$, which shows that (12) does not hold.

We have proved the following

COROLLARY 4.3. (Łojasiewicz [2]).   *If* $k = 2$ *and* h *is of class* C^1, $h(t) \neq 0$ *for* $t \in [0,1]$, *then* $I(U)$ *for* $U(t) = \{0, h(t)\}$ *is R-convex for some* $R < \infty$ *iff* $h(t)$ *and* $h'(t)$ *are linearly independent for each* $t \in [0,1]$.

For $k > 2$ and h absolutely continuous the integral $I(U)$, $U(t) = \{0, h(t)\}$ is not R-convex for any R as also has been proved by Łojasiewicz in [2].

This can be deduced from Corollary 4.2. However we shall prove a slightly more general result.

PROPOSITION 4.1. *Let* $n \in S_{k-1}$ *and* $u_n(t) \in U(t)$ *a.e. in* $[0, 1]$ *be fixed. Let* $\varphi : [0, \infty) \to [0, \infty)$ *be concave and continuous,* $\varphi(0) = 0$. *Assume there exists a measurable function* $n(t) \in S_{k-1}$ *and a measurable selection* $v(t) \in U(t)$ *such that*

(13) $\qquad \varphi\left(\int_0^t |n(\tau) - n| \, d\tau \right) \leqslant \delta(t) t,$

where $\delta(t) \to 0$ *as* $t \to 0$.

(14)
$$\langle u_n(t) - v(t), n(t) \rangle = 0 \qquad and$$
$$0 \leqslant \epsilon \leqslant |v(t) - u_n(t)| \leqslant M < +\infty$$

Then for each $L > 0$ *there is* $x \in I(U)$ *such that*

(15) $\qquad |z_n - x| > L\varphi(\langle z_n - x, n \rangle)$

Proof. Fix $L > 0$. Let $t_0 > 0$ be such that

(16) $\qquad \delta(t_0) < \epsilon / 2k^{3/2} LM$

By (14) for each t there is $i, i = 1, \ldots, k$ such that $|u_n^i(t) - v^i(t)| \geqslant \epsilon/\sqrt{k}$. Put $A_j^i = \{ t \mid 0 \leqslant t \leqslant t_0, (u_n^i(t) - v^i(t))(-1)^j \geqslant \epsilon/\sqrt{k} \}, j = 0, 1$. $m(\bigcup_{i,j} A_j^i) = t_0$. Thus there is i_0 and j_0 such that $m(A_{j_0}^{i_0}) > (t_0/2k)$. Set $A = A_{j_0}^{i_0}$ and consider $u_A = \chi_A \cdot u_n + \chi_A v$. Then for $x_A = \int u_A(t) \, dt$, we have

$$|z_n - x_A| = \left| \int_A (u_n(\tau) - v(\tau)) \, d\tau \right| \geqslant (\epsilon/\sqrt{k}) t_0/2k = \epsilon t_0/2k^{3/2}.$$

On the other hand by (14) we get

$$|\langle u_n(t) - u_A(t), n \rangle| = |\langle u_n(t) - u_A(t), n - n(t) \rangle|$$

$$\leqslant M |n - n(t)| \chi_A(t)$$

Hence by (13) and concavity of φ

$$\varphi(\langle z_n - x_A, n \rangle) \leqslant \varphi\left(M \int_0^{t_0} |n - n(\tau)| \, d\tau\right)$$

$$\leqslant M\varphi\left(\int_0^{t_0} |n - n(\tau)| \, d\tau\right) \leqslant M\delta(t_0)t_0$$

Using (16) we conclude that

$$L\varphi(\langle z_n - x_A, n \rangle) < \epsilon t_0 / 2k^{3/2}.$$

Thus $|z_n - x_A| > L\varphi(\langle z_n - x_A, n \rangle)$, which was to be proved.

If $\varphi(r) = r^{1/2}$ the conclusion of Proposition 4.1 means that $I(U)$ is not R-convex for any R. In this case assumption (14) can be written in the form

$$\int_0^t |n - n(\tau)| \, d\tau \leqslant \delta(t)t^2, \qquad \text{where } \delta(t) \to 0 \quad \text{as } t \to 0$$

and which holds for example if $\dot{n}(0)$ exists and is equal zero. In particular we can deduce from Proposition 4.1 the mentioned result of Łojasiewicz.

COROLLARY 4.4.   *If* $U(t) = \{f_1(t), f_2(t)\}$ *where* $h(t) = f_2(t) - f_1(t)$ *is differentiable for* $t = 0$, $k > 2$, *then* $I(U)$ *is not* R-*convex for any* $R > 0$.

Proof.   There exists $n(t)$ such that $\langle n(t), h(t) \rangle = 0$ and $\dot{n}(0) = 0$. Indeed let us choose $m_1, m_2, \ldots, m_{k-1} \in R^k$ be linearly independent and such that $m_1 = h(0)$ and $m_2 = \dot{h}(0)$ is linearly independent from $h(0)$. Then $n(t)$ is a solution of the system

$$\langle n(t), h(t) \rangle = 0, \langle n(t), m_i \rangle = 0 \qquad i = 2, \ldots, k - 1, |n(t)| = 1$$

Such a solution exists for $0 \leqslant t < t_0$ if t_0 is small enough and the derivative $\dot{n}(0)$ exists and equals zero.

Notice that $\langle \dot{n}(0), n(0) \rangle = 0$ and $\langle \dot{n}(0), h(0) \rangle = \langle \dot{n}(0), m_1 \rangle = 0$ because $\langle n(0), \dot{h}(0) \rangle = 0$. Therefore $\dot{n}(0)$ satisfies the equalities.

$$\langle \dot{n}(0), n(0) \rangle = 0, \qquad \langle \dot{n}(0), m_i \rangle = 0 \qquad i = 1, \ldots, k - 1$$

But $n(0), m_1, \ldots, m_{k-1}$ are linearly independent, therefore we get that $\dot{n}(0) = 0$ and Proposition 4.1 completes the proof.

Example . If $h(t)$ is not regular then $I(U)$, where $U(t) = \{0, h(t)\}$, may be R-convex. An example with $h(t)$ continuous is easily supplied by using Peano's famous construction [4]. Indeed, let $f(t)$ be the continuous map from $[0, 1]$ onto the square $\prod = \{(x_1 x_2) \mid 0 \leqslant x_i \leqslant 1\} \subset R^2$ constructed by Peano. This map has the property that the linear measure of $f^{-1}(A)$ for any measurable $A \subset \prod$ is the same as that of A on the plane. Now, if φ is a measure preserving continuous bijection of $\prod$ into S_2 then $h(t) = \varphi(f(t))$ is the desired function. In fact if $A \subset S_2$ then $h^{-1}A = f^{-1}(\varphi^{-1}A)$, thus $\mu(A) = m_1(h^{-1}A) = m_2(\varphi^{-1}A) \leqslant m(A)$, where m_1, m_2 and m denote, respectively the Lebesgue measure on R^1, R^2 and S_2. Therefore the inequality (12) in Corollary 4.2 holds and $I(U)$ is R-convex by this Corollary.

INSTITUTE OF MATHEMATICS,
POLISH ACADEMY OF SCIENCES, WARSAW

REFERENCES

---

[1] Łojasiewicz, St. (Jr.), Some properties of accessible sets in non linear control systems, *Ann. Polon. Math.* vol. **XXXVII**.2 (1979), 123–127.

[2] Olech, C., Existence Theory in optimal control, *Control Theory and Topics in Functional Analysis*, **I** (1976), Vienna.

[3] Olech, C., Extremal solutions of a control system, *J. of Diff. Eqs.* **2** (1966), 74–101.

[4] Peano, Sur une courbe, qui remplit toute une aire plane, *Math. Ann.* 1890, 157–159.

[5] Pliś, A.; Accessible sets in Control Theory, *Int. Conf. on Diff. Eqs.*, Academic Press (1975), 646–650.

GENERIC PROPERTIES OF AN INTEGRO-DIFFERENTIAL EQUATION*

By Jack K. Hale

Abstract. Consider the functional differential equation

$$\dot{x}(t) = -\int_{-1}^{0} a(-\theta)g(x(t+\theta))\,d\theta$$

where a, g are continuous, $a \geqslant 0$, $a(1) = 0$, $g(0) = 0$, $g'(0) = 1$, $xg(x) > 0$ for $x \neq 0$. The linear function $a_0(s) = 4\pi^2(1-s)$ is such that the characteristic equation

$$\lambda + \int_{-1}^{0} a_0(-\theta)e^{\lambda\theta}\,d\theta = 0$$

has two eigenvalues on the imaginary axis and the remaining ones with negative real parts. In spite of this, it is shown there is no generic Hopf bifurcation for any g. The nature of the bifurcation is characterized under hypotheses which appear to be generic in g.

1. Introduction. Consider the functional differential equation

$$(1.1) \qquad \dot{x}(t) = -\int_{-1}^{0} a(-\theta)g(x(t+\theta))\,d\theta$$

where $a \in C([0,1], \mathbb{R})$, $g \in C^3(\mathbb{R}, \mathbb{R})$, $a \geqslant 0$, $a(1) = 0$, $g(0) = 0$, $g'(0) = 1$. The characteristic equation

$$(1.2) \qquad \lambda = -\int_{-1}^{0} a(-\theta)e^{\lambda\theta}\,d\theta$$

for the linear variational equation around zero for $a = a_0$, $a_0(s) = 4\pi^2(1-s)$, has all solutions with real parts < 0 except two on the

---

*This research was supported in part by the National Science Foundation, under MCS 79-05774, in part by the U.S. Army under AROD-DAAG 29-79-C-0161, and in part by the Air Force Office of Scientific Research under AF-AFOSR 76-3092C.

imaginary axis given by $\pm 2\pi i$. Furthermore, there is a neighborhood U of a_0 and a submanifold Γ of codimension one in $C([0,1], \mathbb{R})$ such that $U \backslash \Gamma = U_1 \cup U_2$ with $U_1 \cap U_2$ the empty set, for $a \in U_1$, all solutions of (1.2) have negative real parts and, for $a \in U_2$, all have negative real parts except two which have positive real parts.

Under such a circumstance, one expects to obtain a generic Hopf bifurcation at a_0 for a residual set of $g \in C^3(\mathbb{R}, \mathbb{R})$. That is, for a residual set of g, a unique hyperbolic periodic orbit should bifurcate from zero as a crosses Γ from U_1 (respectively U_2) to U_2 (respectively U_1). We show this is not the case. Under an assumption which appears to be generic in $g \in C^5(\mathbb{R}, \mathbb{R})$, we characterize the nature of the bifurcation point a_0.

This result shows that the generic properties of Equation (1.1) with the integrand restricted to a product of two functions $a(s)g(x)$ is completely different from the generic properties that would be obtained by considering general functions $h(s, x)$ of two variables.

For notational purposes, we let $C = C([-1, 0], \mathbb{R})$. For any $\phi \in C$, we suppose the solution $x(\phi)(t)$ through ϕ is defined for $t \geq 0$. Let $T_{a,g}(t): C \to C, t \geq 0$, be the semigroup operator by (1.1); that is, $T_{a,g}(t)$ $\phi(\theta) = x(\phi)(t + \theta), -1 \leq \theta \leq 0$.

2. Nongeneric Hopf bifurcation. To prove there can be no generic Hopf bifurcation for Equation (1.1), we need the following proposition from Hale [3, p. 122] or Levin and Nohel [5].

PROPOSITION 2.1. *If $\dot{a} \leq 0, \ddot{a} \geq 0, xg(x) > 0$ for all x, $G(x) = \int_0^x g$ $\to \infty$ as $|x| \to \infty$, then every solution of (1.1) is bounded and*

(i) *if there is an s such that $\ddot{a}(s) > 0$, then every solution approaches zero as $t \to \infty$;*

(ii) *if $\ddot{a}(s) = 0$ for all s (that is, a is linear) then, for any $\phi \in C$, there is either an equilibrium point or a one-periodic solution $p = p(\phi)$ of the ordinary differential equation*

$$(2.1) \qquad \ddot{y} + a(0)g(y) = 0$$

such that the ω-limit set of the orbit through ϕ is $\{p_t, t \in \mathbb{R}\}$, where $p_t \in C, p_t(\theta) = p(t + \theta), -1 \leq \theta \leq 0$.

Since $g'(0) = 1$ in (1.1), we may choose a neighborhood of $x = 0$ and extend g outside this neighborhood so that it satisfies the condition of Proposition 2.1 and so that Equation (2.1) has no one-periodic solutions outside U for any a in a sufficiently small neighborhood V of

$a_0(s) = 4\pi^2(1 - s)$. From the results in Cooperman [1] (see also Hale [14]), the semigroup $T_{a,g}(t)$ has a maximal compact invariant set $A_{a,g}$ which belongs to the set $\tilde{U} = \{\phi \in C : \phi(\theta) \in U, -1 \leqslant \theta \leqslant 0\}$ for each $a \in V$. Also, this set is uniformly asymptotically stable and attracts bounded sets of C. The set $A_{a,g}$ is upper semicontinuous in (a, g).

PROPOSITION 2.2. *For any real μ such that $\mu a_0 \in V$, the set $A_{\mu a_0,g}$ cannot contain a uniformly asymptotically stable periodic orbit.*

Proof. Suppose $A_{\mu a_0,g}$ contains a periodic orbit γ which is uniformly asymptotically stable. Then, for any neighborhood U of γ there is a neighborhood V of γ and a neighborhood W of μa_0 in the C^0-topology such that, for any $a \in W, \phi \in V$, the positive orbit $\gamma^+(\phi)$ of (1.1) belongs to U. One can choose the neighborhood U of γ and W of μa_0 so that U does not contain the equilibrium point zero of (1.1) for any $a \in W$.

In the neighborhood W of μa_0 there exists a strictly convex function a. Thus, $A_{a,g}$ contains no periodic orbits and every solution of (1.1) approaches zero by Proposition 2.1. This contradicts the fact that some positive orbits remain in U and proves that no periodic orbit can be uniformly asymptotically stable.

Let $\omega(b)$ be the period of the solution of (2.1) through $(b,0), b > 0$, for (a_0, g). There is an open dense set of g such that $\omega'(b) \neq 0$. Suppose g is chosen so that this is true. Then there is a neighborhood U of $x = 0$ such that Equation (2.1) for (a_0, g) has no one-periodic solution in U. Thus, we may assume g extended so that $A_{a_0,g} = \{0\}$. Suppose there is a generic Hopf bifurcation at (a_0, g). Since $A_{a,g}$ is upper semicontinuous at (a_0, g), $A_{a,g}$ is uniformly asymptotically stable and the bifurcated periodic orbit is hyperbolic, it follows that the periodic orbit must be uniformly asymptotically stable. This contradicts Proposition 2.2. This proves there cannot be a residual set of g for which there is a generic Hopf bifurcation.

3. The bifurcations at a_0. To understand better the nature of the bifurcations at a_0, we give a more detailed analysis of the equations characterizing the bifurcation of periodic orbits from zero.

We need the following lemma.

LEMMA 3.1. *There is a neighborhood U of a_0 in the C^0-topology, a $\delta > 0$ and an analytic function $\lambda^* : U \to \mathbb{C}$ such that $\lambda^*(a_0) = 2\pi i$ and, for* *every $a \in U$, the equation (1.2) has exactly one solution in each of the circles $|\lambda \pm 2\pi i| < \delta$ given respectively by $\lambda^*(a), \bar{\lambda}^*(a)$ and all other solutions with real parts $\leqslant -\delta$. Furthermore, if*

$$\Gamma^- = \{ a \in U : \text{Re} \lambda^*(a) < 0 \}$$

$$\Gamma^0 = \{ a \in U : \text{Re} \lambda^*(a) = 0 \}$$

$$\Gamma^+ = \{ a \in U : \text{Re} \lambda^*(a) > 0 \}$$

then each of these sets is nonempty and Γ^0 is a submanifold of codimension one.

Proof. If

$$F(\lambda, a) = \lambda + \int_{-1}^{0} a(-\theta) e^{\lambda \theta} \, d\theta$$

then $F(2\pi i, a_0) = 0$ and $\partial f / \partial \lambda = 1 - 8\pi^2$ at $(2\pi i, a_0)$. The Implicit Function Theorem implies the existence of a function $\lambda^*(a) \in \mathbb{C}$ analytic in a neighborhood of a_0 with $\lambda^*(a_0) = 2\pi i$. The other properties of λ^* follow essentially from Rouché's Theorem.

To show Γ^0 has codimension one, consider the family of functions $(a_0 + \nu b_0), b_0(s) = 4\pi^2 s(1 - s), \nu \in \mathbb{R}$. Then the derivative of $\lambda^*(a_0 + \nu b_0)$ with respect to ν at $\nu = 0$ is easily seen

$$\left. \frac{\partial \lambda^*}{\partial \nu} \right|_{\nu = 0} = -4\pi^2 \int_{-1}^{0} \theta (1 + \theta) e^{2\pi i \theta} \, d\theta$$

Thus,

$$\left. \frac{\partial \, \text{Re} \lambda^*}{\partial \nu} \right|_{\nu = 0} = -4\pi^2 \int_{-1}^{0} \theta (1 - \theta) \cos 2\pi \theta \, d\theta > 0.$$

This shows that Γ_0 has codimension one and also that Γ^-, Γ^+ are not empty. This proves the lemma.

Remark 3.2. We knew from the previous section that $\text{Re} \lambda^*(a) < 0$ if $a \in U$ is strictly convex. The above proof shows that $\text{Re} \lambda^*(a) > 0$ if $a \in U$ is of the form $a = a_0 + b$ where b is strictly concave, $b(0) = b(1) = 0$.

For $a = a_0$, the characteristic equation for the linear variational equation around zero has two purely imaginary roots. For a near a_0, and a neighborhood W of zero let $B(r, a, g)$ be the scalar bifurcation function obtained by applying the usual method of Liapunov-Schmidt for the periodic solutions of (1.1) in W which for $a = a_0$ are equal to $r \cos 2\pi t$ (see, for example, deOliveira and Hale [2]). This function has the property that the periodic solutions of the type specified are in one to one correspondence with the nonnegative zeros of $B(r, a, g)$. Furthermore, the stability properties of the periodic solution corresponding to a zero r_0 of $B(r, a, g)$ when restricted to a center manifold at $x = 0$ are the same as the stability properties of the equilibrium point r_0 of the scalar equation

$$(3.1) \qquad \dot{r} = B(r, a, g)$$

(see deOliveira and Hale [2]). The function $B(r, a, g)$ is an odd function of r and has five continuous derivatives if $g \in C^5(\mathbb{R}, \mathbb{R})$. Let

$$(3.2) \qquad B(r, a, g) = \alpha_1(a, g)r + \alpha_3(a, g)r^3 + \alpha_5(a, g)r^5 + o(|r|^5)$$

as $r \to 0$.

If $\lambda^*(a)$ is the function given in Lemma 3.1, the manner in which the bifurcation function is constructed implies that

$$(3.3) \qquad \begin{aligned} &\alpha_1(a, g) = 0 \quad \text{if and only if} \quad \mathrm{Re}\,\lambda^*(a) = 0 \\ &\mathrm{sign}\,\alpha_1(a, g) = \mathrm{sign}\,\mathrm{Re}\,\lambda^*(a). \end{aligned}$$

Thus, $\alpha_1(a_0, g) = 0$.

Let $\alpha_3^0 = \alpha_3(a_0, g)$. We now show that $\alpha_3^0 = 0$. Since the solution $x = 0$ of Equation (1.1) for $a = a_0$ is asymptotically stable, it follows that the zero solution of (3.1) for $a = a_0$ is asymptotically stable. Thus, $\alpha_3^0 \leqslant 0$. A generic Hopf bifurcation corresponds to $\alpha_3^0 < 0$. We have shown in the previous section that $\alpha_3^0 = \alpha_3(a_0, g) = 0$ for an open dense set of g's. Thus, $\alpha_3^0 = 0$ for all g since it is continuous in g. This shows *there is no generic Hopf bifurcation for any g*.

Again, the stability of the zero solution of Equation (1.1) for $a = a_0$ implies that $\alpha_5(a_0, g) \leqslant 0$. We make the hypothesis that

$$(3.4) \qquad \alpha_5^0(g) \overset{\text{def}}{=} \alpha_5(a_0, g) < 0.$$

This implies that

$$(3.5) \qquad B(r, a_0, g) = \alpha_5^0(g)r^5 + o(r^5), \qquad \alpha_5^0 < 0.$$

We have not made the computations (which would be extremely complicated) to obtain the constant $\alpha_5^0(g)$. However, it certainly seems plausible that the set of g for which $\alpha_5^0(g) < 0$ is open in the space $C^5(U, \mathbb{R})$ for a given bounded neighborhood U of $x = 0$.

If $B(r, a, g) = rP(r^2, a, g)$, then

$$(3.6) \qquad P(\rho, a, g) = \alpha_1(a, g) + \alpha_3(a, g)\rho + \alpha_5(a, g)\rho^2 + o(\rho^2)$$

as $\rho \to 0$. This function has a unique maximum $\eta(a, g)$ in a neighborhood U of $a = a_0$ which occurs at a value $\rho^*(a, g)$ and $\eta(a_0, g) = 0$. Let

$$SN^0 = \{a \in U : \eta(a, g) = 0, \rho^*(a, g) \geqslant 0\}$$

(3.7)

$$SN^{+(-)} = \{a \in U : \eta(a, g) > (<)0, \rho^*(a, g) \geqslant 0\}.$$

One can show that every tangent vector to SN^0 is a tangent vector to Γ^0. We suppose that

$$(3.8) \qquad SN^0 \text{ is a submanifold of codimension 1}, \qquad S_{N^+} \neq \phi.$$

It is possible to show that hypothesis (3.8) is satisfied for an open set of $g \in C^5(W, \mathbb{R})$ for a bounded neighborhood W of zero.

We can now prove the following result.

THEOREM 3.3. *With hypotheses* (3.4) *and* (3.8), *there is a neighborhood U of a_0 in the C^0-topology and a neighborhood W of $x = 0$ such that U is subdivided into regions as shown in Figure* 1, *the set $A_{a,g}$ is a disk for each $a \in U$ with boundary being a periodic orbit and the flow on a two dimensional manifold in $A_{a,g}$ is shown in Figure* 1.

Proof. If $a \in \Gamma^+$, then $\alpha_1(a, g) > 0, \alpha_3(a, g) < 0$ implies that $P(p, a, g)$ in (3.6) has a unique positive zero. Thus, there is a unique periodic solution in a small neighborhood of zero and it is asymptotically stable as shown in the flow for Γ^+. This shows that $SN \subseteq \Gamma^-$. By hypothesis (3.8) and the fact that the stability properties of the periodic orbits are determined by (3.1), we have that the flow on SN^0 is the one shown in Figure 1. Also, the flow in the other two regions must be one of those shown in Figure 1. We only need to verify that the regions are ordered as shown. In the proof of Lemma 3.1 we showed that the curve $(a_0 + \nu b_0), b_0(s) = 4\pi^2 s(1 - s)$ was transversal to Γ^0 at $\nu = 0$. For $\nu < 0$ this function is in Γ^- and strictly convex. Thus, the origin is uniformly

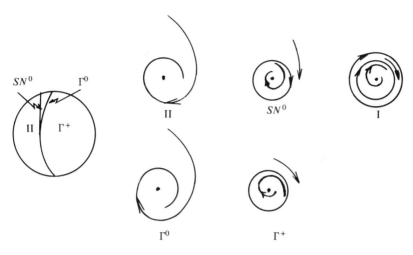

Figure 1.

asymptotically stable. This proves that the flow in Region II in Figure 1 is the one that is depicted. This proves the theorem.

BROWN UNIVERSITY

REFERENCES

[1] Cooperman, G. D., α-condensing maps and dissipative systems. Ph.D. Thesis, Brown University, June, 1978.

[2] deOliveira, J. C. and J. K. Hale, Dynamic behavior from bifurcation equations. *Tohôku Math. J.* **32** (1980), 577–592.

[3] Hale, J. K., *Theory of Functional Differential Equations*, Applied Math. Sci., Vol. 3, 2nd Edition, Springer-Verlag, 1977.

[4] Hale, J. K., Some recent results on dissipative processes. Proc. Symposium on Functional Differential Equations and Dynamical Systems, São Carlos, Brazil, July, 1979. To appear in Springer Lecture Notes in Math.

[5] Levin, J. J. and J. Nohel, On a nonlinear delay equation, *J. Math. Ana. Appl.* **8** (1964), 31–44.

ON THE EFFECT OF THE BOUNDARY CONDITIONS ON THE EIGENVALUES OF ORDINARY DIFFERENTIAL EQUATIONS*

By Evans M. Harrell, II

Abstract. It is shown that the asymptotics of the gap between the eigenvalues of an ordinary differential equation with alternatively Dirichlet or Neumann boundary conditions are given by a simple integral formula involving the standard solutions of the equation. The formula is evaluated in a special case, Mathieu's equation.

1. Despite the persistent and voluminous efforts of Phil Hartman throughout his career attacking the subject of ordinary differential equations, or, perhaps, because of those efforts, it is still very much alive with interesting problems to solve. One such problem is a natural outgrowth of work of Hartman and Putnam thirty years ago [8]. They considered the equation

$$(1) \qquad y''(x) + (\lambda - q(x))y(x) = 0$$

and estimated the intervals of stability and instability, that is, the widths of the bands of the essential spectrum and the gaps between bands. In the classical case, $q(x)$ is a periodic even function, and the bands are demarcated by the eigenvalues of the reduced Sturm-Liouville problem on a finite interval whose length is the periodicity of q, given alternatively Dirichlet or Neumann boundary conditions at the end points. Hochstadt [9] and Trubowitz [18] have proved that the band-gaps decrease rapidly (or exponentially) as functions of their index, assuming that q is smoothly (or real-analytically) periodic, but the problem of the exact asymptotics has remained open. Recent interest in the problem stems from developments in inverse scattering theory [12, 18] and Szegö theorems [19].

Thus consider (1) on an interval $[0, a]$ and with $q(x)$ having an analytic continuation to a strip $S = \{z : -\varepsilon < \mathrm{Re}\, z < a + \varepsilon\}$ for some

---

*Research partially supported by USNSF Grant MCS-7926408.

$\varepsilon > 0$, and suppose, moreover, that, as with Mathieu's equation, $q(z)$ is real on the imaginary axis and for $z = a + ir, r \in \mathbb{R}$. This assumption simplifies things considerably, but the theory can probably be extended in its absence. Let $\beta_1, \beta_2, \ldots$ denote the ordered sequence of eigenvalues λ of (1) given Dirichlet boundary conditions,

$$y(0) = y(a) = 0,$$

and let $\alpha_0, \alpha_1, \ldots$ denote the ordered sequence of Neumann eigenvalues,

$$y'(0) = y'(a) = 0.$$

It is known that α_k and β_k are both asymptotic to $k^2\pi^2/a^2 + C_k + o(1/k)$, where C_k are proportional to the Fourier-cosine coefficients of $q(x)$ (for general references see [7, 10, 11, 17]).

This paper will prove an integral formula for the difference between α_k and β_k and evaluate it for the Mathieu equation in the limit $k \to \infty$. A parallel assault on the problem appears in the physical literature [2, 3], where the key insight is to relate the gaps to the tunneling effect by analytic continuation. The tunneling effect shows up in the context of ordinary differential equations as the exponential growth or shrinking of solutions of (1) between turning points. This is reflected in the form of equations (4) and (5) below, as z_0 will be taken as the complex turning point in practice; these equations are analogous to the dispersion relation of [3], where numerical comparisons with tabulated eigenvalues of Mathieu's equation may also be found. It should be noted that, despite some similarities such as the appearance of a tunneling argument, the problem of estimating the gaps is fundamentally different from the other problem of [3], also treated in [5], which is the complementary situation where $\lambda \ll q(x)$ in some scaling limit.

Define the standard solutions y_1 and y_2 of (1) by

(2) $\qquad y_1(0; \lambda) = y_2'(0; \lambda) = 1 \quad$ and $\quad y_1'(0; \lambda) = y_2(0; \lambda) = 0,$

and, similarly, let

(3) $\qquad y_3(a; \lambda) = y_4'(a; \lambda) = 1 \quad$ and $\quad y_3'(a; \lambda) = y_4(a; \lambda) = 0.$

Note that y_1 is real on the imaginary axis, y_2 is imaginary on the imaginary axis, y_3 is real for $z = a + ir$, and y_4 is imaginary for

$z = a + ir, r \in \mathbb{R}$, owing to the reality conditions on q. The eigenvalues are determined by linear dependence of y_1 and y_3 and, respectively, of y_2 and y_4. For some purposes it may also be relevant to consider the eigenvalues determined by linear dependence of y_1 and y_4 and of y_2 and y_3, for which the theory is quite similar (see Theorem (I')).

2. Connection between the gaps and the standard solutions. It is deceptively easy to derive formulae for the gaps between the eigenvalues in terms of the eigenfunctions—for instance,

$$\beta_k - \alpha_k = \frac{\left(W\{ y_1(z_1, \alpha_k), y_2(z_1, \beta_k) \} - W\{ y_1(z_2, \alpha_k), y_2(z_2, \beta_k) \} \right)}{\int_{z_2}^{z_1} y_2(z, \beta_k) y_1(z, \alpha_k)\, dz},$$

for any $z_{1,2} \in \mathbb{C}$, where the Wronskian $W\{f, g\} \equiv fg' - f'g$ is, of course, not constant here, since $\alpha_k \neq \beta_k$ in general. The extraction of the asymptotics as $k \to \infty$, however, typically reduces to delicate estimates of the ratio of two small quantities, which can seemingly only be estimated accurately enough given knowledge of the gap. The at least as innocuous-looking formula of Theorem (I) is specially adapted to avoid the circularity when the asymptotics are extracted.

(I) *Let* $q(z)$, *etc., be as stipulated above and suppose* z_0 *is an imaginary number. Then, assuming* $y_2 \neq 0$ *for z in the integral,*

$$(4) \qquad \beta_k - \alpha_k = \frac{\mathrm{Re} \int_{z_0}^{z_0+a} (y_2(z, \beta_k))^{-2}\, dz}{\int_0^a (y_1(z, \beta_k))^2\, dz} \left(1 + O(k^{-1}) \right).$$

(*The error is* $O(k^{-1})$ *independently of* z_o.)

The analogous statement for the eigenvalue β_k' for Dirichlet boundary condition at 0 and Neumann at a and the eigenvalue α_k' for Neumann at 0 and Dirichlet at a merely interchanges y_1 and y_2.

(I') *Under the same conditions,*

$$(5) \qquad \beta_k' - \alpha_k' = \frac{\mathrm{Re} \int_{z_0}^{z_0+a} (y_1(z, \alpha_k))^{-2}\, dz}{\int_0^a (y_2(z, \alpha_k))^2\, dz} \left(1 + O(k^{-1}) \right).$$

Because of the uniform asymptotic estimates (as quoted in [17])

$$(6) \qquad y_1(x, \lambda) = \cos\sqrt{\lambda}\, x + \frac{\sin\sqrt{\lambda}\, x}{2\sqrt{\lambda}} \int_0^x q(x')\, dx' + O(\lambda^{-1})$$

and

(7) $y_2(x,\lambda) = \lambda^{-1/2}\sin\sqrt{\lambda}\,x - \dfrac{\cos\sqrt{\lambda}\,x}{2\lambda} \displaystyle\int_0^x q(x')\,dx' + O(\lambda^{-3/2})$

for $0 \leqslant x \leqslant a$ and $\lambda\uparrow\infty$, the denominators of (4) and (5) are respectively $a/2$ and $a/2\lambda$ to leading order in λ, so the leading asymptotics of the gap will require a close analysis of the numerator, i.e., one standard solution at an eigenvalue and near a conveniently chosen complex value (in practice, a complex turning point).

3. Proof of (I). The eigenvalues are determined by the vanishing of the Wronskians of y_1 with y_3 and y_2 with y_4. The Wronskians can obviously be multiplied by any finite, nonvanishing function of λ in the vicinity of α_k and β_k, and it turns out to be convenient to thus rescale the eigenvalue conditions as

(8) $0 = G_D(\beta_k) \quad \text{and} \quad 0 = G_n(\alpha_k),$

where

(9) $G_D(\lambda) = \operatorname{Re} \dfrac{y_1(z_0 + a)y_3(z_0 + a)}{y_1(a)y_3(a)} \cdot \dfrac{W\{y_4, y_2\}}{y_2(z_0 + a)y_4(z_0 + a)}$

$= \operatorname{Re} \dfrac{y_1(z_0 + a)y_3(z_0 + a)}{y_1(a)y_3(a)}$

$\times \left[\dfrac{y_2'(z_0 + a)}{y_2(z_0 + a)} - \dfrac{y_4'(z_0 + a)}{y_4(z_0 + a)} \right]$

and

(10) $G_N(\lambda) = y_1'(a)/y_1(a) = W\{y_3, y_1\}/y_1(a)y_3(a)$

$= \operatorname{Re} \dfrac{y_1(z_0 + a)y_3(z_0 + a)}{y_1(a)y_3(a)}$

$\times \left[\dfrac{y_1'(z_0 + a)}{y_1(z_0 + a)} - \dfrac{y_3'(z_0 + a)}{y_3(z_0 + a)} \right].$

The difference between these two quantities is

$$G_D(\lambda) - G_N(\lambda) = \operatorname{Re} \frac{y_1(z_0 + a)y_3(z_0 + a)}{y_1(a)y_3(a)}$$

$$\times \left[\frac{y_2'(z_0 + a)}{y_2(z_0 + a)} - \frac{y_1'(z_0 + a)}{y_1(z_0 + a)} + \frac{y_3'(z_0 + a)}{y_3(z_0 + a)} \right.$$

$$\left. - \frac{y_4'(z_0 + a)}{y_4(z_0 + a)} \right]$$

$$= \frac{1}{y_1(a)y_3(a)}$$

$$\times \operatorname{Re} \left[\frac{y_3(z_0 + a)}{y_2(z_0 + a)} - \frac{y_1(z_0 + a)}{y_4(z_0 + a)} \right]$$

(because $W\{y_1, y_2\} = W\{y_3, y_4\} = 1$). If $\lambda = \beta_k$, then $y_2(z_0 + a)$ is imaginary, and

(11) $\qquad G_D(\beta_k) - G_N(\beta_k) = \dfrac{-1}{y_1(a)y_3(a)} \operatorname{Re} \dfrac{y_1(z_0 + a)}{y_4(z_0 + a)}$.

According to a well-known formula [7] relating the two solutions of (1),

$$y_1(z) = y_2(z) \int_0^z W\{y_2, y_1\}(y_2(z'))^{-2} \, dz'$$

$$= -y_2(z) \int_0^z (y_2(z'))^{-2} \, dz',$$

and by (7), $y_2(z, \beta_k)/y_4(z, \beta_k) = \pm 1 + O(k^{-1})$ independently of z ($y_2(z, \beta_k)/y_4(z, \beta_k) = \pm 1$ exactly, if q is symmetric as in the application to be made below), so

(12) $\qquad G_D(\beta_k) - G_N(\beta_k) = \pm \dfrac{1 + O(k^{-1})}{y_1(a)y_3(a)} \operatorname{Re} \displaystyle\int_{z_0}^{z_0 + a} (y_2(z'))^{-2} \, dz',$

assuming that y_2 does not vanish on the integration contour (the reality of y_2^2 for imaginary z was used to drop the part of the contour from 0 to z_0).

By Taylor's theorem,

$$0 = G_D(\beta_k) = G_N(\beta_k) + [G_D(\beta_k) - G_N(\beta_k)]$$

$$= G_N(\alpha_k) + (\beta_k - \alpha_k)\frac{\partial}{\partial\lambda}G_N(\gamma) + [G_D(\beta_k) - G_N(\beta_k)]$$

for some γ in the closed interval bounded by α_k and β_k, which thus equals α_k or β_k to all orders in k. Solving for the gap,

$$\beta_k - \alpha_k = -\frac{[G_D(\beta_k) - G_N(\beta_k)]}{(\partial/\partial\lambda)G_N(\lambda)}.$$

The numerator is accounted for in (12), while the denominator is

(13) $$\frac{\partial}{\partial\lambda}\frac{y_1'(a)}{y_1(a)} = -(y_1(a;\lambda))^{-2}\int_0^a y_1^2(x)\,dx$$

(proof: if $u = y'/y$, then $u_\lambda = \partial u/\partial\lambda$ solves the elementary differential equation $u_\lambda' = -1 - 2uu_\lambda$ by substitution into (1) and differentiating. The solution is (13).) From (6) and the closeness of γ to β_k, only a small $o(k^{-1})$ error is incurred by changing γ to β_k, and $y_1(a)/y_3(a) = \pm 1 + O(k^{-1})$, where the sign $\pm$ is the same as in (12), proving (4). The proof of (I') is virtually the same, with the rôles of y_3 and y_4 switched.

4. Application.   The remaining half of the work of obtaining the asymptotics of the gap is the determination of the asymptotics of the function y_2 away from the real axis. The Mathieu equation

(14) $$y''(x) + (\lambda - 2\kappa\cos 2x)y = 0$$

is of the form (1) with $q(x) = 2\kappa\cos 2x$ and $a = \pi$, and the equation obviously continues to the strip. This classical equation is discussed in the general references and [13, 14], and will be treated here as an example. Although it would be possible to write everything in terms of the Mathieu functions $ce_k(z)$ and $se_k(x)$, up to factors identical to $y_1(x;\alpha_k)$ and $y_2(x;\beta_k)$ respectively, the notation and the argument will be as general as conciseness allows. The procedure works, generally speaking, for q of strict exponential type in the sense that its modulus has both upper and lower exponential bounds in S and assuming the growth of q in imaginary directions is maximized on the imaginary axis in such a way that the growth of y_2 in imaginary directions is minimized on the imaginary axis, so the numerator of (4) is dominantly contributed to by z with small real parts or near $z_0 + a$. No attempt will be made to spell out the optimal conditions on q for the asymptotics of y_2 to be ascertainable, as they would be rather gruesome. Instead, a class of functions q will be specified normalized to resemble that of the Mathieu equation, $q(z) = 2\kappa \cos 2z \sim \kappa \exp(-2iz)$, on the imaginary axis and with enough other smooth growth conditions to justify some estimates.

(II) *In addition to the assumptions of sections 1–3, suppose*

a) for z in S with $\operatorname{Im} z$ *greater than some constant, there exist b, $c > 0$ such that* $|q(z)|$, $|q'(z)|$ *and* $|q''(z)|$ *lie between* $b \exp(2|z|)$ *and* $c \exp(2|z|)$;

b) on some strip $0 \leqslant \operatorname{Re} z \leqslant d$, *and on the strip* $0 \leqslant a - \operatorname{Re} z \leqslant d$, *the function*

$$B(z) = q(z) - \kappa \exp(-2iz)$$

is bounded in absolute value by a finite power of $\operatorname{Im} z$ *(actually, weak exponential growth is allowed);*

c) for all z such that $|z - z_0|$ *and* $|z - z_0 + a| \geqslant \lambda^{-\gamma}$ *for some $0 \leqslant \gamma \leqslant 1$, where $z_0(\lambda)$ is the imaginary solution of $q(z_0) - \lambda = 0$, i.e., $z_0 \simeq (i/2)\ln(\lambda/\kappa)$,*

$$|\lambda - q(z)| > \text{const} \cdot \lambda$$

(actually, $\lambda^{4/5+\epsilon}$).
Then, uniformly for $z \in S$ wlth $\operatorname{Im} z \leqslant |z_0|$ *and* $|z - z_0|$ *and* $|z - z_0 + a| \geqslant \lambda^{-\gamma}$,

(15) $$y_2(z) = \lambda^{-1/2}\sin\left(\int_0^z (\lambda - q(z'))^{1/2} dz'\right)\left(1 + 0(\lambda^{-\beta})\right)$$

for some $\beta > 0$. Equation (15) reduces to $\lambda^{-1/2}(i/2)\exp(-i\sqrt{\lambda}\, z)(1 + 0(\lambda^{-\beta}))$ for z imaginary, uniformly for $0 \leqslant |z| \leqslant |z_0|^\alpha$, any $0 < \alpha < 1$. It remains valid when differentiated. Moreover, uniformly on C_λ, the contour extending first from $z_1 = i|z_0|^\alpha$ to z_0 on the imaginary axis and thence *directly to $z_0 + \lambda^{-\gamma}$ with $\operatorname{Im} z = z_0$ fixed,*

$$(16) \quad y_2(z) = i\sqrt{\pi/2}\,\lambda^{-1/4}\left(\frac{2}{e}\sqrt{\frac{\lambda}{\kappa}}\,\right)^{\sqrt{\lambda}} J_{\sqrt{\lambda}}\left(\sqrt{\kappa}\,\exp(-iz)\right)\left(1 + 0(\lambda^{-\beta})\right).$$

The analogous estimates hold for y_4 with $z - a$ imaginary or z near $z_0 + a$.

This has as a corollary the asymptotics of the Mathieu gaps:

(III) *For the Mathieu equation* (14), *as $k \to \infty$,*

$$(17) \quad \alpha_k - \beta_k \sim \frac{4}{\pi} k\left(e^2\kappa/4k^2\right)^k.$$

(The tilde means that the ratio goes to 1.) Since Lemma (II) clearly applies to the Mathieu equation, the corollary is a straightforward computation replacing the denominator of (4) with $a/2 = \pi/2$ and using (16) along with formulae 9.3.4, 9.3.35, 9.3.38, 10.4.9, 10.4.59 of [1]:

$$\alpha_k - \beta_k \sim \frac{4}{\pi^2}\lambda^{1/2}\left(e^2\kappa/4\lambda\right)^{\sqrt{\lambda}}\operatorname{Re}\int_{z_0}^{z_0+\pi} J_{\sqrt{\lambda}}^{-2}\left(\sqrt{\lambda}\,\sqrt{\frac{\kappa}{\lambda}}\,\exp(-iz)\right)dz,$$

$$\lambda \sim k^2 + o(1/k),$$

$$\sim \frac{4}{\pi^2}k\left(e^2\kappa/4k^2\right)^k \cdot 2\operatorname{Re}\int_{z_0}^{z_0+k^{-1/6}} J_k^{-2}\left(k\,\frac{\sqrt{\kappa}}{k}\,\exp(-iz)\right)dz$$

$$\sim \frac{-4}{\pi^2}k\left(e^2\kappa/4k^2\right)^k\operatorname{Re}\int_0^{2^{1/3}ik^{1/2}}\frac{ik^{2/3}}{(k^{2/3}-\zeta/2^{1/3})Ai^2(\zeta)}\,d\zeta,$$

$$\zeta = -2^{1/3}k^{2/3}\left(\frac{\sqrt{\kappa}}{k}e^{-iz}-1\right)$$

$$\sim \frac{4}{\pi^2}k\left(e^2\kappa/4k^2\right)^k\operatorname{Im}e^{2\pi i/3}\int_0^\infty \left(Ai(re^{2\pi i/3})\right)^{-2}dr$$

$$\sim \frac{16}{\pi^2}k\left(e^2\kappa/4k^2\right)^k\operatorname{Im}\int_0^\infty\frac{1}{(Ai(r)-iBi(r))^2}\,dr,$$

which by the variation-of-parameters formula [7] relating two solutions of Airy's equation is

$$\frac{16k}{\pi^2} \left(\frac{e^2\kappa}{4k^2} \right)^k \mathrm{Im} \left. \frac{iAi(x)}{W\{Bi, Ai\}(Ai(x) - iBi(x))} \right|_{x=0}.$$

This equals [1] the expression in (17). In the third to last line of the computation, the contour was extended and distorted without crossing any Stokes lines.

The calculation is only justified when the real part of the integral is larger than the errors in the integral. If z_0 is near the turning point, the real and imaginary parts are of the same order in k, but for other choices of z_0 the integral is highly oscillatory, and the relative magnitudes are not so clear.

5. Proof of (II). Equation (15) is a standard asymptotic estimate; with the methods of Olver [16] it is only necessary to verify the smallness of the total variation of $\int (\lambda - q(z))^{-1/4}(d^2/dz^2)(\lambda - q(z))^{-1/4} dz$ on some contour from 0 to the points in question, which is an easy computation. It would be equally possible to use variation of parameters as in [4,5,7]. Suppose for the moment that y_2 satisfied the differential equation

$$(18) \qquad \eta''(z) + (\lambda - \kappa \exp(-2iz))\eta(z) = 0$$

exactly for $\mathrm{Im}\, z > |z_1|$. Since (18) is a disguised form of Bessel's equation, (16) would follow from matching the asymptotics of (15) to those of the solutions

$$(19) \qquad \begin{aligned} \eta_-(z) &= J_{\sqrt{\lambda}}\left(\sqrt{\kappa}\exp(-iz)\right), \\ \eta_+(z) &= Y_{\sqrt{\lambda}}\left(\sqrt{\kappa}\exp(-iz)\right) \end{aligned}$$

of (18). The function η_+ has the wrong asymptotics in the region between 0 and z_0 by Equation 9.3.36 of [1, see also 15], falling off exponentially along C_λ from z_1 to $z_0 + \lambda^{-\gamma}$, for which reason the correction proportional to it in (16) can be put into the error term $0(\lambda^{-\beta})$.

For (16), it remains to be proved that the exact equation (1) has two solutions that are asymptotically proportional to $\eta_\pm$ as $\lambda \to \infty$, i.e.,

$y_\pm / \eta_\pm \to 1$ uniformly for $z \in C_\lambda$; then the asymptotic matching problem is the same as in the paragraph above. The argument is adapted from [6], Proposition 3.2: Let y solve (1) with the assumptions of (II), and recast the equation by varying parameters,

(20)
$$y(z) = a_+(z)\eta_+(z) + a_-(z)\eta_-(z),$$
$$y'(z) = a_+(z)\eta'_+(z) + a_-(z)\eta'_-(z).$$

Consider $a = (a_+(z), a_-(z))$ an element of the Banach space $\mathscr{B}_\lambda$ of pairs of functions $f = (f_+, f_-)$ on C_λ with the norm $\|f\|_\lambda = \sup_{C_\lambda}|f_-(z)| + \sup_{C_\lambda}|\eta_-^{-2}(z)f_+(z)|$. If (20) is differentiated and solved for $a_\pm$, then

$$\frac{d}{dz}\begin{bmatrix} a_+ \\ a_- \end{bmatrix} = \frac{B(z)}{W\{\eta_-,\eta_+\}}\begin{bmatrix} \eta_+\eta_-(z) & \eta_-^2(z) \\ -\eta_+^2(z) & -\eta_+\eta_-(z) \end{bmatrix}\begin{bmatrix} a_+ \\ a_- \end{bmatrix},$$

which is equivalent to Equation IX.2.2.8 of Hartman [7]. Its integrated version is

$$(1 - M_\lambda)a = a_1,$$

where the operator M_λ is defined by

(21)
$$M_\lambda f = \int_{z_1}^z \frac{B(z')}{W\{\eta_-,\eta_+\}}\begin{bmatrix} \eta_+\eta_-(z') & \eta_-^2(z') \\ -\eta_+^2 & -\eta_+\eta_-(z') \end{bmatrix}\begin{bmatrix} f_+ \\ f_- \end{bmatrix}dz'$$

and $a_1 = \binom{a_+(z_1)}{a_-(z_1)}$ is chosen as $\binom{0}{1}$. Finally, it will be shown that $\|M_\lambda\|_{op,\mathscr{B}_\lambda} \to 0$ as $\lambda \to \infty$, so $\|a(\lambda) - a_1\|_\lambda \to 0$, which implies that $\sup_{C_\lambda}|a_-(z) - 1| \to 1$ and, since the product of the Bessel functions is $o(\lambda^{-1/3})$ uniformly on C_λ [1],

$$\sup_{C_\lambda}|\eta_+(z)a_+(z)/\eta_-(z)| < \mathrm{const}\,\lambda^{-1/3}\sup|\eta_-^{-2}(z)a_+(z)| = o(\lambda^{-1/3}).$$

(Actually, this shows that there is a convergent Neumann series for the solutions of (1) for λ big enough, and that the convergence is more and more rapid as λ increases, and the limits are (19) in some uniform sense.) Therefore $\sup_{C_\lambda}|y(z)/\eta_-(z) - 1| = \sup_{C_\lambda}|a_-(z) - 1 + \eta_+(z)a_+(z)/\eta_-(z)| \to 0$.

The norm of M_λ is bounded by the greatest of the total absolute variations of the components of

$$
\begin{bmatrix} \eta_-^{-2}(z) & 0 \\ 0 & 1 \end{bmatrix} \int_{z_1}^z \frac{B(z')}{W\{\eta_-,\eta_+\}} \begin{bmatrix} \eta_+ \eta_-(z') & \eta_-^2(z') \\ -\eta_+^2(z') & -\eta_+ \eta_-(z') \end{bmatrix}
$$

$$
\times \begin{bmatrix} \eta_-^2(z') & 0 \\ 0 & 1 \end{bmatrix} dz'
$$

$$
= \int_{z_1}^z \frac{B(z')}{W\{\eta_-,\eta_+\}}
$$

$$
\times \begin{bmatrix} n_+(z')\eta_-^3(z')\eta_-^{-2}(z) & \eta_-^2(z')\eta_-^{-2}(z) \\ -\eta_+^2(z')\eta_-^2(z') & -\eta_+(z')\eta_-(z') \end{bmatrix} dz'.
$$

All four terms go to zero as $\lambda \to \infty$, because $|\eta_+ \eta_-|$ is $0(\lambda^{-1/3})$ on C_λ, $|\eta_-(z')/\eta_-(z)| \leqslant 1$ for z and z' in C_λ with z' closer to z_1 than z is, and because of the assumption on $B(z)$. This proves the existence of a solution y approximated uniformly well in ratio by η_-. The existence of a solution y approximated uniformly well in ratio by η_+ follows from the same argument, except that the integral in (21) runs from $z_0 + \lambda^{-\gamma}$ to z, $a_1 = \binom{a_+(z_0+\lambda^{-\gamma})}{a_-(z_0+\lambda^{-\gamma})}$ is chosen as $\binom{1}{0}$, and $\eta_- \leftrightarrow \eta_+$ in $\| \ \|_\lambda$.

THE JOHNS HOPKINS UNIVERSITY

REFERENCES

---

[1] M. Abramowitz and I. A. Stegun, eds., *Handbook of Mathematical Functions*, Applied Mathematics Series **55**, National Bureau of Standards, Washington (1964).

[2] A. M. Dykhne, Quasiclassical particles in a one-dimensional periodic potential, *Soviet Physics JETP* **13** (1960), 999–1001.

[3] N. Fröman, Dispersion relations for energy bands and energy gaps derived by the use of a phase-integral method, with an application to the Mathieu equation, *Journal of Physics A* **13** (1980), 2355–2371.

[4] N. Fröman and P. O. Fröman, *JWKB Method, Contribution to the Theory*, North-Holland, Amsterdam (1965).

[5] E. M. Harrell, The band-structure of a one-dimensional, periodic system in a scaling limit, *Annals of Physics* **119** (1979), 351–369.

[6] E. M. Harrell and B. Simon, The mathematical theory of resonances whose widths are exponentially small, *Duke Mathematical Journal* **47** (1980), 845–902.

[7] P. Hartman, *Ordinary Differential Equations*, Hartman, Baltimore (1973).

[8] P. Hartman and C. R. Putnam, The gaps in the essential spectra of wave equations, *American Journal of Mathematics* **72** (1950), 849–862.

[9] H. Hochstadt, Estimates on the stability intervals for Hill's equation, *Proceedings of the American Mathematical Society* **14** (1963), 930–932.

[10] B. M. Levitan and I. S. Sargstan, *Introduction to Spectral Theory, Translations of Mathematical Monographs* **39**, American Mathematical Society, Providence (1975).

[11] W. Magnus and W. Winkler, *Hill's Equation*, Wiley-Interscience, New York (1966).

[12] H. P. McKean and E. Trubowitz, Hill's surfaces and their theta functions, *Bulletin of the American Mathematical Society* **84** (1978), 1042–1085.

[13] N. W. McLachlan, *Theory and Application of Mathieu Functions*, Clarendon, Oxford (1947).

[14] J. Meixner and F. W. Schäfke, *Mathieusche Funktionen und Sphäroidfunktionen*, Springer, Berlin-Göttingen-Heidelberg (1954).

[15] F. W. J. Olver, The asymptotic expansion of Bessel functions of large order, *Philosophical Transactions of the Royal Society of London A* **247** (1954), 328–368.

[16] F. W. J. Olver, *Introduction to Asymptotics and Special Functions*, Academic Press, New York (1974).

[17] M. Reed and B. Simon, *Methods of Modern Mathematical Physics*, vol. IV: *Analysis of Operators*, Academic Press, New York (1978).

[18] E.Trubowitz, The inverse problem for periodic potentials, *Communications on Pure and Applied Mathematics* **30** (1977), 321–337.

[19] H. Widom, Eigenvalue distribution theorems for certain homogeneous spaces, *Journal of Functional Analysis* **32** (1979), 139–147.

ON THE DEFICIENCY INDEX PROBLEM

By Don Hinton

---

1. Introduction. We discuss here the deficiency index problem for ordinary differential equations and some recent ramifications and open problems. To define the deficiency index problem (DIP), let l be a differential operator

$$(1.1) \qquad l(y) = \sum_{i=0}^{N} P_i y^{(i)}, \qquad a \leqslant x < b,$$

with singularity at b. For simplicity assume the P_i are i-fold continuously differentiable complex-valued functions. Further assume l is formally symmetric, i.e., $l = l^+$ where

$$l^+(y) = \sum_{i=0}^{N} (-1)^i \left(\overline{P}_i y \right)^{(i)}.$$

For a positive continuous function w on $[a, b)$, the operator l determines two operators acting in the Hilbert space $\mathcal{L}^2_w(a, b)$ of all Lebesgue measurable f satisfying $\int_a^b w|f|^2 < \infty$. These operators are defined as follows. Let $D(L)$ be the set of all $f \in \mathcal{L}^2_w(a, b)$ such that f, $f', \ldots, f^{(N-1)}$ are locally absolutely continuous and $w^{-1} l(f) \in \mathcal{L}^2_w(a, b)$. The maximal operator L is defined by:

$$L : D(L) \rightarrow \mathcal{L}^2_w(a, b), L(y) = w^{-1} l(y).$$

Let D'_0 be the set of all $f \in D(L)$ with compact support interior to (a, b), and let L'_0 be L restricted to D'_0. Then L'_0 is a symmetric, densely defined operator; hence L'_0 has a closure L_0. It follows that $L^*_0 = L$. Define the number $n(\lambda)$, $\operatorname{Im} \lambda \neq 0$, by

$$n(\lambda) = \dim\{ f \in D(L) : Lf = \lambda f \}.$$

It is know that $n(\lambda)$ is constant in the upper and lower half planes. We may now define the DIP by:

$$\text{DIP}: \quad \text{Compute } n(i), n(-i).$$

The importance of the numbers $n(i)$, $n(-i)$ is that they determine the number of boundary conditions which must be applied to members of $D(L)$ to determine a self-adjoint operator A. Such an A will exist if and only if $n(i) = n(-i)$ and A will satisfy $L_0 \subset A \subset L$. In terms of partial differential equations, the question is one of a boundary value problem being well-posed, e.g., what boundary conditions associated with $\partial\chi/\partial t = L(\chi)$ yield well-posed problems? For a full development of the above operator theory we refer to [2, 7, 11, 30, 42, 51, 52, 61, 63].

The classification of l according to its deficiency indices was begun in the classical paper of Weyl [65]. The theory developed somewhat slowly until the 1940's. A resurgence in the spectral theory of differential equations began then, particularly with the work of Hartman and Wintner, Titchmarsh, and the Russian school. A historical development may be found in [11, pp. 1581-1628].

Our purpose is not to survey the considerable literature that has evolved in recent years, but to describe a few principal results, state some new problems that have been an outgrowth of the DIP, and discuss some open problems. For additional discussion we refer in particular to the survey articles [10, 22], the lecture notes [42], and their references.

2. Subproblems of the DIP. The first subproblem that arises in the DIP is the *range problem*, i.e., what values of $n(i)$, $n(-i)$ are realized. A rather complete discussion of this problem is given in [22]; we mention here only the general bounds:

$$(2.1) \quad \begin{aligned} N = 2m; \quad & m \leqslant n(i), \quad n(-i) \leqslant 2m \\ N = 2m + 1; \quad & m \leqslant n(i) \leqslant 2m + 1, \\ & m + 1 \leqslant n(-i) \leqslant 2m + 1 \end{aligned}$$

where for $N = 2m + 1$ it is assumed that P_N is of the form $P_N = (-1)^m iP$ with $P(x) > 0$. These general bounds may be deduced from the Von Neumann representation of $D(L)$ [42, pp. 16–17] or from Atkinson's matrix form of self-adjoint systems [2] as shown by Walker [64]. Apparently it is still an open question if all pairs $n(i)$, $n(-i)$ are possible subject only to (2.1); see also Gilbert [26, 27] and Kogan and Rofe-Beketov [46].

In case the minimal values hold (when $n(i) + n(-i) = N$), we say l is in the *limit-point* (LP) case; when the maximal case $n(i) + n(-i) = 2N$ holds, we say l is in the *limit-circle* (LC) case. The terminology arises from Weyl's geometric method of proof of showing that for

$$(2.2) \qquad l(y) = -(py')' + qy,$$

the pair $n(i)$, $n(-i)$ is always either $(1,1)$ or $(2,2)$. (In the case of formally symmetric l with real coefficients one has $N = 2m$ and $n(i) = n(-i)$.)

The LP problem has by far dominated the literature. It is the only case which can be attacked in considerable generality without the use of asymptotic or quasi-asymptotic methods. For asymptotic methods in the DIP we refer to [9, 46]. In a physical problem the LP case means that the self-adjoint operators generated by l are determined by placing all the boundary conditions at the regular point a, i.e., no boundary conditions at the singular point b are needed. In numerical work this often means that truncation near b with zero boundary conditions can be expected to yield good results. An interpretation of the LP case in quantum mechanics may be found in [53].

For our discussion of the LP case we take l to be of the form,

$$(2.3) \qquad l(y) = \sum_{i=0}^{m} (-1)^i \left(q_i y^{(i)} \right)^{(i)},$$

where the q_i are real with $q_m > 0$. Associated with l is Green's formula

$$\int_a^c \left[\bar{z} l(y) - y \overline{l(z)} \right] dx = \left[y, z \right] \Big|_a^c$$

where $[\, , \,]$ is the Lagrange bilinear form

$$[y, z] = D(y, \bar{z}) - D(\bar{z}, y)$$

with

$$(2.4) \qquad D(y, z) = \sum_{i=0}^{m-1} y^{(i)} z^{[2m-i-1]}$$

See [34] for definition of the quasi-derivatives $z^{[2m-i-1]}$.

In the second order case some early general LP criteria were given by Hartman [28], Hartman and Winter [29], Levinson [50], and Titchmarsh [62]. For the special case of (2.2),

$$(2.5) \qquad l(y) = -y'' + q(x)y, \qquad a \leqslant x < \infty,$$

one result of [29] is that (2.5) is LP at ∞ if

$$\int_a^x q_-(t)\,dt = 0(x^3)$$

as $x \to \infty$ where $q_-(t) = \max\{0, -q(t)\}$; a result of [28] is that (2.5) is LP at ∞ if

$$(2.6) \qquad q(x) \geqslant k, \qquad k \text{ a constant},$$

for all x in a sequence of non-overlapping intervals of constant length.

Condition (2.6) is remarkable in that only the behavior of q on a subset of $[a, \infty)$ is considered (for the Schrodinger equation, this has a quantum-mechanical interpretation [16]). More general 2nd order interval-type *LP* criteria were subsequently given by Atkinson and Evans [4], Eastham [12], Evans [17], Knowles [44], and Read [57]. The most general results seem to be those of Evans [17] and Read [57] (see also [54, 55]). The generalization of Hartman's interval criteria now seems to have ceased. We indicate in section 4 how to apply the ideas to an equation with matrix coefficients. We note however that the interval criteria for (2.2) do not apply easily when p is "large"; for $p(x) = x^n$ and $w(x) = 1$, "large" means $n > 2$.

Early LP criteria for equations of order greater than 2 were derived by asymptotic methods (cf. [9, 52]). The first general results that avoided asymptotic methods were given by Everitt [19] in the fourth order case. Everitt's work was extended to equations of even order by Hinton [31, 32]; for the two term operator $l(y) = (-1)^n y^{(2n)} + q(x)y$ with $w \equiv 1$ this LP condition is that for some $k > 0$, $q(x) \geqslant -kx^{2n/(2n-1)}$. Interval-type criteria have been given by Atkinson [3] and Evans and Zettl [18]. Most of the higher-order results apply only to even order equations and frequently require $w \equiv 1$.

An interesting aspect of the higher order equations is that many of them require that, as applied to (2.3), q_i for $i > 0$ not be too large. For positive polynomial coefficients this may be interpreted as

$\int_a^\infty [w/q_i]^{1/2i} dx = \infty$, i.e., degree q_i − degree $w \leqslant 2i$. This distinction shows up already in

(2.7) $l(y) = -(x^\alpha y')' + q(x)y,$ $1 \leqslant x < \infty,$

with $w(x) = 1$. For $\alpha \leqslant 2$, (2.7) is LP if $q(x) \geqslant -kx^{2-\alpha}$ for some $k > 0$; for $\alpha > 2$, (2.7) is LP if $q(x) \geqslant -(2\alpha - 3)x^{\alpha-2}/4$ and the constant is sharp.

A notable exception to the above remark is the work of Kauffman [39, 41]. In [39] polynomial-like coefficients are considered (with $w \equiv 1$), and a detailed analysis of the structure of l is given. A comparison of l with a certain Euler operator is made, and it is proved that (2.3) with positive, polynomial-like coefficients is "almost always" LP at infinity, but that certain exceptional cases arise. An example of [39] disproved the positive coefficient conjecture, i.e., that (2.3) is always of LP type at infinity if the coefficients are positive. The proof of the counterexample does not apply to the 4th order case; hence a remaining open question is

$$l(y) = (ry'')'' - (qy')' + py,   a \leqslant x < \infty,$$

with $w \equiv 1$ LP at infinity if r, q, and p are positive? The paper [41] considers other classes of coefficients.

The work of Kauffman shows that many open problems remain in establishing LP criteria for q_i "large", $i > 0$. In particular what results can be obtained by the inclusion of an arbitrary weight w? The transformation theory of [32] shows that inclusion of a weight allows finite singularities to be transformed to infinity.

As shown by Eastham and Thompson [15] and Read [56], LC differential expressions are extremely sensitive to changes in the coefficients, e.g., changing a coefficient on an open set of arbitrarily small measure may alter the deficiency index. It is not surprising then that LC criteria generally require somewhat stringent growth conditions on the coefficients. A general second-order result has been given by Kupcov [47]; a simple proof of this result is given in Knowles [43] as well as many references for second-order LC results. There seem to be few LC results for higher order equations; however, some methods have been developed by Eastham [13, 14], Read [58], and Zettl [66].

We conclude this section with an open question. For (2.3) and an integer r, $m \leqslant r \leqslant 2m$, does there exist a weight w such that $n(i) = n(-i) = r$ with respect to the weight w? For $r = 2m$ the answer is clearly yes since one need only choose w so small that all solutions y of $l(y) = 0$ satisfy $\int_a^b w|y|^2\, dx < \infty$.

3. Recent ramifications. During the last decade there have been a number of new concepts arise from studying the deficiency index problem. The operator (2.3) is said to satisfy the *Dirichlet condition* (D) if

$$|q_i|^{1/2}|y| \in \ell^2(a, \infty), \qquad i = 0, \ldots, m \quad (y \in D(L))$$

A closely related concept is the *strong limit point* (SLP) condition defined by

$$\lim_{x \to b} D(y, z)(x) = 0 \qquad (y, z \in D(L))$$

where $D(y, z)$ is given by (2.4). This property was introduced by Everitt, Giertz, and Weidmann; cf. [20]. That SLP implies LP is an immediate consequence of the fact that LP is equivalent to

$$\lim_{x \to b} [y, z](x) = 0 \qquad (y, z \in D(L)).$$

The relations LP, SLP, and D are closely related, but independent; cf. [21, 24, 37, 48, 49]. Criteria for SLP and D may be found in [3, 6, 17, 20, 25, 33, 34, 37].

Consider now (2.3) in $\ell_w^2\,(a, b)$ with each $q_i \geqslant 0$. Define now the *Dirichlet index* Di(L) by:

$$\text{Di(L)} = \dim\left\{ y \in D(L) : L(y) + wy = 0, \right.$$

$$\left. \sum_{i=0}^{m} \int_a^b q_i|y^{(i)}|^2 < \infty \right\}$$

The Dirichlet index has proved to be quite useful in considering the convergence of eigenfunction expansions in the energy norm [35, 36, 38, 40] given by

$$\|y\|_E^2 = \int_a^b \left[w|y|^2 + \sum_{i=0}^{m} q_i|y^{(i)}|^2 \right] dx.$$

The Dirichlet index has also proved to play an important role in connection with the minimization problem associated with (2.3) [5], i.e.,

$$(3.1) \qquad \min_{y \in Q} \int_a^b \sum_{i=0}^m q_i |y^{(i)}|^2 \, dx$$

where Q is the set of all sufficiently differentiable y with (3.1) finite and $\int_a^b w |y|^2 = 1$.

For $w \equiv 1$, criteria for $Di(L) = m$ are given in [59]. The generality of these conditions has suggested the following conjecture by Kauffman: Does there exist an l in (2.3) with $q_i \geqslant 0$ such that $Di(L) > m$?

Two other recent outgrowths of the DIP are (i) the *separation problem*, i.e., when does $f \in D(L)$ in (1.1) (with $w \equiv 1$) imply $P_i^{(j)} f^{(i-j)} \in \mathcal{L}^2(a, b)$ for all $0 \leqslant j \leqslant i \leqslant m$, and (ii) the *problem of powers* defined as the study of the relationship of the deficiency indices of powers of l with those of l. The first problem was introduced by Everitt and Giertz in [23]; the second problem is the principal content of [42].

An interesting recent development is the investigation of J-symmetry which is defined as follows. Let J be a conjugation operator in $\mathcal{L}^2_w(a, b)$, e.g., $J(f) = \bar{f}$, and define a closed, densely defined linear operator T in $\mathcal{L}^2_w(a, b)$ to be *J-symmetric* if $JTJ \subset T^*$ (operator inclusion); T is called *J-selfadjoint* if $JTJ = T^*$. These concepts have application to differential operators with complex coefficients. As with the real coefficient case, the problem arises of describing, via boundary conditions, all J-selfadjoint restrictions of formally J-symmetric differential operators. We refer to the basic paper of Knowles [45].

4. Matrix coefficients. Equations with matrix coefficients have not been as extensively studied as the scalar case; we refer to Anderson [1], Eastham [13, 14] Rofe-Beketov [60], and their references. We state in this section how a theorem of Read [57], which represents a generalization of Hartman's theorem, may be extended to the matrix case. The proof, which we omit, is a combination of the methods of [32] and [57].

Let K, P, S, T be continuous $n \times n$ hermitian matrices on $a \leqslant x < \infty$ with $K(x) > 0$, $P(x) > 0$, and either $T \equiv 0$ or $T(x)$ is of constant sign ($T(x) \geqslant 0$ or $T(x) \leqslant 0$) but $T \not\equiv 0$. Define

$$\tau(Y) = K^{-1} \left[-(PY')' + (S + iT)Y \right]$$

Then τ acts in the Hilbert space $\mathscr{L}_K^2(a, \infty)$ of all measurable, n-vector functions Y satisfying $\int_a^\infty Y^* KY\, dx < \infty$. Define

$$V = \left\{ Y \in \mathscr{L}_K^2(a, \infty) : Y, PY' \text{ are } Ac_{loc}, \tau(Y) = 0 \right\}$$

where AC_{loc} means locally absolutely continuous. Note that dim $V \leqslant 2n$.

THEOREM 4.1. *Suppose* $S = S_1 + S_2$ *where* S_1 *and* S_2 *are hermitian and continuous and there is a real function* $w \geqslant 0$ *with* $w \in AC_{loc}$ *such that for some constant* $k \geqslant 0$ *and* $a \leqslant x < \infty$:
(i) $\| K^{-1/2}(x) P^{1/2}(x) \| \quad |w'(x)| \leqslant k$.
(ii) $\| K^{-1/2}(x) P^{1/2}(x) \| w(x) \leqslant kx$.
(iii) $-S_1(x) w(x)^2 \leqslant k K(x)$.
(iv) $|\int_a^x s_2 w^{1-b}| w(x)^b \| P^{-1/2}(x) K^{-1/2}(x) \| \leqslant k$ *for some* $b, 0 \leqslant b \leqslant 1$, *where* $s_2(x) = \| S_2(x) \|$.
Then if $Y \in V$,
(v) $\int_a^\infty w^2 Y'^* PY'\, dx < \infty, \int_a^\infty w^2 Y^* S_1 Y\, dx < \infty$.
(vi) dim $V \leqslant n$ *if also* $\int_a^\infty w \| K^{-1/2} P^{1/2} \|^{-1}\, dx = \infty$.

We have necessarily omitted many important topics of recent years, e.g., integral inequalities, Titchmarsh-Weyl $m(\lambda)$ functions, pairs of differential operators (cf. Coddington and Snoo [8]). The topics chosen reflect the interests of the author.

UNIVERSITY OF TENNESSEE

REFERENCES

[1] R.L. Anderson, " Limit-point and limit circle criteria for a class of singular symmetric differential operators," *Can. J. Math.* **28** (1976), 905–914.

[2] F.V. Atkinson, *Discrete and continuous boundary problems* (Academic Press, New York, 1964).

[3] ———, "Limit-n criteria of integral type", *Proc. Royal Soc. Edinburgh* (A) **73** (1975), 167–198.

[4] F.V. Atkinson and W.D. Evans, "Solutions of a differential equation wich are not of integrable square," *Math. Z.* **127** (1972), 323–332.

[5] J. Bradley, D. Hinton, R. Kauffman, "On the minimization of singular quadratic functions," *Proc. Royal Soc. Edinburgh* **87A** (1981), 193–208.

[6] B. Brown and W. Evans, "On the limit-point and strong limit-point classification of 2nth order differential expressions with wildly oscillating coefficients," *Math. Z.* **134** (1973), 351–368.

[7] E. Coddington and N. Levinson, *Theory of ordinary differential equations* (McGraw-Hill, New York, 1955).

[8] E. Codding and H.S.V. Snoo, "Differential subspaces associated with pairs of ordinary differential expressions," *J. Differential Eqs.*, to appear.

[9] A. Devinatz, "The deficiency index of a certain class of ordinary self-adjoint differential operators," *Advances in Math.* **8** (1972), 434–473.

[10] ———, "The deficiency index problem for ordinary self-adjoint differential operators", *Bull. Amer. Math. Soc.* **79** (1973), 1109–1127.

[11] N. Dunford and J.T. Schwartz, *Linear operators: II* (Interscience, New York, (1963).

[12] M.S.P. Eastham, "On a limit-point method of Hartman", *Bull. London Math. Soc.* **4** (1972), 340–344.

[13] ———, "Self-adjoint differential expressions with all solutions $\ell^2(0, \infty)$, *Differential Equations*. Proc. 1977 Uppsala International Conference (Almqirst and Wiksell International, Stockholm, Stockholm, 1977).

[14] ———, "The limit-4 case of fourth-order self-adjoint differential operators," *Proc. Roy. Soc. Edinburgh* **79A** (1977), 51–60.

[15] M.S.P. Eastham and M.L. Thompson, "On the limit-point, limit-circle classification of second-order ordinary differential equations", *Quart. J. Math.* **24** (1973), 531–35.

[16] M.S.P. Eastham, W.D. Evans, and J.B. McLeod, "The essential self-adjointness of Schrodinger-type operators", *Arch. Rat. Mech. Analysis* **60** (1976), 185–204.

[17] W.D. Evans, "On limit-point and Dirichlet-type results for second-order differential expressions," *Lecture Notes in Mathematics* (Springer-Verlag, Berlin) **564** (1976), 78–92.

[18] W.D.Evans and A. Zettl, "Interval limit-point criteria for differential expressions and their powers," *J. London Math Soc.* (2), **13** (1976), 543–556.

[19] W.N. Everitt, "On the limit-point classification of fourth-order differential operators," *J. London Math. Soc.* **44** (1969), 273–281.

[20] ———, "On the strong limit-point condition of second-order differential expressions," *Proc. of the International Conference on Differential Equations, Los Angeles* (1974), 287–307. (Academic Press, Inc., New York).

[21] ———, "A note on the Dirichlet condition for second-order differential expressions," *Canad. J. Math.* **28** (1976), 312–320.

[22] ———, "On the deficiency index problem for ordinary differential operators, 1910–1970," *Differential Equations*. Proc. 1977 Uppsala International Conference (Almqirst and Wiksell International, Stockholm, 1977).

[23] W.N. Everitt and M. Giertz, "Some properties of the domains of certain differential operators," *Proc. London Math. Soc.* (3), **23** (1971), 301–324.

[24] W.N. Everitt, M. Giertz and J.B. McLeod, "On the strong and limit-point classification of second-order differential expressions," *Proc. London Math. Soc.* **28** (1974), 142–158.

[25] W.N. Everitt, D.B. Hinton and J.S. Wong, "On the strong limit-n classification of linear ordinary differential expressions of order $2n$," *Proc. London Math. Soc.* **29** (1974), 351–367.

[26] R.C. Gilbert, "Asymptotic formulas for solutions of a singular linear ordinary equations," *Proc. Roy. Soc. Edinburgh* **81A** (1978), 57–70.

[27] ——, "A class of symmetric ordinary differential operators whose deficiency numbers differ by an integer," *Proc. Roy. Soc. Edinburgh* **82A** (1979), 117–134.

[28] P. Hartman, "The number of L^2-solutions of $x'' + q(t)x = 0$," *Amer. J. of Math.* **73** (1951), 635–645.

[29] P. Hartman and A. Winter, "A criterion for the non-degeneracy of the wave equation," *Amer. J. Math.* **71** (1949), 206–213.

[30] E. Hille, *Lectures on ordinary differential equations* (Addison-Wesley, London, 1969).

[31] D. Hinton, "Limit-point criteria for differential equations," *Can. J. Math.* **24** (1972), 293–305.

[32] ——, "Limit-point criteria for differential equations, II," *Can. J. Math.* **26** (1974), 340–341.

[33] ——, "Limit point-limit circle criteria for $(py')' + qy = \lambda ky$," *Lecture Notes in Mathematics* (Springer Verlag, Berlin), **415** (1974), 173–183.

[34] ——, "Strong limit-point and Dirichlet criteria for ordinary differential expressions of order $2n$," *Proc. Royal Soc. Edinburgh* **76A** (1977), 301–310.

[35] ——, "On the eigenfunction expansions of singular ordinary differential equations," *J. Differential Equations* **24** (1977), 282–308.

[36] ——, "Eigenfunction expansions and spectral matrices of singular differential operators," *Proc. Royal Soc. Edinburgh* **80A** (1978), 289–308.

[37] H. Kalf, "Remarks on some Dirichlet type results for semi-bounded Sturm-Liouville operators," *Math. Ann.* **210** (1974), 197–205.

[38] ——, "A characterization of the Friedrichs extension of Sturm-Liouville operators," *J. London Math. Soc.* (2) **17** (1978), 511–521.

[39] R.M. Kauffman, "On the limit-n classification of ordinary differential operators with positive coefficients," *Proc. London Math. Soc.* **35** (1977), 496–526.

[40] ——, "The number of dirichlet solutions to a class of linear ordinary differential equations," *J. Differential Eqs.* **31** (1979), 117–129.

[41] ——, "On the limit-n classification of ordinary differential operators with positive coefficients, II," *Proc. London Math. Soc.* **41** (1980), 499–515.

[42] R.M. Kauffman, T.T. Read, and A. Zettl, "The deficiency index problem for powers of ordinary differential expressions," *Lecture Notes in Mathematics* (Springer-Verlag, Berlin) **621**, 1977.

[43] I. Knowles, "On the number of $\mathcal{L}^2$-solutions of second order linear differential equations," *Proc. Royal Soc. Edinburgh* **80A** (1978), 1–13.

[44] ——, "Note on a limit-point criterion," *Proc. Amer. Math. Soc.* **41** (1973), 117–119.

[45] ——, "Symmetric conjugate-linear operators in Hilbert space," to appear.

[46] V.I. Kogan and F.S. Rofe-Beketov, "On the question of deficiency indices of differential operators with complex coefficients," *Proc. Roy. Soc. Edinburgh* **72A** (1973/74), 281–298.

[47] N.P. Kupcov, "Conditions of non-self-adjointness of a second order linear differential operator," (Russian) *Dokl. Akad. Nauk.* **138** (1961), 767–770.

[48] M.K. Kwong, "Note on the strong limit point condition of second-order differential expressions," *Quart. J. Math. Oxford* (2), **28** (1977), 201–208.

[49] ——, "A second-order Dirichlet differential expression that is not bounded below," *Proc. Royal Soc. Edinburgh* **83A** (1979), 39–43.

[50] N. Levinson, "Criteria for the limit-point case for second-order linear differential operators," *Casopis pro pesto vanyi matematiky a fysiky'*, **74** (1949), 17–20.

[51] B.M. Levitan and I.S. Sargsjan, *Self-adjoint ordinary differential operators* (English translation from the Russian in Translations of Mathematical Monographs, **39**, Amer. Math. Soc. 1975.)

[52] M.A. Naimark, *Linear differential operators*, **II** (New York: Ungar, 1968).

[53] J. Rauch and M. Reed, "Two examples illustrating the differences between classical and quantum mechanics," *Commun, Math. Phys.* **29** (1973), 105–111.

[54] T.T. Read, "A limit-point criterion for expressions with oscillatory coefficients," *Pac. J. Math.* **66** (1976), 243–255.

[55] ———, "A limit-point criterion for expressions with intermittently positive coefficients," *J. London Math. Soc.* (2) **15** (1977), 271–276.

[56] ———, "Perturbations of limit-circle expressions," *Proc. Amer. Math. Soc.* **56** (1976), 108–110.

[57] ———, "A limit-point criterion for $-(py')' + qy$," *Lecture Notes in Mathematics* (Springer-Verlag, Berlin) **564** (1976), 383–390.

[58] ———, "Higher order differential equations with small solutions," to appear.

[59] J. Robinette, "On the Dirichlet problem for ordinary differential equations," M.S. Thesis, University of Tennessee, 1979.

[60] F.S. Rofe-Beketov, "Square-integrable solutions, self-adjoint extensions and spectrum of differential systems," *Differential equations* Proc. 1977 Uppsala International Conference (Almqirst and Wiksell International, Stockholm, 1977).

[61] M.H. Stone, "Linear transformations in Hilbert space and their applications to analysis," (Amer. Math. Soc. Coll. Publ. 15, New York, 1932).

[62] E.C. Titchmarsh, "On the uniqueness of the Green's function associated with a second-order differential equation," *Can. J. Math.* **1** (1949), 191–198.

[63] E.C. Titchmarsh, *Eigenfunction expansions*: I(Oxford University Press, 1962).

[64] P.W. Walker, "A vector-matrix formulation for formally symmetric ordinary differential equations with applications to solutions of integrable square," *J. London Math. Soc.* (2) **9** (1974), 151–159.

[65] H. Weyl, "Uber gewohnliche Differentialgleichungen mit Singularitaten und die zugehorigen Entwicklungen willkurlicher Funktionen," *Math. Ann.* **68** (1910), 220–269.

[66] A. Zettl, "An algorithm for the construction of limit-circle expressions," *Proc. Royal Soc. Edinburgh* **75A** (1975–76), 1–3.

ISOMETRY OF COMPACT RIEMANNIAN MANIFOLDS TO SPHERES

By Chuan-Chih Hsiung

1. Let M^n be a Riemannian manifold of dimension $n \geq 2$ and class C^3, (g_{ij}) the symmetric matrix of the positive definite metric of M^n, and (g^{ij}) the inverse matrix of (g_{ij}), and denote by ∇_i, R_{hijk}, $R_{ij} = R^k_{ijk}$ and $R = g^{ij}R_{ij}$ the operator of covariant differentiation with respect to g_{ij}, the Riemann tensor, the Ricci tensor and the scalar curvature of M^n respectively, where all indices take the values $1, \ldots, n$, and repeated indices indicate summation. Let d be the operator of exterior differentiation, δ the operator of codifferentiation, and $\Delta = d\delta + \delta d$ the Laplace-Beltrami operator.

Let v be a vector field defining an infinitesimal conformal transformation of M^n, and L_v the Lie derivative with respect to v. Then we have

$$(1.1) \qquad L_v g_{ij} = \nabla_i v_j + \nabla_j v_i = 2\rho g_{ij}.$$

The infinitesimal transformation v is said to be homothetic or an infinitesimal isometry according as the scalar function ρ is constant or zero. We also denote by $L_{d\rho}$ the Lie derivative with respect to the vector field ρ^i defined by

$$(1.2) \qquad \rho^i = g^{ij}\rho_j, \qquad \rho_j = \nabla_j\rho.$$

Let $\xi_{I(p)}$ and $\eta_{I(p)}$ be two tensor fields of the same order $p \leq n$ on a compact orientable manifold M^n, where $I(p)$ denotes an ordered subset $\{i_1, \ldots, i_p\}$ of the set $\{1, \ldots, n\}$ of positive integers less than or equal to n. Then the local and global scalar products $\langle \xi, \eta \rangle$ and (ξ, η) of the tensor fields ξ and η are defined by

$$(1.3) \qquad \langle \xi, \eta \rangle = \frac{1}{p!} \xi^{I(p)}\eta_{I(p)},$$

$$(1.4) \qquad (\xi, \eta) = \int_{M^n} \langle \xi, \eta \rangle \, dV,$$

where dV is the element of volume of the manifold M^n at a point.

The following is a long standing conjecture.

Conjecture. If a compact Riemannian manifold M^n of dimension $n > 2$ with positive scalar curvature R admits an infinitesimal non-isometric conformal transformation, then M^n is isometric to an n-sphere.

It will be seen below that the conjecture in this form is false. A weakened formulation is proposed at the end of this paper.

2. In the last two decades or so, related to the conjecture of §1 various authors have obtained various results among which the following two theorems jointly obtained by the author and N. H. Ackerman [2] are most general.

Throughout this section, M^n will denote a compact Riemannian manifold of dimension $n > 2$ with metric g_{ij}, which admits an infinitesimal nonisometric conformal transformation v satisfying (1.1) with $\rho \neq 0$.

THEOREM 2.1. *An oriented manifold M^n is isometric to an n-sphere if it satisfies one of the following three equivalent conditions:*

$$\left(P + \frac{c}{n} \left[nR\rho_i\rho^i - (L_v R + nR\rho)\Delta\rho \right], 1 \right) \geq 0,$$

(1.5) $$\left(P - \frac{c}{n} \rho(nL_{d\rho}R + \Delta L_v R), 1 \right) \geq 0,$$

$$\left(P + \frac{c}{n} \left[L_v, L_{d\rho} \right] R, 1 \right) \geq 0,$$

where

(1.6)
$$P = \rho L_v \left[a^2 A + \frac{c - 4a^2}{n - 2} B - \frac{1}{n} \left(\frac{2a^2}{n - 1} + \frac{c - 4a^2}{n - 2} \right) R^2 \right],$$

$$\left[L_v, L_{d\rho} \right] = L_v L_{d\rho} - L_{d\rho} L_v,$$

A and B are defined by

(1.7) $$A = R^{hijk} R_{hijk}, \qquad B = R^{ij} R_{ij},$$

and a, c are constants such that

$$(1.8) \qquad c \equiv 4a^2 + (n-2)\left[2a \sum_{i=1}^{4} b_i + \left(\sum_{i=1}^{6} (-1)^{i-1} b_i \right)^2 \right.$$

$$\left. - 2(b_1 b_3 + b_2 b_4 - b_5 b_6) + (n-1) \sum_{i=1}^{6} b_i^2 \right]$$

$$> 0,$$

b's being arbitrary constants.

An elementary calculation shows that $c \geqslant 0$ where equality holds if and only if $b_1 = \cdots = b_4, b_5 = b_6 = 0, a = -(n-2)b_1$.

For $L_t R = 0$, Theorem 2.1 (referred to the first inequality of (1.5) for $P = 0$ and $(nR\rho_i\rho^i - (L_t R + nR\rho)\Delta\rho, 1) \geqslant 0)$ with "isometric" replaced by "conformal" is due to Yano [11] for either $a \neq 0$, $c - 4a^2 = 0$ or $a = 0$, $c - 4a^2 \neq 0$, and due to Hsiung and Stern [8] for general a and b's. Theorem 2.1 (referred to the first inequality of (1.5) for $P = 0$ and $(nR\rho_i\rho^i - (L_t R + nR_\rho)\Delta\rho, 1) \geqslant 0)$ is due to Yano and Hiramatu [12] for $a \neq 0$, $c - 4a^2 = 0$ or $a = 0$, $c - 4a^2 \neq 0$.

For constant R, Theorem 2.1 (referred to the second inequality of (1.5) for $P = 0$) is due to Lichnerowicz [9] for $a = 0$, $c \neq 0$, B = constant, due to Hsiung [5] for $a \neq 0$, $c - 4a^2 = 0$, A = constant, due to Yano [10] for either $a = 0$, $c \neq 0$, or $a \neq 0$, $c - 4a^2 = 0$, due to Hsiung [6] for $b_2 = \cdots = b_6 = 0$, due to Yano and Sawaki [13] for $b_1 = \cdots = b_4 = b/(n-2), b_5 = b_6 = 0$, and due to Hsiung [7] for general a and b's. For $L_t R = 0, L_{d\rho} R = 0$, Theorem 2.1 (referred to the second inequality of (1.5) for $P = 0$) is due to Ackler and Hsiung [2].

THEOREM 2.2. *A manifold M^n is isometric to an n-sphere if it satisfies*

$$(1.9) \qquad L_t (A^a B^b) = 0,$$

$$(1.10) \qquad c\left(\frac{2a}{A} + \frac{(n-1)b}{B} \right) = \frac{2^a(a+b)R^{2(a+b-1)}}{n^{a+b-1}(n-1)^{a-1}},$$

$$(1.11) \qquad \left(\frac{b}{2(n-1)} A^a B^{b-1} R L_t R - A^a B^b \left(\frac{4a}{A} + \frac{(n-2)b}{B} \right) \right.$$

$$\left. \times \left(R^{ij} \nabla_i \nabla_j \rho + \frac{R^2 \rho}{n(n-1)} \right), \rho \right) \leqslant 0,$$

where A, B are given by (1.7), *and a, b are nonnegative integers and not both zero.*

For constant R and $A^a B^b$, Theorem 2.2 is due to Lichnerowicz [9] for $a = 0$, $b = 1$, and due to Hsiung [5] for general a and b. For constant $A^a B^b$ and $L_c R = 0$, $L_{d\rho} R = 0$, Theorem 2.2 is due to Hsiung and Stern [8]; in this case condition (1.11) is satisfied automatically since

$$(1.12) \quad \left(R^{ij} \nabla_i \nabla_j \rho + \frac{R^2 \rho}{n(n-1)} , \rho \right) \geqslant 0,$$

which is due to Hsiung and Stern [8], and due to Lichnerowicz [9] for constant R.

3. Let B and F be Riemannian manifolds, and f a positive C^∞ function on B. Consider the product manifold $B \times F$ with projections $\pi : B \times F \to B$ and $\eta : B \times F \to F$. The warped product $M = B \times_f F$ defined by Bishop and O'Neill [3] is the manifold $B \times F$ with the Riemannian structure such that $\|v\|^2 = \|\pi_* v\|^2 + (f(\pi m))^2 \|\eta_* v\|^2$ for all tangent vectors $v \in T_m M$, where $\| \ \|$ denotes the length of a vector in $T_m M$, and $T_m M$ the tangent space of M at a point $m \in M$.

Very recently, N. Ejiri [4] established the following theorem showing the conjecture of §1 to be false.

THEOREM 3.1. *Let r be a positive number, E the set of real numbers, and M^n a complete Riemannian n-manifold with positive constant scalar curvature R. If $f(t), t \in E$, is a positive periodic function satisfying*

$$\frac{d^2 f}{dt^2} = - \frac{n-1}{2} \frac{1}{f} \left(\frac{df}{dt} \right)^2 + \frac{R}{2n} \frac{1}{f} - \frac{r}{2n} f,$$

then $E \times_f M^n$ is a complete Riemannian manifold with positive constant scalar curvature R admitting an infinitesimal nonisometric conformal transformation. In particular, if M^n is compact, then there exist compact Riemannian manifolds with constant scalar curvature admitting an infinitesimal nonisometric conformal transformation.

Remark 1. Because of Theorem 3.1, the conditions in Theorems 2.1 and 2.2 are not extraneous.

Remark 2. Now it is natural to ask whether the following conjecture holds or not.

Conjecture. If a compact Riemannian manifold M^n of dimension $n > 2$ with scalar curvature R admits an infinitesimal nonisometric conformal transformation v satisfying (1.1) with $\rho \neq 0$ such that $L_v R = 0$ and $L_{d\rho} R = 0$, then M^n is isometric to an n-sphere.

LEHIGH UNIVERSITY

REFERENCES

[1] N. H. Ackerman and C. C. Hsiung, Isometry of Riemannian manifolds to spheres. II, *Canad. J. Math.* **28** (1976), 63–72.

[2] L. L. Ackler and C. C. Hsiung, Isometry of Riemannian manifolds to spheres, *Ann. Mat. Pura Appl.* **99** (1974), 53–64.

[3] R. L. Bishop and B. O'Neill, Manifolds of negative curvature, *Trans. Amer. Math. Soc.* **145** (1969), 1–49.

[4] N. Ejiri, A negative answer to a conjecture of conformal transformations of Riemannian manifolds, *J. Math. Soc. Japan* **33** (1981), 261–266.

[5] C. C. Hsiung, On the group of conformal transformations of a compact Riemannian manifold, *Proc. Nat. Acad. Sci. U.S.A.* **54** (1965), 1509–1513.

[6] ———, On the group of conformal transformations of a compact Riemannian manifold. II, *Duke Math. J.* **34** (1967), 337–341.

[7] ———, On the group of conformal transformations of a compact Riemannian manifold. III, *J. Differential Geometry* **2** (1968), 185–190.

[8] C. C. Hsiung and L. W. Stern, Conformality and isometry of Riemannian manifolds to spheres, *Trans. Amer. Math. Soc.* **161** (1972), 65–73.

[9] A. Lichnerowicz, Sur les transformations conformes d'une variété riemannienne compacte, *C. R. Acad. Sci. Paris* **259** (1964), 697–700.

[10] K. Yano, On Riemannian manifolds with constant scalar curvature admitting a conformal transformation group, *Proc. Nat. Acad. Sci. U.S.A.* **55** (1966), 472–476.

[11] ———, On Riemannian manifolds admitting an infinitesimal conformal transformation, *Math. Z.* **113** (1970), 205–214.

[12] K. Yano and H. Hiramatu, Riemannian manifolds admitting an infinitesimal conformal transformation, *J. Differential Geometry* **10** (1975), 23–38.

[13] K. Yano and S. Sawaki, Riemannian manifolds admitting a conformal transformation group, *J. Differential Geometry* **2** (1968), 161–184.

ON A THEOREM OF LIAPOUNOFF ON HOLOMORPHIC
SOLUTIONS OF SINGULAR FIRST ORDER PARTIAL
DIFFERENTIAL EQUATIONS

By Stanley Kaplan

---

1. Introduction. Let $Z_i(z,u)(= Z_i(z_1, \ldots, z_d, u_1, \ldots, u_p))$ and $F_j(z,u)$ be holomorphic functions in a neighborhood of the origin in $\mathbb{C}^d \times \mathbb{C}^p$ with $Z_i(0,0) = 0$ and $F_j(0,0) = 0$ for $i = 1, \ldots, d$ and $j = 1, \ldots, p$. Under hypotheses to be given below, the theorem of our title asserts the existence of a unique p-tuple of functions $u(z) = (u_1(z), \ldots, u_p(z))$, holomorphic in a neighborhood of the origin in $\mathbb{C}^d$, and vanishing at the origin, satisfying

$$\sum_{i=1}^{d} Z_i(z, u(z)) \frac{\partial u_j}{\partial z_i}(z) = F_j(z, u(z)) \qquad j = 1, \ldots, p$$

which we rewrite, in matrix form, as

(1) $$D_z u(z) Z(z, u(z)) = F(z, u(z)).$$

We denote the $d \times d$ matrix of complex numbers

$$\left(\frac{\partial Z_i}{\partial z_j}(0,0) \right)_{i,j=1,\ldots,d} \qquad \text{by } (D_z Z)_0$$

and the $p \times d$ matrix

$$\left(\frac{\partial u_i}{\partial z_j}(0) \right)_{\substack{i=1,\ldots,p \\ j=1,\ldots,d}} \qquad \text{by } (D_z u)_0, \text{ etc.}$$

Given square matrices of complex numbers $\mathcal{B}(d \times d)$ and $\mathcal{C}(p \times p)$, if $\lambda_1, \ldots, \lambda_d$ are the eigenvalues of $\mathcal{B}$, repeated according to algebraic multiplicity, then, for l a non-negative integer we use the abbreviation $E(l; \mathcal{B}, \mathcal{C})$ for the following statement:

no eigenvalue of $\mathcal{C}$ is of the form $\sum_{i=1}^{d} \alpha_i \lambda_i$ where

$\alpha = \langle \alpha_1, \ldots, \alpha_d \rangle \in \mathbb{N}^d$ and $|\alpha| \equiv \alpha_1 + \cdots + \alpha_d = l.$

169

Liapounoff's hypotheses are these:

i) the eigenvalues $\lambda_1, \ldots, \lambda_d$ of $(D_z Z)_0$ satisfy

$$0 \notin CH\{\lambda_1, \ldots, \lambda_d\} \quad (CH \equiv \text{convex hull}).$$

ii) $E(l; (D_z Z)_0, (D_u F)_0)$ holds for all $l = 1, 2, 3, \ldots$ and
iii) $(D_u Z)_0 = 0$.

Liapounoff further observed that if iii) is replaced by
iii') $(D_z F)_0 = 0$
then a unique solution $u(z)$ is still obtained, subject to the additional requirement $(D_z u)_0 = 0$; in this case $E(l; (D_z Z)_0, (D_u F)_0)$ need be assumed to hold only for $l \geq 2$ (see [3], pp. 310–315.)

In a previous work ([2]) we considered the special case of (1) in which Z does not depend on u. We assumed that $Z(z) = (Z_1(z), \ldots, Z_d(z))$ has as its zero variety near the origin the germ S of an analytic sub-manifold through the origin of dimension $s < d$. Thus (at least) s of the eigenvalues $\lambda_1, \ldots, \lambda_d$ of $(D_z Z)_0$ are zero; we denote by $\lambda_1, \ldots, \lambda_n$ $(n = d - s)$ the remaining ones. Under the further hypothesis that

$$(2) \qquad 0 \notin CH\{\lambda_1, \ldots, \lambda_n\}$$

we showed that every formal power series solution $u(z) = (u_1(z), \ldots, u_p(z))$ of (1), such that each $u_i(z)$ has constant term zero, has what we shall call here the Artin property (see [1]) with respect to (1); in general, a p-tuple $u(z) = (u_1(z), \ldots, u_p(z))$ of formal power series which satisfies a system of equations $(\mathcal{E})$ will be said to have the Artin property with respect to $(\mathcal{E})$ if, for every positive integer ν there exists a p-tuple $u^*(z) = (u_1^*(z), \ldots, u_p^*(z))$ of convergent power series (that is, convergent in some neighborhood of the origin) satisfying $(\mathcal{E})$, and such that

$$\text{order}(u^*(z) - u(z)) \equiv \min_{i=1, \ldots, p} \text{order}(u_i^*(z) - u_i(z)) \geq \nu.$$

This result extends to the general equation (1) as follows:

THEOREM 1. *Let* $Z_i(z, u)$ *and* $F_j(z, u)$ *be holomorphic in a neighborhood of the origin in* $\mathbb{C}^d \times \mathbb{C}^p$, *with* $Z_i(0, 0) = 0$ *and* $F_j(0, 0) = 0$ *for*

$i = 1, \ldots, d$ and $j = 1, \ldots, p$. Let $\bar{u}(z) = (\bar{u}_1(z)), \ldots, \bar{u}_p(z))$ be a p-tuple of formal power series, with constant term zero, satisfying (1). Suppose there exists the germ S of an analytic sub-manifold of dimension s through the origin in $\mathbb{C}^d$, such that

(3) $\qquad Z_i(z, \bar{u}(z)) = 0 \quad$ on S for $i = 1, \ldots, d.$

Let $\lambda_1, \ldots, \lambda_n$, with $n = d - s$, be the non-trivial eigenvalues of

$$\mathcal{B} \equiv (D_z Z)_0 + (D_u Z)_0 (D_z \bar{u})_0$$

(the other s eigenvalues are easily seen to be zero, as a result of (3).) If $\lambda_1, \ldots, \lambda_n$ satisfy (2) then $\bar{u}(z)$ has the Artin property with respect to (1) and (3).

What follows in this section is a proof of Theorem 1 with one major technical detail, a generalization of Liapounoff's theorem, appearing in the next section as Theorem 2. First, a word about the meaning of (3) is in order: S may be defined as the image of the s-dimensional plane

$$\{ z' = (z'_1, \ldots, z'_d) \in \mathbb{C}^d : z'_1 = z'_2 = \cdots = z'_n = 0 \}$$

under an invertible holomorphic map. In different notation, with $x = (x_1, \ldots, x_n)$ replacing $(z'_1, \ldots, z'_n)$ and $y = (y_1, \ldots, y_s)$ replacing $(z'_{n+1}, \ldots, z'_d)$, we have $S = \{ z = \Phi(0, y) : y \in \mathbb{C}^s \}$ near 0, where $\Phi(x, y)$ is an invertible holomorphic map of a neighborhood of $(0,0)$ in $\mathbb{C}^n \times \mathbb{C}^s$ onto a neighborhood of 0 in $\mathbb{C}^d$, with $\Phi(0,0) = 0$. Thus (3) is understood to mean

(3) $\qquad Z_i(\Phi(0, y), \bar{u}(\Phi(0, y))) = 0 \qquad$ for $i = 1, \ldots, d.$

With $v(x, y) \equiv u(\Phi(x, y))$, (1) becomes, in the new variables (x, y)

(1') $\qquad D_x v(x, y) X(x, y, v(x, y)) + D_y v(x, y) Y(x, y, v(x, y))$

$$= G(x, y, v(x, y))$$

where $X(x, y, v) = (X_1(x, y, v), \ldots, X_n(x, y, v))$ and $Y(x, y, v) = (Y_1(x, y, v), \ldots, Y_s(x, y, v))$ are related to $Z(x, u)$ by $D_x \Phi(x, y) X(x, y, v) + D_y \Phi(x, y) Y(x, y, v) = Z(\Phi(x, y), v)$, and

$G(x, y, v) \equiv F(\Phi(x, y), v)$. (3) becomes now

(3')
$$X(0, y, v(0, y)) = 0$$
$$Y(0, y, v(0, y)) = 0$$

It is easy to see that the Jacobian matrix at $(0,0)$ in $\mathbb{C}^n \times \mathbb{C}^s$ of the map $(x, y) \rightarrow (X(x, y, 0), Y(x, y, 0))$ is given by

$$J^{-1}(D_z Z)_0 J$$

where J is the Jacobian matrix of Φ at the origin. Since $\bar{v}(x, y) \equiv \bar{u}(\Phi(x, y))$ (where $\bar{u}(z)$ is the solution given in Theorem 1) satisfies (3'), it follows that

$$(D_x X)_0 + (D_v X)_0 (D_x \bar{v})_0 = 0$$

and

(4) $$(D_y Y)_0 + (D_v Y)_0 (D_y \bar{v})_0 = 0$$

Consequently, the eigenvalues of the $n \times n$ matrix

$$\mathcal{B}' \equiv (D_x X)_0 + (D_v X)_0 (D_x \bar{v})_0$$

are precisely $\lambda_1, \ldots, \lambda_n$.

Let us agree to say that a power series $\varphi(x, \omega)$ in $(x_1, \ldots, x_n)$ and $\omega = (\omega_1, \ldots, \omega_m)$ vanishes to order k in x $(k = 0, 1, 2, \ldots)$ if we can write

$$\varphi(x, \omega) = \sum_{\alpha \in \mathbb{N}^n} \varphi_\alpha(\omega) x^\alpha$$

where each $\varphi_\alpha(\omega)$ is a power series in ω, and $\varphi_\alpha(\omega) = 0$ for all α with $|\alpha| < k$. A system of equations $(\mathcal{E})$ involving power series in x and ω will be said to hold to order k in x if the difference of the two sides of each equation in $(\mathcal{E})$ vanishes to order k in x. We shall say that $\varphi(x, \omega) = (\varphi_1(x, \omega), \ldots, \varphi_p(x, \omega))$ satisfies $(\mathcal{E}_k)$ if $\varphi(x, \omega)$ satisfies $(\mathcal{E})$ to order k in x. Then the proof of Theorem 1 consists of two steps.

Step 1.   If $\bar{v}(x, y)$ is as above, for every $k = 1, 2, \ldots$ $\bar{v}(x, y)$ has the Artin property with respect to $(1'_k)$ and (3').

Step 2. For any sufficiently large k, if $v^*(x, y)$ is a convergent power series solution of $(1'_k)$ and $(3')$ with $v^*(0,0) = 0$ and $(D_x v^*)_0 = (D_x \bar{v})_0$, then there exists a unique convergent power series solution $v(x, y)$ of $(1')$ satisfying $v(x, y) = v^*(x, y)$ to order k in x.

Step 2 is accomplished using Theorem 2 below along with the observation that (2) implies that $|\sum_{i=1}^n \alpha_i \lambda_i| \to \infty$ as $|\alpha| \to \infty (\alpha \in \mathbb{N}^n)$. Thus, given any $\mathcal{C}$, $E(l; \mathcal{B}', \mathcal{C})$ holds for all sufficiently large l, so that Theorem 2 is easily seen to apply to $(1')$.

As for Step 1: $(1'_k)$ and $(3')$ are a finite system of equations for the "undetermined coefficients" $\{v_\alpha(y) : \alpha \in \mathbb{N}^n, |\alpha| < k\}$ in the expansion $v(x, y) = \sum_{\alpha \in \mathbb{N}^n} v_\alpha(y) x^\alpha$. Were it not for the fact that derivatives of the $v_\alpha(y)$ occur in these equations, a straightforward application of Artin's theorem [1] would suffice for Step 1. In [2], where we treated the special case $X = X(x, y)$ and $Y = Y(x, y)$ (independent of v) this difficulty is eliminated by means of a change of variables $x' = x, y = \eta(x, y)$ where $\eta(x, y)$ is the unique s-tuple of convergent power series satisfying

$$D_x \eta(x, y) X(x, y) + D_x \eta(x, y) Y(x, y) = 0$$

and $\eta(0, y) = y$. It is then easy to see that $(1')$ is transformed into a new equation of the same form in the variables x' and y', but one in which no derivatives in y' occur. In the present context we modify this technique as follows: With $\bar{v}(x, y)$ as above, let $\bar{\eta}(x, y)$ be the unique s-tuple of formal power series satisfying

(5) $$D_x \bar{\eta}(x, y) X\big(x, y, \bar{v}(x, y)\big) + D_y \bar{\eta}(x, y) Y\big(x, y, \bar{v}(x, y)\big) = 0$$

and

(6) $$\bar{\eta}(0, y) = y.$$

(The existence of a unique such $\bar{\eta}$ follows easily from (2); in fact, it is enough that no non-trivial relation $\sum_{i=1}^n \alpha_i \lambda_i = 0$ holds, with $\alpha = \langle \alpha_1, \ldots, \alpha_n \rangle \in \mathbb{N}^n$. See, e.g., the proof of Theorem 2 below or the proof of Lemma 1 of [2]). Then $x' = x, y' = \bar{\eta}(x, y)$ defines a formal invertible change of variables whose inverse is of the form $x = x'$, $y = \bar{v}(x', y') = \bar{v}(x, y')$ where $\bar{\eta}$ and $\bar{v}$ are related by the equations

(7)
$$\bar{\eta}\big(x, \bar{v}(x, y')\big) = y'$$
$$\bar{v}\big(x, \bar{\eta}(x, y)\big) = y$$

Let $\bar{w}(x, y') \equiv \bar{v}(x, \bar{v}(x, y'))$; an easy calculation shows that the pair $(\bar{v}(x, y'), \bar{w}(x, y'))$ is a formal power solution of the system

(8)
$$D_x v(x, y')X(x, v(x, y'), w(x, y')) = Y(x, v(x, y'), w(x, y'))$$

$$D_x w(x, y')X(x, v(x, y'), w(x, y')) = G(x, v(x, y'), w(x, y'))$$

(9) $v(0, y') = y'$

and

(10) $X(0, y', w(0, y')) = 0.$

On the other hand, if $(v^*(x, y'), w^*(x, y'))$ is any solution of (8), (9), and (10), and if we define $\eta^*(x, y)$ so that $\eta^*(x, y)$ and $v^*(x, y')$ satisfy the equations (7) and then define $v^*(x, y) \equiv w^*(x, \eta^*(x, y))$, we thereby obtain a solution of (1') and (3'). It is easy to see (but very important for our argument) that if $(\bar{v}(x, y), \bar{w}(x, y'))$ has the Artin property with respect to (8), (9), and (10), then $\bar{v}(x, y)$ has the Artin property with respect to (1') and (3'). (The converse is true, less easy to see, and not needed here.) Observe that

$$(D_x \bar{v})_0 = (D_x \bar{w})_0 + (D_y \bar{w})_0 (D_x \bar{\eta})_0;$$

from (7) we deduce $(D_x \bar{v})_0 = -(D_x \bar{\eta})_0$ and, since $\bar{w}(0, y') = \bar{v}(0, y')$, (4) enables us to conclude that

$$\mathcal{B} \equiv (D_x X)_0 + (D_v X)_0 (D_x \bar{v})_0$$
$$= (D_x X)_0 + (D_y X)_0 (D_x \bar{v})_0 + (D_v X)_0 (D_x \bar{w})_0.$$

Thus, for the system (8), (9), and (10) for which Step 1 no longer presents any problem, Step 2 works just as well using Theorem 2 below and our hypothesis (2) on the eigenvalues of $\mathcal{B}'$.

2. A generalization of Liapounoff's theorem.

We present here the technical result (Theorem 2) omitted in the previous section. It is a generalization of Liapounoff's theorem (see also the remark at the end of this section). Theorem 2 is stated in greater generality than is needed; proving a less general result by the methods used here would not seem to save much effort. The methods are essentially those used in the proof of

Lemma 5 of [2], which is, in fact, a special case of Theorem 2, so we present the proof in outline and refer the reader to [2] for details.

THEOREM 2. *Let k be an integer ≥ 2. Suppose $v^*(x, y) = (v_1^*(x, y), \ldots, v_p^*(x, y))$ is a p-tuple of convergent power series which satisfies*

$$(11) \qquad D_x v(x, y) X(x, y, v(x, y), D_y v(x, y))$$
$$= G(x, y, v(x, y), D_y v(x, y))$$

to order k in x, along with

$$(12) \qquad v^*(0, 0) = 0$$

$$(13) \qquad X(0, y, v^*(0, y), D_y v^*(0, y)) = 0$$

and

$$(14) \qquad D_q G(0, y, v^*(0, y), D_y v^*(0, y))$$
$$= D_x v^*(0, y) D_q X(0, y, v^*(0, y), D_y v^*(0, y))$$

where $X = (X_1, \ldots, X_n)$, $G = (G_1, \ldots, G_p)$ with $X_i(x, y, v, q)$ and $G_j(x, y, v, q)$ holomorphic functions in a neighborhood of $(0, 0, 0, q^{(0)})$ in $\mathbb{C}^n \times \mathbb{C}^s \times \mathbb{C}^p \times \mathbb{C}^{ps}$; here $q^{(0)} = (D_y v^*)_0$ and, in general, $q = (\{q_{ij}\} : i = 1, \ldots, p; j = 1, \ldots, s) = (q_{11}, \ldots, q_{1s}, q_{21}, \ldots, q_{ps})$. Suppose further that the eigenvalues $\lambda_1, \ldots, \lambda_n$ of $\mathfrak{B} = D_x X(0, 0, 0, q^{(0)}) + D_v X(0, 0, 0, q^{(0)}) D_x v^*(0, 0)$ satisfy (2), and that

$$(15) \qquad E(l; \mathfrak{B}, \mathcal{C}) \text{ holds for all } l \geq k,$$

where $\mathcal{C} \equiv D_v G(0, 0, 0, q^{(0)}) - D_x v^*(0, 0) D_v X(0, 0, 0, q^{(0)})$. It then follows that there exists a unique convergent power series solution $v(x, y)$ of (11) such that

$$(16) \qquad v(x, y) = v^*(x, y) \qquad \text{to order } k \text{ in } x.$$

Proof. $v(x, y)$ satisfies (11) if and only if $v'(x, y) \equiv v(x, y) - v^*(x, y)$ satisfies

$$(11') \qquad D_x v'(x, y) X'(x, y, v'(x, y), D_y v'(x, y))$$
$$= G'(x, y, v'(x, y), D_y v'(x, y))$$

where $X'(x, y, v, q) \equiv X(x, y, v^*(x, y) + v, D_y v^*(x, y) + q)$ and

$$G'(x, y, v, q) = G(x, y, v^*(x, y) + v, D_y v^*(x, y) + q)$$

$$- D_x v^*(x, y) X'(x, y, v, q).$$

Thus, we may assume that

$$v^*(x, y) = 0, \qquad q^{(0)} = 0$$

and that

(11$_k$) $G(x, y, 0, 0) = 0$ to order k

(13) $X(0, y, 0, 0) = 0$

(14) $D_q G(0, y, 0, 0) = 0$

and that (2) and (15) hold, with $\lambda_1, \ldots, \lambda_n$ the eigenvalues of $\mathcal{B}$ $= (D_v X)_0$ and with $\mathcal{C} = (D_v G)_0$. $v(x, y)$ is now sought to satisfy (11) and

(16) $v(x, y) = 0$ to order k in x.

A linear change of variables $x = Tx'$ has the effect of replacing $\mathcal{B}$ by $T^{-1} \mathcal{B} T$, so we may assume $\mathcal{B}$ is in suitable normal form, say upper triangular, with $\lambda_1, \ldots, \lambda_n$ along the main diagonal (in our discussion of the convergence of our solution below we require a more specialized normal form for $\mathcal{B}$). If we write $v(x, y) = \sum_{|\alpha| \geq k} v_\alpha(y) x^\alpha$ we see from (13) and (14) that the coefficient of x^α in

$$D_x v(x, y) X(x, y, v(x, y), D_y v(x, y))$$

$$- G(x, y, v(x, y), D_y v(x, y))$$

is given by an expression of the form

(17) $c_\alpha(y) \equiv \left(\sum_{i=1}^{n} \lambda_i \alpha_i \right) v_\alpha(y) - \mathcal{C} v_\alpha(y) + \sum \Gamma_{\alpha\beta}(y) v_\beta(y) + R_\alpha(y)$

where, i) the sum is over these β for which $|\beta| = |\alpha|$ and each $\Gamma_{\alpha\beta}(y)$ is a $p \times p$ matrix of convergent power series in y; moreover, from the fact that $\mathcal{B}$ is upper triangular, it follows that $\Gamma_{\alpha\beta}(0) = 0$ unless $\beta > \alpha$ in the sense that $\beta_i > \alpha_i$ at the first place i where $\beta_i \neq \alpha_i$, and ii) $R_\alpha(y)$ is a p-tuple of polynomials in the components of $\{v_\beta(y), D_y v_\gamma(y) : |\beta|, |\gamma| < |\alpha|\}$ with coefficients which are convergent power series in y. (15) then implies that the equation $\{c_\alpha(y) = 0 : \alpha \in \mathbb{N}^n, |\alpha| = l\}$ may be solved for the coefficients $\{v_\alpha(y) : \alpha \in \mathbb{N}^n, |\alpha| = l\}$ successively for $l = k, k + 1, \ldots$. The coefficients $\{v_\alpha(y)\}$ so obtained are clearly p-tuples of convergent power series in y. It remains to show that $v(x, y)$ so constructed is a convergent power series in x and y.

If $u(z)$ is any power series with complex coefficients in any set of variables z, say $u(z) = \sum u_\alpha z^\alpha$, we write $|u|(z) \equiv \sum |u_\alpha| z^\alpha$; if $u(z) = \sum u_\alpha z^\alpha$ and $v(z) = \sum v_\alpha z^\alpha$, then $u(z) \prec v(z)$ means $|u_\alpha| \leqslant v_\alpha$ for all α. For p-tuples $u(z) = (u_1(z), \ldots, u_p(z)), v(z) = (v_1(z), \ldots, v_p(z))$, $u(z) \prec v(z)$ means $u_i(z) \prec v_i(z)$ for all $i = 1, \ldots, p$ and $|u|(z) \equiv (|u_1|(z), \ldots, |u_p|(z))$. If $u(x, \omega)$ is a p-tuple of power series in the variables $x = (x_1, \ldots, x_n)$ and any other set of variables ω, $u^\#(t, \omega)$ is the p-tuple of power series in t and ω obtained by substituting $x_1 = x_2 = \cdots = x_n = t$ in $u(x, \omega)$. We shall accomplish our proof by showing that

$$\varphi(t, y) \equiv |v|^\#(t, y)$$

is convergent.

We define the $p \times p$ matrix $M = (M_{ij})$ by

$$M_{ij} = \sup\{|M_{ij}^{(\alpha)}| : \alpha \in \mathbb{N}^n, |\alpha| \geqslant k\}$$

where

$$M^{(\alpha)} = \left(M_{ij}^{(\alpha)} \right) = \left[I - \left(\sum_{i=1}^{n} \lambda_i \alpha_i \right)^{-1} \mathcal{C} \right]^{-1}.$$

With

$$c_0 \equiv \inf\left\{ \frac{|\sum_{i=1}^{n} \lambda_i \alpha_i|}{|\alpha|} : |\alpha| \geqslant k \right\} > 0$$

we choose $\epsilon > 0$ so small that

(18)   the largest eigenvalue of $\dfrac{2(n-1)\epsilon}{c_0} M$ is less than 1,

and take for the normal form of $\mathfrak{B}$

$$
\mathfrak{B} = \begin{bmatrix}
\lambda_1 & \mu_1 & 0 & . & . & 0 \\
0 & \lambda_2 & \mu_2 & . & . & 0 \\
0 & 0 & \lambda_3 & . & . & 0 \\
. & . & . & . & . & . \\
. & . & . & . & \lambda_{n-1} & \mu_{n-1} \\
0 & . & . & . & 0 & \lambda_n
\end{bmatrix}
$$

where $0 \leqslant \mu_i \leqslant \epsilon$ for $i = 1, \ldots, n-1$. If we now define

$$
X'(x, y, v, q) \equiv X(x, y, v, q) - \mathfrak{B} x,
$$

$$
G'(x, y, v, q) \equiv G(x, y, v, q) - \mathcal{C} v
$$

and $w(x, y) = \sum_{\alpha \in \mathbb{N}^n} w_\alpha(y) x^\alpha \equiv G'(x, y, v(x, y), D_y v(x, y))$ $- D_x v(x, y) X'(x, y, v(x, y), D_y v(x, y))$ then we may rewrite the equation $c_\alpha(y) = 0$ (see (17)), as

$$
v_\alpha(y) = \left(\sum_{i=1}^{n} \lambda_i \alpha_i \right)^{-1} M^{(\alpha)} \left\{ w_\alpha(y) - \sum_{i=1}^{n-1} \mu_i(\alpha_{i+1}) v_{\alpha^{(i)}}(y) \right\}
$$

where $\alpha^{(i)}$ is the multi-index obtained from α by increasing α_i by 1 and decreasing α_{i+1} by 1 (unless $\alpha_{i+1} = 0$ in which case $\alpha^{(i)} \equiv 0$, and $v_{\alpha^{(i)}}(y) = 0$). From this we obtain, for $l \geqslant k$,

$$
\sum_{|\alpha| = l} |\alpha| |v_\alpha|(y) \prec c_0^{-1} M \left(\sum_{|\alpha| = l} |w_\alpha|(y) \right)
$$

$$
+ \left(2\epsilon(n-1)c_0^{-1} \right) M \left(\sum_{|\alpha| = l} |\alpha| |v_\alpha|(y) \right).
$$

From this, and (18) we deduce, using Lemma 3 of [2], that for $l \geqslant k$,

(19) $\quad \displaystyle\sum_{|\alpha| = l} |\alpha| |v_\alpha(y)| \prec N \left(\sum_{|\alpha| = l} |w_\alpha|(y) \right)$

where

$$
N \equiv c_0^{-1} \left[I - \left(2\epsilon(n-1)c_0^{-1} \right) M \right]^{-1} M.
$$

We rewrite (19) in the form

$$tD_t\varphi(t, y) \prec N|w|^{\#}(t, y).$$

Since

$$|w|^{\#}(t, y) \prec \sum_{i=1}^{n} |X_i'|^{\#}(t, y, \varphi(t, y), D_y\varphi(t, y))D_t\varphi(t, y)$$
$$+ |G'|^{\#}(t, y, \varphi(t, y), D_y\varphi(t, y)),$$

the rest of our proof is furnished by the following variant of [2], Lemma 4.

LEMMA . *Let* $\mathfrak{A}$ *denote the set of all convergent power series with nonnegative coefficients in the variables* (t, y, v, q) *and let* $\mathsf{M}_p(\mathfrak{A})$ *denote the set of all* $p \times p$ *matrices with entries in* $\mathfrak{A}$. *Suppose*

$$g(t, y, v, q) \in \mathfrak{A}^p \qquad and \qquad A(t, y, v, q) \in \mathsf{M}_p(\mathfrak{A})$$

are given, and suppose $\varphi(t, y)$ *is a p-tuple of formal power series with non-negative coefficients such that*

(20) $tD_t\varphi(t, y) \prec A(t, y, \varphi(t, y), D_y\varphi(t, y))D_t\varphi(t, y)$

$$+ g(t, y, \varphi(t, y), D_y\varphi(t, y))$$

and

(21) $\varphi(t, y) = 0$ *to order 2 in* t.

Suppose further that

(22) $A(0, y, 0, 0) = 0$

(23) $D_t A(0, 0, 0, 0) = 0$

(24) $D_v g(0, 0, 0, 0) = 0$

and

(25) $D_q g(0, y, 0, 0) = 0.$

Then $\varphi(t, y)$ *is convergent.*

Proof.   Without loss of generality, we assume that $g(t, y, 0, 0) = 0$ to order 2 in t. We write

$$A(t, y, v, q) = A(0, y, v, q) + tA^*(t, y, v, q)$$

and

$$g(t, y, v, q) = g(t, y, 0, q) + B^*(t, y, v, q)v$$

with A^* and $B^* \in M_p(\mathfrak{A})$, and

$$A^*(0, 0, 0, 0) = D_t A(0, 0, 0, 0) = 0$$

and

$$B^*(0, 0, 0, 0) = D_v g(0, 0, 0, 0) = 0.$$

Since $\varphi(t, y) \prec tD_t\varphi(t, y)$ follows from (21), another application of [2], Lemma 3 allows us to replace $A(t, y, v, q)$ and $g(t, y, v, q)$ in (20) by

$$A'(t, y, v, q) \equiv \left[I - C^*(t, y, v, q) \right]^{-1} A(0, y, v, q)$$

and

$$g'(t, y, v, q) = \left[I - C^*(t, y, v, q) \right]^{-1} g(t, y, 0, q)$$

where $C^* \equiv A^* + B^*$. Thus, we may as well assume that

$$A(t, y, 0, 0) = 0$$

$$g(t, y, v, 0) = 0 \qquad \text{to order 2 in } t,$$

and

$$D_q g(0, y, v, 0) = 0.$$

Therefore, $A(t, y, v, q)$ may be expanded

$$(26) \qquad A(t, y, v, q) = \sum_{i=1}^{p} v_i A_i^{(1)}(t, y, v, q) + \sum_{\substack{1 \leqslant j \leqslant p \\ 1 \leqslant l \leqslant s}} q_{jl} A_{jl}^{(2)}(t, y, v, q)$$

where the various A's are all in $\mathbb{M}_p(\mathfrak{A})$, and $g(t, y, v, q)$ may be written

(27) $\qquad g(t, y, v, q) = tg_0(y) + \sum t^j v^\alpha q^\beta g_{j\alpha\beta}(t, y, v, q)$

where $g_0(y)$ is a p-tuple of convergent power series in y with non-negative coefficients, the $g_{j\alpha\beta}$ are all in $\mathfrak{A}^p$, and the sum is taken over all $\alpha \in \mathbb{N}^p$, $\beta \in \mathbb{N}^s$ and $j \in \mathbb{N}$ satisfying

    i) $|\alpha| = 0, |\beta| = 1$ and $j = 1$ or
    ii) $|\alpha| = 0, |\beta| = 2$ and $j = 0$ or
    iii) $|\alpha| = 1, |\beta| = 0$ and $j = 2$.

Taking

(28) $\qquad \varphi(t, y) = t^2 \psi(t, y)$

and substituting (26), (27) and (28) in (20) enables us to conclude that

(29) $\qquad \psi(0, y) \prec \frac{1}{2} g_0(y)$

and

(30) $\qquad D_t\psi(t, y) \prec \left\{ \sum_{i=1}^p \psi_i(t, y)A_i^{(1)} + \sum_{\substack{1 \leqslant j \leqslant p \\ 1 \leqslant l \leqslant s}} \frac{\partial \psi_j}{\partial yl}(t, y)A_{jl}^{(2)} \right\}$

$$\times \left[tD_t\psi(t, y) + 2\psi(t, y) \right]$$

$$+ \sum t^{j-1+2(|\alpha|+|\beta|-1)}g_{j\alpha\beta}(\psi(t, y))^\alpha (D_y\psi(t, y))^\beta$$

where we have written $A_i^{(1)}$ for $A_i^{(1)}(t, y, t^2\psi(t, y), t^2 D_y\psi(t, y))$ and similarly for $A_{jl}^{(2)}$ and $g_{j\alpha\beta}$. Because the term containing $D_t\psi(t, y)$ on the right hand side of (3) contains a factor of t, yet another application of [2], Lemma 3 enables us, in effect, to discard this term, so as to obtain a majorization of the form

(31) $\qquad D_t\psi(t, y) \prec F(t, y, \psi(t, y), D_y\psi(t, y))$

with

$$F(t, y, v, q) \in \mathfrak{A}^p.$$

We may then apply the Cauchy-Kowalewski theorem exactly as in the proof of [2], Lemma 4 to conclude our proof.

Remark. The case $k = 1$ of Theorem 2 may be treated as follows (we take $v^*(x, y) = 0$ to begin with, for convenience). Suppose

$$X(0, y, 0, 0) = 0$$

$$G(0, y, 0, 0) = 0$$

$$D_q G(0, y, 0, 0) = 0$$

and

$$(D_v X)_0 \equiv D_v X(0, 0, 0, 0) = 0.$$

Suppose $\lambda_1, \ldots, \lambda_n$, the eigenvalues of $\mathfrak{B} = (D_x X)_0$, satisfy (2) and suppose further that $E(l; \mathfrak{B}, \mathcal{C})$ holds for all $l \geqslant 1$, where $\mathcal{C} = (D_v G)_0$. Then there exists a unique convergent power series solution $v(x, y)$ of (11) such that $v(0, y) = 0$.

Proof. To reduce this assertion to that of Theorem 2 we need only show that there exists a unique $p \times p$ matrix of convergent power series $\mathcal{C}(y)$ such that $v(x, y) \equiv \mathcal{C}(y)x$ satisfies (11_2). Under the present hypotheses, (11_2) reduces to the equation

$$(11_2) \quad \begin{aligned} \mathcal{C}(y)\{D_x X(0, y, 0, 0) + D_v X(0, y, 0, 0)\mathcal{C}(y)\} \\ = D_x G(0, y, 0, 0) + D_v G(0, y, 0, 0)\mathcal{C}(y) \end{aligned}$$

The hypotheses $E(1; \mathfrak{B}, \mathcal{C})$ is equivalent to the assertion that the linear map L of the vector space of $p \times n$ matrices of complex numbers into itself given by $L(A) \equiv A\mathfrak{B} - \mathcal{C}A$ is an isomorphism. We then define, as we must,

$$(32) \quad \mathcal{C}(y) = \mathcal{C}_0 + \mathcal{C}'(y)$$

where $\mathcal{C}_0$ is the unique solution of

$$L(\mathcal{C}_0) = (D_x G)_0$$

and $\mathcal{C}'(0) = 0$; an easy application of the implicit function theorem now gives us a unique $p \times n$ matrix $\mathcal{C}'(y)$ of convergent power series, with $\mathcal{C}'(0) = 0$, such that $\mathcal{C}(y)$ defined by (32) satisfies (11_2).

THE CITY COLLEGE OF NEW YORK

THE CITY UNIVERSITY OF NEW YORK

REFERENCES

[1] M. Artin, On the solution of analytic equations, *Invent. Math.* **5** (1968), 277–291.

[2] S. Kaplan, Formal and convergent power series solutions of singular partial differential equations, *Trans. Amer. Math. Soc.*, **256** (1979), 163–183.

[3] A. Liapounoff, Problème général de la stabilité du mouvement, *Annals of Math. Studies* 17, Princeton Univ. Press, Princeton (1949), (translation of Russian version of 1892).

REMARKS ON TAUBERIAN CURVES*

By J. J. Levin

Abstract. A lemma concerning tauberian curves is obtained. Its relationship to earlier results, including related matters for integral equations, is discussed.

Let $\mathcal{Y} = \{ y : \mathbb{R}^1 \to \mathbb{C}^N \}$. For brevity, all the elements of $\mathcal{Y}$, not just the continuous ones, will be called curves here. We consider the tauberian curves, $T \subset \mathcal{Y}$, and its subset $\tilde{T} \subset T$, where

$$x \in T = T(\mathbb{R}^1, \mathbb{C}^N) \quad \text{if} \quad \lim_{t \to \infty, \tau \to 0} |x(t + \tau) - x(t)| = 0,$$

$$x \in \tilde{T} \quad \text{if} \quad x \in T \quad \text{and} \quad \limsup_{t \to \infty} |x(t)| < \infty.$$

Here: $\mathbb{R}^1 = (-\infty, \infty)$ and $|z| = \sum_{k=1}^{N} |z_k|$, where $z = (z_1 \ldots z_N) \in \mathbb{C}^N$ (complex N-space). Convergence in $\mathbb{C}^N$ is denoted by: $x_n \to x(x_n, x \in \mathbb{C}^N)$ if $|x_n - x| \to 0$ ($n \to \infty$ is understood). C, C_u, L^∞, and BV denote, repsectively, the continuous, uniformly continuous, essentially bounded, and totally bounded variation curves in $\mathcal{Y}$.

$\tilde{T}$ plays a key role in, among other places, the study of the asymptotic behavior as $t \to \infty$ of the bounded solutions of each of the equations

(E$_1$) $\quad \dfrac{d}{dt} x(t) + \displaystyle\int_{-\infty}^{\infty} g(y(t - \xi)) \, dA(\xi) = f(t) \quad (-\infty < t < \infty),$

(E$_2$) $\quad x(t) + \displaystyle\int_{-\infty}^{\infty} g(x(t - \xi)) \, dA(\xi) = f(t) \quad (-\infty < t < \infty),$

where $x \in \tilde{T}$, $g \in C(\mathbb{C}^N, \mathbb{C}^N)$, $A_{ij} \in BV(A = (A_{ij}))$, $f \in L^\infty$, and $f(t) \to f(\infty)(t \to \infty)$ exists. See, for example, [1],[2], and [3] for references and for a discussion of the hypothesis $x \in \tilde{T}$. In [1] it was shown how a

---

*This research was supported by NSF Grant No. MCS-8001524.

185

portion of the analysis relating the bounded solutions of (E_1) and of (E_2) to solutions of their limit equations, which are defined, respectively, by

(E_1^*) $\dfrac{d}{dt} y(t) + \displaystyle\int_{-\infty}^{\infty} g(y(t - \xi))\, dA(\xi) = f(\infty)$ $(-\infty < t < \infty)$,

(E_2^*) $y(t) + \displaystyle\int_{-\infty}^{\infty} g(y(t - \xi))\, dA(\xi) = f(\infty)$ $(-\infty < t < \infty)$,

holds for arbitrary curves in $\tilde{T}$, without the integral equation setting. One of the corollaries of [1], which is meaningful apart from integral equations, is generalized here and some consequences drawn.

Corollary 1 of [1] is

LEMMA 1.   *If $x \in \tilde{T}$, then*

(1) $x = \varphi + \eta$,

where

(2) $\varphi \in L^{\infty} \cap C_u$,

(3) $\displaystyle\lim_{t \to \infty} \eta(t) = 0$.

The proof of Lemma 1 given in [1] involved the formula

(4) $\varphi = \displaystyle\sum_{m=1}^{\infty} \psi_m y_m$,

where: $\{\psi_m\} \subset C^{\infty}(\mathbb{R}^1, [0, 1])$ is (uniformly) equicontinuous with $\|\psi'_m\|_{\infty} \to 0$, $\{y_m\} \subset \Gamma(x) \subset \mathcal{Y}$ is uniformly bounded and (uniformly) equicontinuous ($\Gamma(x)$ is defined below), and, for each t, at most two terms of (4) are nonvanishing.

In the context of integral equations: if $x \in \tilde{T}$ is a bounded solution of (E_1) (or (E_2)), then the y_m of (4) are solutions of (E_1^*) (or (E_2^*)). Formulas (1), (3), and (4) show how solutions of integral equations "tend" to solutions of their limit equations as $t \to \infty$. Thus, the asymptotic behavior of solutions of the (E_i) may sometimes be deduced from the behavior of solutions of (E_i^*).

Here we show that the hypothesis $x \in \tilde{T}$ of Lemma 1 can be weakened to $x \in T$, if L^{∞} is deleted from (2).

LEMMA 2. *Let* $x \in T$ *and*

(5) $s_1 < s_2 < \ldots, s_n \to \infty, \qquad s_{n+1} - s_n \to 0,$

(6) $\varphi(t) = \begin{cases} x(s_1) & (t \in (-\infty, s_1]) \\ x(s_n) + \dfrac{x(s_{n+1}) - x(s_n)}{s_{n+1} - s_n}(t - s_n) & \\ & (t \in [s_n, s_{n+1}], \qquad n = 1, 2, \ldots), \end{cases}$

(7) $\eta = x - \varphi.$

Then (3) *holds and*

(8) $\varphi \in C_u.$

The generality of the polygonal construction of φ in (5) and (6) is at the expense of (4). The latter does not generally hold in T, as is shown by an example below. In $\tilde{T}$, (4) is of considerable importance for integral equation interpretations, as noted above.

Proof of Lemma 2. Clearly, $\varphi \in C_u((-\infty, t_0], \mathbb{C}^N)$ for all $t_0 \in \mathbb{R}^1$. Hence, if (8) does not hold, there exist sequences $\{t_m\}$ and $\{\tau_m\}$, with

(9) $s_1 \leqslant t_1 < t_2 < \ldots, \qquad t_m \to \infty, \quad \tau_m \to 0,$

and a constant $\delta > 0$ such that

(10) $|\varphi(t_m) - \varphi(t_m + \tau_m)| \geqslant \delta \qquad (m = 1, 2, \ldots).$

Define $u(m)$ and $v(m)$ for $m = 1, 2, \ldots$ by

(11) $\{s_{u(m)}\} \subset \{s_n\}, \qquad \{s_{v(m)}\} \subset \{s_n\},$

(12) $s_{u(m)} \leqslant t_m < s_{u(m)+1}, \qquad s_{v(m)} \leqslant t_m + \tau_m < s_{v(m)+1}$

$(m = 1, 2, \ldots).$

Clearly, (5), (9), (11), and (12) imply that

(13) $s_{u(m)} \to \infty, \qquad |s_{u(m)} - s_{v(m)}| \to 0.$

From (6) and (12) it follows that

$$\left|\varphi(t_m) - x(s_{u(m)})\right| = \frac{t_m - s_{u(m)}}{s_{u(m)+1} - s_{u(m)}} \left|x(s_{u(m)+1}) - x(s_{u(m)})\right|$$

$$\leqslant \left|x(s_{u(m)+1}) - x(s_{u(m)})\right|$$

and, similarly,

$$\left|\varphi(t_m + \tau_m) - x(s_{v(m)})\right| \leqslant \left|x(s_{v(m)+1}) - x(s_{v(m)})\right|.$$

Hence,

$$\left|\varphi(t_m) - \varphi(t_m + \tau_m)\right|$$

$$\leqslant \left|\varphi(t_m) - x(s_{u(m)})\right| + \left|x(s_{u(m)}) - x(s_{v(m)})\right|$$

$$+ \left|x(s_{v(m)}) - \varphi(t_m + \tau_m)\right|$$

$$\leqslant \left|x(s_{u(m)+1}) - x(s_{u(m)})\right| + \left|x(s_{u(m)}) - x(s_{v(m)})\right|$$

$$+ \left|x(s_{v(m)+1}) - x(s_{v(m)})\right|,$$

which together with (5), (11), (13), and $x \in T$ implies that $|\varphi(t_m) - \varphi(t_m + \tau_m)| \to 0$. This contradicts (10) and, thus, establishes (8).

For $t \in [s_n, s_{n+1}]$, (6) and (7) imply

$$\eta(t) = \frac{s_{n+1} - t}{s_{n+1} - s_n}\left[x(t) - x(s_n)\right] + \frac{t - s_n}{s_{n+1} - s_n}\left[x(t) - x(s_{n+1})\right].$$

Hence,

$$|\eta(t)| \leqslant |x(t) - x(s_n)| + |x(t) - x(s_{n+1})| \qquad (t \in [s_n, s_{n+1}]),$$

which together with (5) and $x \in T$ establishes (3) and completes the proof.

Lemma 2 has implications for some of the results in [1] which dealt with T and $\tilde{T}$ and not specifically with applications to integral equations. Before indicating these it is necessary to recall some facts concerning $\mathcal{Y}$.

Convergence in the compact open topology (c.o.) on $\mathcal{Y}$ is defined by

(14) $\quad\begin{cases} x_n \to x \text{ c.o.} \, (x_n, x \in \mathcal{Y}) \text{ if } |x_n(t) - x(t)| \to 0 \\ \text{uniformly on every compact subset of } \mathbb{R}^1. \end{cases}$

It is well known that (14) is equivalent to convergence in the complete metric defined on $\mathcal{Y}$ by

(15) $\qquad d(x, y) = \sum_{n=1}^{\infty} \frac{1}{2^n} \sup_{|t| \leqslant n} \frac{|x(t) - y(t)|}{1 + |x(t) - y(t)|}.$

For $y \in \mathcal{Y}$, let $R(y)$ denote the range of y and, for $Y \subset \mathcal{Y}$, let

$$R(Y) = \bigcup_{y \in Y} R(y).$$

For $y \in \mathcal{Y}$, $\tau \in \mathbb{R}^1$, let $y_\tau : t \mapsto y(t + \tau)$ denote the τ translate of y. Associated with each $x \in \mathcal{Y}$ are two positive limit sets, $\Omega(x) \subset \mathbb{C}^N$ and $\Gamma(x) \subset \mathcal{Y}$, where

$$\Omega(x) = \left\{ \omega \in \mathbb{C}^N \,|\, x(t_n) \to \omega \text{ for some } t_n \to \infty \right\},$$

$$\Gamma(x) = \left\{ y \in \mathcal{Y} \,|\, x_{t_n} \to y \text{ c.o. for some } t_n \to \infty \right\}.$$

An easy consequence of Lemma 2 and the Ascoli-Arzelà theorem is:

LEMMA 3. *If $x \in T$ and if there exist $t_n \to \infty$ such that $\sup_n |x(t_n)| < \infty$, then there exist a subsequence $\{t_{n_k}\}$ of $\{t_n\}$ and a $y \in \Gamma(x)$ such that $x_{t_{n_k}} \to y$ c.o.*

In [1, Lemma 2.4], Lemma 3 was obtained under the more stringent requirement that $x \in \tilde{T}$, and in [3, Lemma 3.2], it was obtained under the still stricter hypothesis that $x \in L^\infty \cap T$. Although the proof of [1, Lemma 2.4] may be trivially modified to yield Lemma 3 (that of [3, Lemma 3.2] may not be), it is considerably simpler to obtain it as a corollary of Lemma 2. Lemmas 2.4 of [1] and 3.2 of [3] are used repeatedly in those papers.

The following lemma was established in $\tilde{T}$ in [1,(2.22)]. As a consequence of Lemma 3 above, rather than Lemma 2.4 of [1], the proof of [1,(2.22)] now yields the more general:

LEMMA 4. *If $x \in T$, then $R(\Gamma(x)) = \Omega(x)$.*

The example $x(t) \equiv t \in T$, in which $\Omega(x) = \varnothing$, shows that Lemma 4 may hold trivially. If $x \in \tilde{T}$, then, clearly (also, [1(2.21)]), $\Omega(x) \neq \varnothing$. We now illustrate some properties of $\tilde{T}$, besides $\Omega(x) \neq \varnothing$, that do not generally hold in T.

If $x \in \tilde{T}$, then, by [1, Lemma 2.5], $\Omega(x)$ and $\Gamma(x)$ are compact and connected sets (in their respective spaces). Also, $d(x_t, \Gamma(x)) \to 0$ as $t \to \infty$, where d is defined by (15). That none of this, nor (4), need hold in T is easily seen from the following example in which $x \in C_u(\mathbb{R}^1, \mathbb{R}^2)$ and $\Omega(x) \neq \varnothing$: Let $x(t) = (x_1(t), x_2(t)) = (0,0)$ on $(-\infty, 0]$ and let $R(x)$ be a rectangular spiral starting at the origin and consisting of horizontal and vertical line segments; half of the horizontal line segments tend toward 1, the other half toward -1; half of the vertical segments tend toward ∞, the other half toward $-\infty$; on $[0, \infty)$, $x(t)$ moves along $R(x)$ at constant speed. Clearly $\Omega(x)$ is the union of the horizontal lines $x_2 = \pm 1$, and the assertion is evident. Note also that (1), (3), (4), with $y_m \in \Gamma(x)$ can not hold simultaneously here.

It should be noted that Lemma 2 may be used to simplify some proofs of [1] which dealt with T. An example, in which the proof is now evident, is [1, Lemma 2.3]: If $x \in T$, then $\Gamma(x)$ is (uniformly) equicontinuous.

UNIVERSITY OF WISCONSIN

REFERENCES

[1] J.J. Levin, Tauberian curves and integral equations, *J. Integral Equations* **2** (1980), 57–91.

[2] ———, On some geometric structures for integrodifferential equations, *Adv. Math.* **22** (1976), 146–186.

[3] J.J. Levin and D.F. Shea, On the asymptotic behavior of the bounded solutions of some integral equations, I, II, III, *J. Math. Anal. Appl.* **37** (1972), 42–82, 288–326, 537–575.

EVERY SPHERE EVERSION HAS A QUADRUPLE POINT

By Nelson Max* and Tom Banchoff**

---

§0. Let $i : S^2 \to R^3$ be the inclusion of the sphere, and let $\chi : R^3 \to R^3$ be the reflection in the XY plane. An eversion of the sphere is a regular homotopy of immersions, starting with i and ending at χi. Since Smale [4] first showed that eversions must exist, several examples have been given. The eversions described in [1], [2], [3], [6], and §6 below are all variants of one created by Froissart and improved by Morin, and have one quadruple point.

Using this Froissart-Morin eversion $F : S^2 \times [0, 1] \to R^3$, we show here that every eversion must have at least one quadruple point. In §1, we define generic eversions and show that the number of quadruple points is preserved mod 2 under a homotopy between generic eversions. In §2 we define Γ_c, the space of *standard* immersions, which are fixed in a neighborhood of the north pole. Smale [4] has shown that the differential defines an isomorphism between $\pi_1(\Gamma_c)$ and $\pi_3(V_{3\,2})$, which is known to be the group of the integers. Let $L = F \cdot \chi F$ be the loop in Γ_c which turns the sphere inside out by F and back again by the mirror image χF. Lemma 2 of §2 states that $[L]$ is a generator of $\pi_1(\Gamma_c)$.

Since L has two quadruple points, we can compose F with a power of L, to get, in every homotopy class, a generic eversion with an odd number of quadruple points. This, together with the result of §1, shows that every generic eversion has a quadruple point, and in §3 we apply a limiting argument to prove the general case.

The geometry behind Smale's isomorphism is explained in §4, together with a further isomorphism between $\pi_3(V_{3\,2})$ and $\pi_3(S^3)$. In §5,

---

*The first author worked under the auspices of the U.S. Department of Energy, on contract number W-7405-ENG-48 to Lawrence Livermore National Laboratory.

**We gratefully acknowledge the help of Bernard Morin, who first conjectured the main theorem, and the fact that $F \cdot \chi F$ was a generator of $\pi_1(\Gamma_c)$. He has suggested that the degree of h could be calculated by integration from the equations in [3], but so far this has not been done. He has also had several helpful discussions and correspondences with us during the preparation of this paper. Finally, by describing his eversion to us in terms of its 14 singularities [2], he made it possible for us to visualize it as described in §6 below, and thereby compute the degree of h. We are also grateful for the suggestions of the referee.

we develop a method of evaluating $[L]$ in terms of the degree of the resulting map from S^3 to S^3, and in §6 we apply the method to the Froissart-Morin eversion, in a "proof by pictures."

In this paper, an n-tuple point will mean a point whose inverse image has cardinality at least n. Thus a sextuple point is also a quadruple point. The non-generic eversion given in [7] has a sextuple point at its "halfway" stage.

§1. A *generic* immersion from M^m to N^k is one for which the several tangent planes to the surfaces at an n-tuple point are in general position. By transversality, the set of n-tuple points will be an immersed submanifold of dimension $k - n(k - m)$.

Since we are using the results of Smale [4], using curvature in §3, and also using second derivatives in §5, we will assume that all our immersions are C^2. By the Morse-Sard Theorem (see [8]), the set of generic immersions is open and dense in the space of C^2 immersions with the C^2 topology, so every immersion can be approximated by a generic one. If $G: S^2 \times [0, 1] \to R^3$ is an eversion, G need not be a differentiable function of its deformation parameter $t \in [0, 1]$. However, every regular homotopy in the space of C^2 regular homotopies can be approximated by a C^2 differentiable homotopy (see [5]). If G is differentiable, then the level preserving function $\overline{G}: S^2 \times [0, 1] \to R^4$ defined by $\overline{G}(p, t) = (G(p, t), t)$ is an immersion. If, in addition the immersion $\overline{G}$ is a generic immersion, then G will be called a *generic eversion*. Again by the Morse-Sard theorem, every differentiable eversion can be approximated by a generic eversion in the space of C^2 differentiable eversions.

LEMMA 1. *If two generic eversions are homotopic, then they have the same mod 2 number of quadruple points.*

Proof. Let H be a homotopy of eversions between H_0 and H_1, that is, a two-parameter regular homotopy $H: S^2 \times [0, 1] \times [0, 1] \to R^3$, such that

    i.) $H(p, t, 0) = H_0(p, t)$

    ii.) $H(p, t, 1) = H_1(p, t)$

    iii.) $H(p, 0, u) = i(p)$

    iv.) $H(p, 1, u) = \chi i(p)$.

If H is not differentiable and generic, we can use a relative version of the Morse-Sard theorem to approximate it by a generic homotopy between the same two eversions, which we again call H. Let $\overline{H}(p,t,u)$ $= (H(p,t,u),t,u)$. The set of quadruple points of $\overline{H}$ will then be a properly immersed one dimensional manifold M with boundary.

The boundary of M lies in $R^3 \times [0,1] \times (\{0\} \cup \{1\})$, since $H \mid S^2 \times (\{0\} \cup \{1\}) \times [0,1]$ is an embedding. Since ∂M contains an even number of points, the number lying in $R^3 \times [0,1] \times \{0\}$, which are the quadruple points of H_0, must be the same, mod 2, as the number lying in $R^3 \times [0,1] \times \{1\}$, the quadruple points of H_1.

Note that five arcs of quadruple points may intersect at a generic quintuple point, as in Figure 1.

§2.   Let Γ be the space of C^2 immersions of S^2 into S^3, with the C^2 topology. An arc in Γ between the immersions i and χi is then an eversion of the sphere.

Let ρ be the projection of R^3 onto the XY plane, let N be the north pole of S^2, and let K be a small round neighborhood centered at N. An immersion f will be called *standard* if $f \mid K = \rho \mid K$. Let Γ_c be the subspace of Γ consisting of standard immersions. If we identify $S^2 - K$ with the unit disk D^2, we may also consider Γ_c as the space of immersions of D^2 into R^3 which agree with ρ, together with their first and second derivatives, along ∂D^2.

Any immersion can be made standard by a general affine motion which brings N to the origin, with the correct differential, followed by a slight flattening near N. For example, a translation along the negative Z

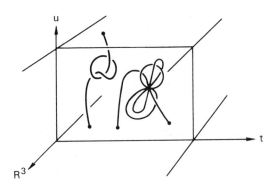

Figure 1.

axis and a slight flattening with a C^2 "fillet" will turn the inclusion i into the standard inclusion i'. This process can be carried out continuously in any finite number of parameters (see Smale [4]). So if we have an eversion G, we can modify it to get a *standard eversion*, i.e., a regular homotopy through standard immersions, from i' to $\chi i'$. If in addition G is generic, and we keep the quadruple points of G away from N, we get a standard generic eversion with the same number of quadruple points.

Two standard eversions E and G are homotopic through standard eversions if and only if their concatenation, the loop $E \cdot G^{-1}$, defines the zero element of $\pi_1(\Gamma_c, i')$. Let $L = F \cdot \chi F$, where F is the Froissart-Morin eversion, so that L has two quadruple points.

LEMMA 2. *The fundamental group $\pi_1(\Gamma_c, i')$ is equal to the integers, generated by the class of L.*

This lemma will be proved in §4 through §6.

THEOREM 1. *Every generic eversion has an odd number of quadruple points.*

To prove theorem 1 from lemma 2, let G be a generic eversion. By the argument above, we may assume G is standard. By lemma 2, we can find an integer n so that $[G \cdot F^{-1}] = [L]^n$ in $\pi_1(\Gamma_c, i')$. Thus G is homotopic to the generic eversion $L^n \cdot F$, which has an odd number of quadruple points. Therefore, by lemma 1, G also has an odd number.

§3. We now want to prove the general case.

THEOREM 2. *Every sphere eversion G has a quadruple point.*

Proof. Since each immersed surface in the eversion is assumed to be C^2, it inherits a riemannian metric from R^3, and the concepts of curvature and geodesics make sense. Let Y be the maximum of the principal curvatures during the eversion, and let Z be the maximum of the length of $dG_t(V)$, for any unit tangent vector V to S^2. Then there is a neighborhood A of the eversion G in the space of C^2 regular homotopies (not necessarily differentiable homotopies), such that if H is any eversion in A, then any principal curvature at any point is less than $2Y$, and dH_t takes any unit vector tangent to S^2 into a vector of length less than $2Z$.

Suppose H is such an eversion in A, and P and Q are two different points on S^2 such that $H_t(P) = H_t(Q)$. Let $d(P, Q)$ represent the length of the shortest great circle arc Σ on S^2 between P and Q, in the standard metric dq on S^2. Also, let Γ be the shortest geodesic between P and Q, in the metric ds pulled back by the immersion H_t from the standard metric dr on R^3. Then

$$d(P, Q) = \int_\Sigma dq > \frac{1}{2Z} \int_\Sigma ds \geqslant \frac{1}{2Z} \int_\Gamma ds = \frac{1}{2Z} \int_{H_t(\Gamma)} dr$$

$$> \frac{1}{4YZ} \int_{H_t(\Gamma)} (\text{normal curvature}) \, dr$$

$$= \frac{1}{4YZ} \int_{H_t(\Gamma)} (\text{total curvature}) \, dr > \frac{\pi}{4YZ} .$$

Here the first inequality follows from the bound on the stretching of tangent vectors by dH_t, the second inequality follows because Γ is the shortest geodesic, the third inequality follows from the upper bound on principal curvatures, the next equality follows because the tangential curvature for a geodesic is zero, and the last inequality follows because the curve $H_t(\Gamma)$ is closed. Therefore $d(P, Q)$ is greater than the constant $K = \pi/(4YZ)$.

Since generic eversions are dense in the space of all C^2 eversions, we may find a sequence of generic eversions $H_1, H_2, H_3, \ldots$ in A, whose limit is G. Each of these must have at least one quadruple point, so let a_n, b_n, c_n, d_n, be four distinct points on S^2, and t_n be in $[0, 1]$, such that $H_n(a_n, t_n) = H_n(b_n, t_n) = H_n(c_n, t_n) = H_n(d_n, t_n)$. Since S^2 and $[0, 1]$ are compact, we may choose five successive subsequences, ending at the index sequence $n1, n2, n3, \ldots, ni, \ldots$, so that $a_{ni}, b_{ni}, c_{ni}, d_{ni}$, and t_{ni} converge respectively to points a, b, c, d, and t. Since the points a_n, b_n, c_n, and d_n maintain a minimum distance K from each other for each fixed n, the points a, b, c, and d are distinct. But $G(a, t) = G(b, t) = G(c, t) = G(d, t)$, since $\lim_{i \to \infty} H_{ni} = G$. Therefore G has a quadruple point. As we noted in §0, this may actually turn out to also be an n-tuple point for some $n > 4$.

§4.   We now turn to the proof of Lemma 2.

Let V_{32} be the space of pairs of linearly independent vectors in R^3. We construct a map $j: \pi_1(\Gamma_c, i') \to \pi_3(V_{32})$, which Smale [4] has shown to be an isomorphism. We then identify $\pi_3(V_{32})$ with $\pi_3(S^3) = Z$. In §5 and §6, we will verify that $j([L])$ is a generator.

A loop J in Γ_c based at i' is specified by a map $J: D^2 \times [0, 1] \to R^3$ which agrees with $\rho \times$ identity on $\partial D^2 \times [0, 1]$.

If u and v are stereographic coordinates on D^2, via projection from the north pole N, we may define $dJ : D^2 \times [0, 1] \to V_{32}$ by

$$dJ(u, v, t) = \left(\frac{\partial J}{\partial u}(u, v, t), \frac{\partial J}{\partial v}(u, v, t) \right).$$

By the definition of a regular homotopy, dJ will be a continuous map into V_{32}. Since J is a loop based at i', dJ respects the identification $(u, v, 0) \sim (u, v, 1)$ for all (u, v) in D^2, making its domain a solid torus, $D^2 \times S^1$. Also, since J is standard, dJ respects the identifications $(u, v, t) \sim (u, v, t')$ for all (u, v) in ∂D^2, and t, t' in $[0, 1]$. These further identifications make the domain of dJ into the join of two circles (see [10]) which is homeomorphic to S^3. It is clear that the homotopy class of dJ in $\pi_3(V_{32})$ depends only on the homotopy class of J in $\pi_1(\Gamma_c, i')$, so we may define $j : \pi_1(\Gamma_c, i') \to \pi_3(V_{32})$ by $j[J] = [dJ]$. In [4] (theorem 3.9 for $k = 1$ and $n = 3$) Smale proves that j is an isomorphism.

For use in the following section, we now describe a specific isomorphism $\pi_3(V_{32}) \approx Z$. Let F_{32} be the flag manifold of pairs of oriented linear subspaces $V^1 \subset V^2 \subset R^3$ where V^i has dimension i. Let $\sigma : V_{32} \to F_{32}$ be the projection which maps a pair of vectors (A, B) to the oriented subspaces generated by A, and by A and B together. The fibre $R_+^1 \times R_+^2$ of σ is contractible, so σ is a homotopy equivalence.

We now show that the universal covering space of F_{32} is homeomorphic to S^3. Let I be the standard flag $\langle(-1, 0, 0)\rangle \subset \langle(-1, 0, 0), (0, 1, 0)\rangle \subset R^3$, where "$\langle \cdots \rangle$" means "the oriented linear subspace spanned by." We define a map $\lambda : R^3 \to F_{32}$ by

$$\lambda(V) = \begin{cases} I & \text{if } |V| = 0 \\ R(V, |V|)I & \text{if } |V| > 0, \end{cases}$$

where $R(V, \theta)$ is the rotation of θ radians about an axis through V, counterclockwise when viewed in the direction from V to origin.

The image of the closed ball $\bar{B}_\pi$ of radius π about the origin in R^3 is all of F_{32}. The points V in the open ball B_π are mapped to unique flags, but the two points V and $-V$ determine the same flag if $|V| = \pi$. Therefore F_{32} is homeomorphic to the B_π with antipodal points on its boundary identified, i.e., to three dimensional projective space P^3.

Similarly, the open ball $B_{2\pi}$ gives a double covering of all points of F_{32} except I, which is covered once. Points a distance 2π apart on the same ray through the origin in R^3 map into the same flag. Also, points on the boundary of $\bar{B}_{2\pi}$ all map to I. Thus the sphere S^3, formed from $\bar{B}_{2\pi}$ by identifying its boundary to a point, is a double covering space of F_{32}, with covering map λ. But S^3 is simply connected, so it is the universal covering space of F_{32}. Since the higher homotopy groups of a space and its covering space are isomorphic, we get the sequence of isomorphisms

$$\pi_3(V_{32}) \approx \pi_3(F_{32}) \approx \pi_3(S^3) \approx Z.$$

§5. A more general way to construct the universal covering space $\tilde{X}$ of a space X is to let $\tilde{X}$ be the space of homotopy classes of based paths in X, and to let the covering projection take a path to its free endpoint. (See [9].) Since any two universal covering spaces of the same space are homeomorphic (see [9]), we can also view $\tilde{F}_{32}$ as the space of homotopy classes of paths of flag in R^3, based at I. Let L be the loop $F \cdot \chi F$ in Γ_c. Using this construction of $\tilde{F}_{32}$, it is now possible to define an explicit lifting h of $\sigma \circ dL$ to S^3.

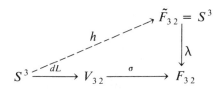

Let (u, v, t) be a point in the domain $D^2 \times [0,1]$ of dL before its identification to form S^3, and let ζ be a smooth simple arc on $D^2 \times \{t\}$ from $(1,0,t)$ to (u,v,t). Then $\zeta' = \sigma \circ dL \circ \zeta$ is a path of frames, based at I. Since D^2 is simply connected, the homotopy lifting property can be used to show that $[\zeta']$ in $\tilde{F}_{32}$ is independent of the choice of ζ. Thus we define $h(u, v, t) = [\zeta']$. Note that the points $(1,0,t)$ identify to the base point B of S^3, and that $(\partial L/\partial u)(1,0,t) = (-c,0)$ and $(\partial L/\partial v)(1,0,t) = (0,c)$, where c is a positive constant depending on the size of the fixed neighborhood K of the north pole N. Thus $\lambda \circ h(B) = I$, explaining our backwards choice of the "standard" flag I.

This construction is easy to visualize on the surfaces $L_t(S^2)$. Let R be a strip on $D^2 \times \{t\}$ with ζ as a centerline, and $L(R)$ be its image on $L_t(S^2)$, as shown in Figure 2a. As the surface moves, the strip $L(R)$ will move and twist, showing the changes in the tangent frame along its center line $L(\zeta)$.

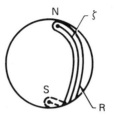

Figure 2a.

Let $C \subset F_{32}$ be the great circle containing those paths of flags with
positively oriented horizontal planes at their free ends. In the ball $\overline{B}_{2\pi}$, C
corresponds to the Z axis. It contains 4π worth of rotations, and doubly
covers a generator of $\pi_1(F_{32}, I)$. A complementary great circle D, which
links C once, projects to the flags with negatively oriented horizontal
tangent planes.

On the surfaces $L_t(S^2)$, $Lh^{-1}(C)$ consists of those points where the
tangent plane is horizontal and the original "outside" normal points
towards the negative Z axis.

Since L is C^2, the lifting h of its differential is C^1. By changing L
slightly within its homotopy class if necessary, we may find a point Q on
C such that the jacobian matrix dh is non-singular at the finitely many
points of $h^{-1}(Q)$. Then, to compute the degree of h, we need only add
up the members of $h^{-1}(Q)$, with signs from the determinant $|dh|$.

By the choice of Q, $h^{-1}(Q)$ consists of non-degenerate critical
points for the height function Z on the surfaces $L_t(S^2)$. It is straightfor-
ward to verify that the sign of $|dh|$ is the product of the two factors f_1
and f_2, where

$$f_1(u,v,t) = \begin{cases} +1 & \text{at a local maximum or minimum} \\ -1 & \text{at a saddle point} \end{cases}$$

and

$$f_2(u,v,t) = \begin{cases} +1 & \text{if } \dfrac{\partial L(u,v,t)}{\partial u} \quad \text{turns counterclockwise for} \\ & \hspace{4.5em} \text{increasing } t, \\ -1 & \text{if } \dfrac{\partial L(u,v,t)}{\partial u} \quad \text{turns clockwise for} \\ & \hspace{4.5em} \text{increasing } t. \end{cases}$$

The immersions in the Froissart-Morin eversion F all have two-fold rotational symmetry about the Z axis, so the tangent plane at the south pole S remains horizontal. We take ζ to be the segment in the negative u direction in the stereographic coordinate plane, joining the B to the south pole S. In the figures, B is identified with the nearby north pole N. The images $F_t(N)$ and $F_t(S)$ of the poles are again labeled N and S for simplicity.

The eversion starts by pulling the south pole through and above the north pole to the position of Figure 2b.

The tangent vector at S has not rotated, but the strip now intersects itself. To remove the resulting twist so that the sphere ends up reflected as in Figure 2c, the rest of the eversion F must apply an odd number of $360°$ rotations to the tangent plane at S. In §6, we verify that during the eversion F, the tangent vector $(\partial L/\partial u)$ at S rotates exactly $360°$ and brings the strip R to the position shown in Figure 2c.

Note that an eversion is defined by a specific parametrization, not just a set of images, and the parametrization affects the final position of the strip. If the strip had ended up as in Figure 2d, the regular homotopy would not satisfy our definition of an eversion, and the rotation count for $(\partial L/\partial u)$ at S would be invalid. However, if F brings Figure 2a to

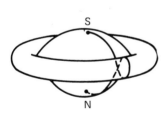

Figure 2b.

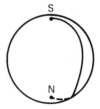

Figure 2c.

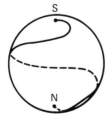

Figure 2d.

Figure 2c, then χF will bring Figure 2c back to Figure 2a, rotating the tangent vector an additional 360° counterclockwise.

The total rotation of 720° during L means that $h\,|\,(S \times [0,1])$ traverses the great circle C exactly once, up to homotopy, so the algebraic sum of those points in $h^{-1}(Q)$ corresponding to S will be 1. In §6 we also show that the other points in $h^{-1}(Q)$ cancel in pairs, so that the total degree of h is 1. To verify this cancellation, we will need the following lemma.

LEMMA 3.   *Let $J: D^2 \times [0,1] \to R^3$ be a generic loop in Γ_c, and let h be the lifting of $\sigma \circ dJ$ as described above, transversal to $Q \in C \subset \tilde{F}_{32}$. Suppose a saddle a_t and minimum b_t emerge from a degenerate critical point at time t_0, and rejoin at another degenerate critical point at time t_1. Suppose further that a continuous family of arcs γ_t (e.g. arcs of steepest descent from a_t to b_t) exists on the surfaces $J_t(S^2)$, for $t_0 \leqslant t \leqslant t_1$, connecting a_t to b_t, such that the tangent plane to $J_t(S^2)$ is never vertical at any point of γ_t. Then the algebraic sum of those points a_t and b_t belonging to $h^{-1}(Q)$ is zero.*

Proof of Lemma 3.   Let $\alpha(t): [t_0, t_1] \to D^2$ be the preimage of a_t, so that $J(\alpha(t), t) = a_t$, and let $\beta(t)$ be the preimage of b_t. Similarly, let $\Gamma: [0,1] \times [t_0, t_1] \to D^2$ be the preimage of the curves γ_t so that for fixed t, $\Gamma(s,t)$ connects $\alpha(t)$ and $\beta(t)$. Note that $\alpha(t_0) = \beta(t_0) = \Gamma(s, t_0)$ and $\alpha(t_1) = \beta(t_1) = \Gamma(s, t_1)$, for all s in $[0,1]$.

Suppose $\zeta(s,t)$ is a continuous family of paths on the surfaces $D^2 \times t$, connecting $(1,0,t)$ with $\alpha(t)$. Then for fixed t in $[t_0, t_1]$, $\phi_t(s) = \sigma \circ dJ \circ \zeta(s,t)$, considered as a path in s, is a representative for $h(\alpha(t))$ in $\tilde{F}_{32}$. Similarly

$$
\psi_t(s) = \begin{cases} \sigma \circ dJ \circ \zeta(2s, t); & 0 \leqslant s \leqslant 1/2 \\ \sigma \circ dJ \circ \Gamma(2s - 1, t); & 1/2 \leqslant s \leqslant 1 \end{cases}
$$

is a representative for $h(\beta(t))$.

The equivalence classes of ϕ_t and ψ_t lie in $C \subset \tilde{F}_{32}$, and join the same pair of endpoints. Since $J(\alpha(t))$ consists of saddle points, the contribution of $h\,|\,\alpha([t_0, t_1])$ to the degree of h is the negative of the algebraic number of times the curve ϕ_t crosses Q in a counterclockwise direction. Similarly, the contribution of $h\,|\,\beta([t_0, t_1])$ is equal to the number of times ψ_t crosses Q. These contributions will cancel if ϕ_t and ψ_t are homotopic, relative to their endpoints, as paths from $[t_0, t_1]$ to C. We construct the required homotopy using the tangent flags along the curves γ_t, projected to become horizontal.

Let ρ be the projection of R^3 onto the XY plane. If P lies in the domain of J, let

$$U_x(P) = (1 - x)\frac{\partial J}{\partial u}(P) + x\rho\left(\frac{\partial J}{\partial u}(P)\right) \qquad \text{and}$$

$$V_x(P) = (1 - x)\frac{\partial J}{\partial v}(P) + x\rho\left(\frac{\partial J}{\partial v}(P)\right).$$

Since the tangent planes to J are never vertical along the curves γ_t, the two vectors $U_x(\Gamma(s, t))$ and $V_x(\Gamma(s, t))$ are linearly independent, so we may define the flag

$$W_x(s, t) \colon \langle U_x(\Gamma(s, t)) \rangle \subset \langle U_x(\Gamma(s, t)), V_x(\Gamma(s, t)) \rangle \subset R^3.$$

Then

$$\omega_{xt}(s) = \begin{cases} \sigma \circ dJ \circ \zeta(2s, t); & 0 \leqslant s \leqslant 1/2 \\ W_x(2s - 1, t); & 1/2 \leqslant s \leqslant 1 \end{cases}$$

agrees with $\psi_t(s)$ when x is 0, and when x is 1, $\omega_{1t}(s)$ has horizontal flags for $1/2 \leqslant s \leqslant 1$. The paths $\omega_{1t}(s)$ can also serve as representatives of $h(\beta(t))$ in F_{32}. The homotopy $\omega_{1t}((1 + u)s)$ in the parameter $u \in [0, 1]$ then shrinks these paths back along the arcs γ_t, ending up at ϕ_t, and remaining inside $C \subset \tilde{F}_{32}$.

§6. We will now describe a variant of the Froissart-Morin eversion, using a finite series of images $F_t(S^2)$ through which the reader can imagine the continuous regular homotopy. (The film [1] has many more closely spaced computer drawn images, giving the illusion of continuous motion.)

We will also draw an arc ζ_t on each image, connecting the north pole N to the south pole S. By imagining a continuous family of strips about these arcs, the reader can verify that the tangent vector $(\partial L/\partial u)$ at S turns $360°$ counterclockwise when viewed from above, as required by §5. At the same time, we will verify that the other points in $h^{-1}(Q)$ cancel by drawing the arcs γ_t of lemma 3.

The diagrams serve two functions: a) to define the surfaces and convince the reader that they are stages of a continuous eversion, and b) to show that the degree of h is 1. For each position of the surface, we offer two figures, adorned to serve these two functions. Both figures have continuous black lines, indicating either the visible profile edges, where the tangent plane contains the line of sight, or visible portions of the self-intersection curves. In the figures a), the self-intersection curves are continued by dashed lines showing their hidden portions. In the figures b) the arc ζ_t between N and S is indicated, as well as other arcs γ_t and δ_t to which lemma 3 will be applied. These arcs are drawn with a continuous line where they are visible, and dashed where hidden.

At time t_0 the south pole S has been pushed flat, into a degenerate critical point. (This means that the point Q of §5 must be chosen away from $h(S, t_0)$.) For $t > t_0$, S becomes a local maximum. If continuous rotational symmetry about the Z axis were retained, the resulting dimple would be bounded by a circle of degenerate critical points, as in Figure 3. Instead, as suggested by Pugh [6], we immediately start twisting the surface, so that it has only two-fold rotational symmetry. The degenerate critical point $h(S, t_0)$ will then give birth to five non-degenerate ones. In Figure 4, these are a local maximum at S, two saddle points at a and c, and two local minima at b and d. There is an arc of steepest decent γ_t from a to b, and a similar arc δ_t from c to d, related to γ_t by two-fold symmetry.

The south pole is next pushed above the north pole creating a circle of self-intersection. The result, Figure 5, is a twisted version of Figure 2b. Next the "east" and "west" poles are similarly pushed through each other (Figure 6). This creates a second self-intersection circle, intersecting the first in two triple points. The first circle now intersects itself in these two triple points, so the self-intersection set consists of four arcs connecting one triple point to the other, and two loops, one at each triple point.

In Figure 7, the twisting continues, so that the lowest of the intersection arcs connecting the two triple points makes a zigzag, and its projection in Figure 7a crosses itself. Additional twisting contorts this arc still further, as the surface approaches two points of "saddle" tangency (Figure 8). After these tangencies, the topology of the self-intersection set changes (Figure 9). There are now two arcs joining the two triple points, and two loops at each triple point.

During these motions, the tangent to ζ_t at S twists 90° counterclockwise with respect to the tangent at N, when viewed from above. In Figures 10 and 11, the point N has started to rise, and by Figure 12, it

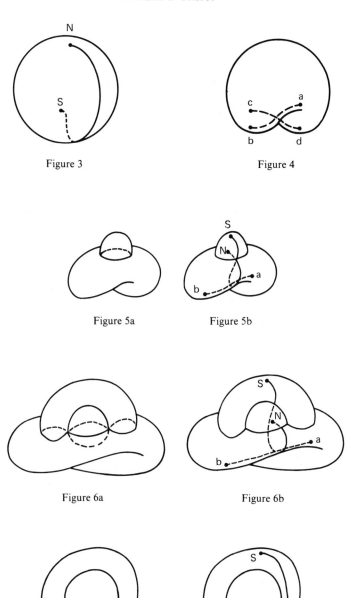

Figure 3

Figure 4

Figure 5a

Figure 5b

Figure 6a

Figure 6b

Figure 7a

Figure 7b

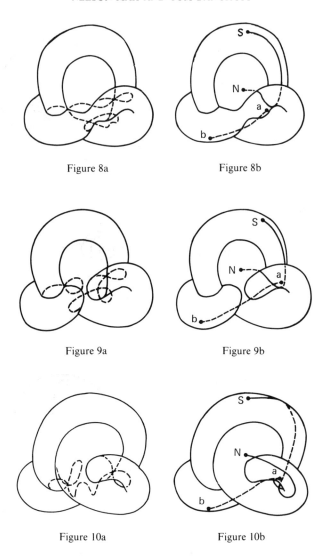

Figure 8a   Figure 8b

Figure 9a   Figure 9b

Figure 10a   Figure 10b

becomes tangent to a line of double points, creating a singular triple point, which separates into two transversal triple points in Figure 13. The self-intersection set for Figure 13 consists of four triple points joined by short arcs into a tetrahedron, and also by six longer arcs. By Figure 12, the tangent to ζ_t at S has rotated further counterclockwise, so that it makes an angle of 180° with the negative of the tangent at N. The saddle point at a_t has shifted across an arc of double points, and it is now

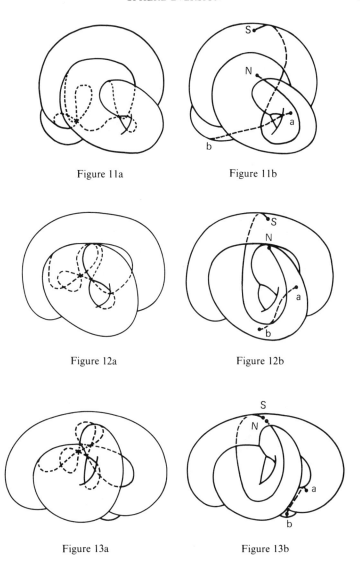

Figure 11a   Figure 11b

Figure 12a   Figure 12b

Figure 13a   Figure 13b

visible from above in Figure 13. It is still connected by γ_t to the minimum at b_t, which has now shifted so that it is directly behind a_t.

In Figure 14, N has risen to the same level as S forming a "saddle" tangency, and the image momentarily has four-fold symmetry. The four triple points have merged into the promised transversal quadruple point T. The double point set consists of four arcs joining T to N (which

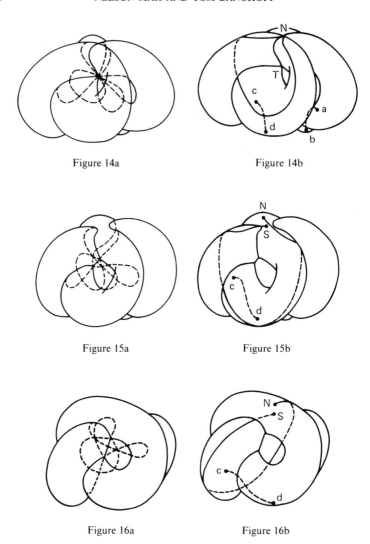

Figure 14a   Figure 14b

Figure 15a   Figure 15b

Figure 16a   Figure 16b

coincides with S) and four loops joining T to itself. The arc ζ_t becomes a closed loop. The two arcs γ_j joining a to b, and δ_t, joining c to d, are still related by two-fold symmetry. In subsequent pictures δ_t will be more visible than γ_t, so we will concentrate on it instead. Due to the four-fold symmetry, there are two other pairs of minima and saddle points, but their tangent planes have the opposite orientation, so they are in $h^{-1}(D)$ rather than $h^{-1}(C)$. By examining the figures, one can verify that these exhaust the places where the tangent plane to the surface is horizontal.

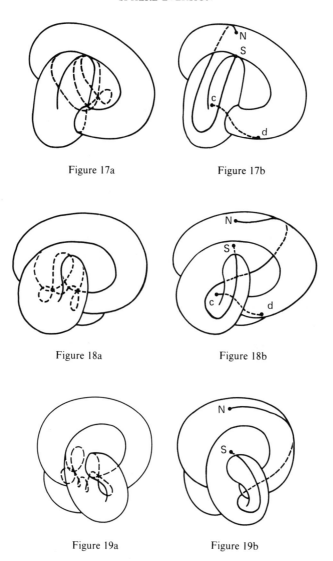

Figure 17a

Figure 17b

Figure 18a

Figure 18b

Figure 19a

Figure 19b

In Figure 15 the surface containing N has risen above the surface containing S, and the two have switched roles, so that if Figure 15 were rotated 90°, it would look like Figure 13. We partially carry out this rotation in Figures 16 through 19, while following the previous homotopy in reverse. It should be evident that Figure 19 is a slightly rotated version of Figure 10, with the roles of N and S switched. The angle between the tangent to ζ_t at S and the negative of the tangent at N is

270° in Figure 19, since the arc ζ_t is in the same position, but its direction is reversed. As the steps in Figures 3 through 9 are reversed, this angle will increase to 360°. When the sphere again becomes round, the arc ζ_t between N and S will take the position of Figure 2c, verifying that the eversion is correctly parametrized.

Between Figures 18 and 19, the surface containing the arc δ_t tilts until c and d join at a singular critical point, and δ_t disappears. The arc γ_t disappears symmetrically at the rear of the surface. Therefore, the conditions for lemma 3 are satisfied, and the contributions of the points a_t, b_t, c_t, and d_t to the degree of h cancel.

The second part of the loop L consists of the mirror image, with respect to the XY plane, of the eversion F just described. Since this reflection keeps the XY plane fixed, the rest of the set $h^{-1}(C)$ of points with horizontal positively oriented tangent planes will consist of the mirror images of the points S, a_t, b_t, c_t, and d_t whose course has just been exhaustively followed. The tangent vector at S will again turn 360° counterclockwise, giving a total of 720°, or one complete transversal of the circle C, up to homotopy. Thus the algebraic number of points in $h^{-1}(C)$ lying at the image of the south pole will be exactly 1, and all other contributions cancel. This completes the proof of lemma 2.

LAWRENCE LIVERMORE NATIONAL LABORATORY
BROWN UNIVERSITY

REFERENCES

---

[1] Max, Nelson, Turning a Sphere Inside Out, 23 minute color sound 16 mm film, and 12 page accompanying pamphlet. International Film Bureau, Chicago, Illinois (1976). The pamphlet is also available as Unit 289, UMAP project, E.D.C. Newton, Mass. (1978).

[2] Morin, Bernard, and Petit, Jean Pierre, Le Retournement de la sphere, Pour la Science, **15** (1979), 34–41.

[3] Morin, Bernard, Equations du retournement de la sphere, C. R. Acad. Sc. Paris, **287** (1978), 879–882.

[4] Smale, Steven, A classification of immersions of the two-sphere, Transactions A.M.S. **90** (1959), 281–290.

[5] Munkres, James, Elementary differential topology. Annals of Mathematics Studies no. 54, (1966), Princeton University Press.

[6] Pugh, Charles, Chicken wire models, now stolen from the University of California, Berkeley, but recorded in the film [1].

[7] Phillips, Anthony, Turning a surface inside out, *Scientific American* **214** (May 1966), 112–120.

[8] Hirsh, Morris, Differential Topology (1976), Springer Verlag, New York.

[9] Murkres, James, Topology, a first course (1975), Prentice-Hall, Englewood Cliffs, New Jersey.

[10] Milnor, John, Construction of universal bundles, II, *Annals of Mathematics* **63** (1956), 430–436.

CLASSIFYING HARMONIC MAPS OF A SURFACE IN E^n

By TILLA KLOTZ MILNOR

The space of all non-congruent minimal immersions $X : R \to E^n$ of a Riemann surface R which are isometric to a fixed holomorphic curve $X : R \to C^m$ was studied by Calabi in [3]. (Note that n is free to vary.) In this paper we allow X to have branch points, so that its induced metric ds^2 may be degenerate at isolated points. Using Calabi's methods, we describe the larger class of non-congruent harmonic maps $X : R \to E^n$ whose energy 1 metric Γ coincides with ds^2, a condition which reduces to isometry in case X yields a generalized minimal surface. (See [13].)

At the heart of Calabi's work is the fact that there is exactly one holomorphic curve among all non-congruent, minimal immersions X isometric to X, namely X itself. Analogously, we have the fact that there is exactly one holomorphic curve among all non-congruent, harmonic maps X with energy 1 metric $\Gamma = ds^2$.

1. Preliminary Notions. While our results apply only to maps of R into E^n, it seems best to present some basic definitions for maps $X : R \to M^n$ into an arbitrary Riemannian manifold M^n. This allows a proper indication of the central role of the energy 1 metric which seems generally useful in the study of harmonic immersions. (See [9] or [12].) We assume C^∞ smoothness throughout.

The induced metric of a map $X : R \to M^n$ is denoted by I, except that the symbol ds^2 is used for the induced metric of a particular holomorphic curve $X : R \to C^m$. If X is an immersion, then I is a Riemannian metric, the second fundamental form B of X is defined as in [7], and the mean curvature vector field $\mathcal{H}$ of X is given by

$$\mathcal{H} = \operatorname{tr}_I B.$$

A (possibly degenerate) metric g on R is conformal provided that $g = \lambda \, dz \, d\bar{z}$ with $\lambda \geq 0$ a function, for any conformal parameter z on R. We call a map $X : R \to M^n$ conformal if I is a conformal metric on R.

Henceforth, we reserve the symbol g to denote a conformal Riemannian metric on R. Once g is fixed, we associate to any map

211

$X : (R, g) \to M^n$ the energy function

$$e = e(X) = \tfrac{1}{2} \operatorname{tr}_g I.$$

When X is an immersion, e is everywhere positive.

Definition. If $X : (R, g) \to M^n$ is a map with energy function e, then the conformal metric $\Gamma = eg$ on R is called the *energy 1 metric* of X. When X is an immersion, Γ is the only conformal metric on R for which $X : (R, \Gamma) \to M^n$ has energy function identically equal to 1. If $z = x + iy$ is a conformal parameter on R, and $I = E \, dx^2 + 2F \, dx \, dy + G \, dy^2$, then Γ can be computed without reference to g by the formula

$$(1) \qquad 2\Gamma = (E + G) \, dz \, d\bar{z} = (E + G)(dx^2 + dy^2).$$

In particular, Γ is independent of the choice of g on R. We therefore use (1) to define the energy 1 metric of any map $X : R \to M^n$. Note that when X is conformal, Γ coincides with I.

The definition of a harmonic map $X : R \to M^n$ can be found in [4], [5] or [6]. Roughly speaking, X is harmonic if it is extremal for the Γ-area integral, and minimal if it is conformal, and extremal for the ordinary I-area integral. Thus X is minimal if and only if it is conformal and harmonic. An immersion X is minimal if and only if it is conformal with $\mathcal{H} \equiv 0$. The following facts help to explain the usefulness of the energy 1 metric Γ in the study of harmonic immersions.

Remark 1.   (See [10] and [11].) Suppose $X : R \to M^n$ is an immersion. Then the energy 1 metric Γ is complete if I is. If X is also harmonic, then I satisfies the Codazzi-Mainardi equations with respect to Γ as metric, a property we denote by $\operatorname{Cod}(\Gamma, I)$. Indeed, one can characterize harmonic immersions $X : R \to M^n$ by the two conditions $\operatorname{Cod}(\Gamma, I)$ and $\mathcal{H}^\Gamma \equiv 0$, where $\mathcal{H}^\Gamma$ is the Γ-mean curvature vector $\operatorname{tr}_\Gamma B$. However, even if just $\operatorname{Cod}(\Gamma, I)$ holds for X, the intrinsic curvatures of Γ and I satisfy the inequality

$$K(\Gamma) \leq \mu K(I)$$

where $0 < \mu \equiv \det I / \det \Gamma \leq 1$.

A map $X : R \to E^n$ is harmonic if and only if $\Delta X = \partial^2 X / \partial z \partial \bar{z} \equiv 0$ for any conformal parameter z on R. When X is an immersion, the component of ΔX tangent to $X(R)$ vanishes if and only if $\mathrm{Cod}(\Gamma, I)$ holds, while the component of ΔX normal to $X(R)$ vanishes if and only if $\mathcal{K}^\Gamma \equiv 0$.

The exposition below follows very closely the discussion of Calabi's work in Lawson's book [8]. Adaptation of the arguments there to cover the present situation is remarkably straightforward.

2. Results. Given a harmonic map $X : R \to E^n$, the fact that $\partial^2 X / \partial z \partial \bar{z} \equiv 0$ for any conformal parameter z on R indicates that

$$(2) \qquad \phi = 2 \partial X / \partial z = (\phi_j)$$

is holomorphic. Moreover, since $\phi \, dz$ is invariant, X generates n holomorphic differentials $\phi_j \, dz$ on R. If we normalize X so that $X(p_0) = 0$ for some p_0 on R, it is easy to check that

$$(3) \qquad X = \mathrm{Re} \int_{p_0}^{P} \phi \, dz.$$

The induced metric is given by

$$(4) \qquad I = |\mathrm{Re}\, \phi|^2 \, dx^2 - 2 \langle \mathrm{Re}\, \phi, \mathrm{Im}\, \phi \rangle \, dx \, dy + |\mathrm{Im}\, \phi|^2 \, dy^2,$$

so that (1) provides the expression

$$(5) \qquad 2\Gamma = |\phi|^2 \, dz \, d\bar{z}$$

for the energy 1 metric Γ, where $|\phi|^2 = \sum |\phi_j|^2$. The map X is minimal if and only if

$$(6) \qquad \phi^2 = \sum \phi_j^2 = |\mathrm{Re}\, \phi|^2 - |\mathrm{Im}\, \phi|^2 + 2i \langle \mathrm{Re}\, \phi, \mathrm{Im}\, \phi \rangle \equiv 0,$$

that is, if and only if $I \equiv \Gamma$. It is customary to call $X(R)$ a minimal surface when X is a minimal immersion, and a generalized minimal surface if X is a non-constant minimal map. In the latter case, isolated branch points occur wherever $\phi = 0$. (See [13].)

Remark 2. Formula (3) provides a way to generate all harmonic maps of R into E^n. Specifically, given any n holomorphic differentials

$\phi_j dz$ on R with no real periods, the map X given by (3) for $\phi = (\phi_j)$ is harmonic with I given by (4) and Γ by (5). The map X is an immersion provided that $\mathrm{Re}\,\phi$ and $\mathrm{Im}\,\phi$ are linearly independent. The map X is minimal if and only if (6) holds.

Remark 3. If R is simply connected and $X : R \to E^n$ is harmonic, there are many examples of non-congruent harmonic maps of R into E^n which have the same energy 1 metric Γ as X. Simply take the *associate harmonic maps* $X_\theta : R \to E^n$ given by

$$X_\theta = \mathrm{Re} \int_{p_0}^{P} e^{i\theta} \phi\, dz,$$

so that $X_0 = X$. When X is minimal, it is well known that the X_θ are also minimal, and have the same induced metric I as X. When X is harmonic, the X_θ need not be isometric, but they do share the same energy 1 metric Γ as X.

When R is simply connected and $X : R \to E^n$ is harmonic, it makes sense to speak of the conjugate harmonic immersion $\tilde{X} : R \to E^n$ given by $\tilde{X} = -X_{\pi/2}$. Then we can associate with X the holomorphic curve $\hat{X} : R \to C^n = E^n \times iE^n$ defined by

$$(7) \qquad \sqrt{2}\,\hat{X} = (X, \tilde{X}).$$

Because $\hat{X}$ is holomorphic, it is minimal, and its induced metric $\hat{I}$ is the energy 1 metric. Moreover, since $\hat{I} = |d\hat{X}/dz|^2\, dz\, d\bar{z}$, (5) shows that $\hat{I}$ coincides with Γ. Thus $\hat{X}$ is singularity free if and only if Γ is a Riemannian metric on R. We will need the following result due to Calabi. (See [1] or [8].) The possibility of isolated branch points does not cause any problems.

LEMMA 1. *Let* $Z : R \to C^k$ *and* $\hat{Z} : R \to C^j \subseteq C^k$ *be holomorphic curves with the same induced metric. Then there is a holomorphic isometry* $T : C^k \to C^k$ *(that is, a unitary transformation plus a translation) such that* $Z = T \circ \hat{Z}$.

We have thus established the following result.

THEOREM 1. *Among all non-congruent non-constant harmonic maps of R into Euclidean space of any dimension which share the same energy 1* *metric* Γ, *there exists at most one holomorphic curve, and exactly one if R is simply connected.*

In view of Theorem 1, we fix a non-constant, holomorphic map $\mathbf{X} : R \to C^m$ and let ds^2 denote its induced metric. Assume that $\mathbf{X}(p_0) = 0$ and that $\mathbf{X}(R)$ is contained in no linear subspace of complex dimension less than m. We seek to describe the space $\mathcal{S}(\mathbf{X})$ of harmonic maps $X : R \to E^n$ for any n with energy 1 metric $\Gamma = ds^2$, $X(p_0) = 0$ and $X(R)$ contained in no linear subspace of dimension less than n. The remark which follows shows that these normalizations make m an invariant of $\Gamma = ds^2$ which can be computed given any X in $\mathcal{S}(\mathbf{X})$.

Remark 4. For any X in $\mathcal{S}(\mathbf{X})$ the holomorphic curve $\hat{X}$ given by (7) must be well defined even if R is not simply connected. To see this, remove curves from R so as to obtain a connected, simply connected subsurface R_0 with closure equal to R. Then $\hat{X}$ is defined on R_0 and isometric to the restriction of $\mathbf{X}$ to R_0. By Lemma 1, these maps must be congruent, and since $\mathbf{X}$ extends to R, so must $\hat{X}$.

LEMMA 2. *Any X in $\mathcal{S}(\mathbf{X})$ is given by*

(8) $\qquad X = \sqrt{2}\, \mathrm{Re}\{ S\mathbf{X} \}$

where S is an $n \times m$ complex matrix such that
  (a) $\bar{S}'S = 1_m$
  (b) $(S, \bar{S})$ has rank n
  (c) $m \leqslant n \leqslant 2m$.
Moreover, X is minimal if and only if the corresponding matrix S satisfies

(9) $\qquad (d\mathbf{X}/dz)'S'S(d\mathbf{X}/dz) \equiv 0.$

Finally, any X given by (8) lies in $\mathcal{S}(\mathbf{X})$.

Proof. We follow the argument in [8]. Suppose X is in $\mathcal{S}(\mathbf{X})$. Since $\Gamma = ds^2 = |d\mathbf{X}/dz|^2\, dz\, d\bar{z}$, (2) and (5) yield

(10) $\qquad \sqrt{2}\, |\partial X/\partial z|^2 = |d\mathbf{X}/dz|^2.$

Because $\partial X/\partial z$ is holomorphic, Lemma 1 and (10) provide an $N \times N$ unitary matrix U so that

(11) $\qquad \sqrt{2}\, \partial X/\partial z = U(d\mathbf{X}/dz)$

for $N = \max(n, m)$. Here zero components may have been added to the vector on one side of (11) to make it into an N-vector. If this were done to $\partial X/\partial z$, it would mean that $\mathbf{X}(R)$ is contained in a linear subspace of complex dimension less than m. Hence $m \leqslant n = N$, so that U is an $n \times n$ matrix. Similarly, by (2), (3), (11) and $X(p_0) = \mathbf{X}(p_0) = 0$, we have $X = \sqrt{2}\,\mathrm{Re}(U\mathbf{X})$ or

(12) $X = (U\mathbf{X} + \overline{U}\overline{\mathbf{X}})/\sqrt{2}\,.$

Again, zero components may have been added to the vector on one side of (12). If this were done to the right side, it would mean that $X(R)$ is contained in a linear subspace of dimension less than n. Thus we have $n \leqslant 2m$. To verify the main claim of the lemma, let S be composed of the first m columns of U. The next claim follows since X is minimal if and only if $(\partial X/\partial z)^2 \equiv (d\mathbf{X}/dz)'S'S(d\mathbf{X}/dz) \equiv 0$. To check the last claim, note that any $n \times m$ matrix S satisfying (a), (b) and (c) can be formed by the first m rows of an $n \times n$ unitary matrix U. Thus, if (8) gives X, (12), (11) and (10) hold, so that (2) and (5) give the rest.

Remark 5. Two elements X and X' of $\mathcal{S}(\mathbf{X})$ are congruent if and only if there is an orthogonal $n \times n$ matrix 0 such that $X = 0X'$, that is, if and only if the corresponding matrices satisfy $S = 0S'$.

Remark 6. One can think of $\mathbf{X} = (\mathrm{Re}\,\mathbf{X}, \mathrm{Im}\,\mathbf{X}): R \to E^{2m}$ as an element of $\mathcal{S}(\mathbf{X})$. For $\mathbf{X}$, $n = 2m$ and the corresponding $2m \times m$ matrix $S^{\mathbf{X}}$ is given in $m \times m$ block form by

$$\sqrt{2}\,S^{\mathbf{X}} = \begin{pmatrix} 1 \\ -i \end{pmatrix}.$$

By $\mathcal{P}(\mathbf{X})$ we denote some maximal collection of non-congruent elements from $\mathcal{S}(\mathbf{X})$. Note that $\mathcal{P}(\mathbf{X})$ can be thought of as the space of all non-congruent harmonic immersions of R into Euclidean space of any dimension which have energy 1 metric $\Gamma = ds^2$, and which fit into no lower dimensional Euclidean space. As an extension of a theorem of Calabi (see [3]) we have the following.

THEOREM 2. *Suppose that $\mathbf{X}: R \to C^m$ is a holomorphic curve (with or without branch points) which fits into no lower dimensional affine subspace of C^m. If $\mathbf{X}$ has induced metric ds^2, then the space $\mathcal{P}(\mathbf{X})$ of all non-congruent harmonic maps $X: R \to E^n$ which have energy 1 metric*

$\Gamma = ds^2$ and which fit into no lower dimensional affine subspace of E^n is naturally described as the set of all complex, symmetric $m \times m$ matrices P such that $1_m - P\bar{P} \geqslant 0$. Moreover

$$n - m = \mathrm{rank}\left(1_m - P\bar{P}\right),$$

so that $m \leqslant n \leqslant 2m$ and $n = 2m$ if and only if $1_m - P\bar{P} > 0$. Finally, an X in $\mathscr{P}(\mathbf{X})$ is minimal if and only if the corresponding P satisfies

$$(d\mathbf{X}/dz)^t P(d\mathbf{X}/dz) \equiv 0.$$

Proof. Let Σ be the set of all $n \times m$ matrices S satisfying conditions (a), (b) and (c) of Lemma 2, and let Π be the set of all complex, symmetric $n \times m$ matrices P such that $1_m - P\bar{P} \geqslant 0$ has rank $n - m$. Lawson shows on pp. 156–7 [8] that for any $n \times m$ matrix S in Σ, $P = S'S$ is in Π with $n - m = \mathrm{rank}(1_m - P\bar{P})$. Moreover, S and S' in Σ yield the same P if and only if $S = 0S'$ for an orthogonal matrix 0, so that by Remark 5, congruent elements of $\mathcal{S}(\mathbf{X})$ yield the same P in Π. Since Lawson's procedure allows one to retrieve an S from Σ for any P in Π so that $P = S'S$, the theorem follows from Lemma 2.

Remark 7. If $\mathcal{I}(\mathbf{X})$ is the space of minimal immersions in $\mathscr{P}(\mathbf{X})$, Calabi's work shows that $\mathcal{I}(\mathbf{X})$ can be naturally imbedded as a linear variety in the closure of the Siegel domain $\mathcal{D}$ of complex, symmetric $m \times m$ matrices P with $1_m - P\bar{P} > 0$. The point $P = 0$ is in the variety and corresponds to the unique holomorphic curve $\mathbf{X}$. Here we see that $\mathscr{P}(\mathbf{X})$ can be naturally imbedded as the closure of $\mathcal{D}$, with $\mathcal{I}(\mathbf{X}) \subset \mathscr{P}(\mathbf{X})$ going to the variety just described.

In [2], Calabi gives necessary and sufficient conditions that a real analytic conformal metric g on a simply connected Riemann surface R is achieved as the induced metric I for a minimal immersion of R into some Euclidean space. The considerations above show that the same necessary and sufficient conditions hold if g is achieved as the energy 1 metric Γ for a harmonic immersion of R into some Euclidean space.

RUTGERS UNIVERSITY

REFERENCES

[1] E. Calabi, Isometric imbeddings of complex manifolds, *Annals of Math.* **58** (1953), 1–23.

[2] ———, Metric Riemann surfaces, *Contributions to the Theory of Riemann Surfaces*, Princeton University Press, Princeton, N.J., 1953, 77–85.

[3] ———, Quelques applications l'analyse complexe aux surfaces d'aire minima, with *Topics in Complex Manifolds* (Hugo Rossi), Les Presses de l'Universite de Montreal, 1968.

[4] J. Eells and J. H. Sampson, Harmonic mappings of Reimannian manifolds, *American Journal of Math.* **86** (1964), 109–160.

[5] F. B. Fuller, Harmonic mappings, *Proceedings of the National Academy of Sciences* **40** (1954), 987–991.

[6] P. Hartman, On homotopic harmonic maps, *Canadian Journal of Math.* **19** (1967), 673–687.

[7] S. Kobayashi and K. Nomizu, *Foundations of Differential Geometry* II, Interscience Publishers, New York, N.Y., 1969.

[8] H. B. Lawson, Jr., *Lectures on Minimal Submanifolds*, Notas de Math. Inst. de Mat. Pure e Aplicada, Rio de Janeiro, 1974.

[9] T. K. Milnor, Mapping surfaces harmonically into E^n, *Proceedings of the American Math. Society* **78** (1980), 269–275.

[10] ———, Harmonically immersed surfaces, *Journal of Differential Geometry*, **14** (1979), 205–214.

[11] ———, Abstract Weingarten surfaces, *Journal of Differential Geometry*, **15** (1980), 365–380.

[12] ———, The energy 1 metric on harmonically immersed surfaces, *Michigan Math. Journal*, to appear.

[13] R. Osserman, *A Survey of Minimal Surfaces*, Van Nostrand Math Studies 25, Princeton (1969).

ON THE THEORY OF INFINITESIMAL DEFORMATIONS
OF SUBMANIFOLDS IN EUCLIDEAN SPACE

By V. I. Oliker*

Introduction. In this paper we investigate deformations of compact submanifolds in Euclidean space of codimension $\geqslant 1$ which admit a global nondegenerate parallel section of the normal bundle. The deformations we consider are assumed to preserve this section in the sense that it remains in the normal bundle throughout the deformation.

Existence of such a section (call it ξ) is a property modeled from the case of hypersurfaces, and the condition on deformations means that the spherical image of the submanifold via ξ on a unit hypersphere remains fixed. In this setting we are able to obtain various generalizations of rigidity results on deformations previously known only for hypersurfaces.

More specifically, we study infinitesimal deformations with respect to the metric and with respect to symmetric functions of principal radii of curvature in the direction ξ. The latter is motivated by an attempt to find an analogue of the Minkowski problem for submanifolds of codimension $\geqslant 1$, whose analytic solution involves, as a first step, a study of infinitesimal deformations with respect to the symmetric functions of principal radii of curvature (see [13], and also [4]). The next step is the global solubility of a certain nonlinear problem, which in case of codimension > 1 is also underdetermined. We hope to present this part of the problem in our subsequent paper.

It should be noted that the theory of infinitesimal deformations of submanifolds of codimension > 1 does not seem to be well developed. Actually, we know of only a few papers in this area, some of which are quite recent [5], [7], [14] (see also further references given there), [15].

This paper is organized as follows.

In §1 we describe the notations and recall a few basic formulas and definitions.

---

*Partially supported by the National Science Foundation Grant #MCS 80-02779, The University of Iowa, and by the Sonderforschungsbereich Theoretische Mathematik at the Universität Bonn.

In §2 we introduce deformations preserving a nondegenerate parallel section ξ of the normal bundle and give several of their basic properties. Then we discuss the concept of parallel deformations introduced by Yano [15]. For submanifolds, which admit a section ξ as above, parallel deformations form a subset of the set of deformations preserving ξ. Later, in §5 we prove a rigidity result related to parallel deformations (Theorem 5.2).

The partial differential equations of infinitesimal deformations are derived in §3.

Except for one special case, in order to assure that these equations are elliptic we have to impose one further restriction. Namely, it is required that the second fundamental form of the submanifold in the direction ξ is definite at least at one point. This condition together with the assumed compactness and nondegeneracy of ξ imply that the second fundamental form in the direction ξ is definite everywhere. Correspondingly, submanifolds satisfying those conditions are said to be elliptically immersed. For the case of codimension one the ellipticity of immersion is equivalent to local convexity; in codimension > 1 the class of elliptically immersed submanifolds is much wider. The elliptic immersions are introduced in §4, where we also give examples illustrating this notion. Certain results on the corresponding differential equations are also presented in this section.

Finally, in §5, we prove several results establishing global rigidity "in the direction ξ." Roughly speaking, these results show that the freedom of an infinitesimal deformation (say, isometric) is completely determined by the component of the deformation field which lies in the normal bundle and orthogonal to ξ. In particular, if this component vanishes identically, and some other hypotheses are satisfied, then the submanifold does not admit any (isometric) infinitesimal deformations.

In rather preliminary form, a part of the information contained in §§2, 3 has been discussed in [10], where the main subject was the deformations of convex hypersurfaces. This paper can be read independently of [10].

1. Preliminaries.

1. Unless it is stated otherwise the following convention on the ranges of indices is assumed to be in effect throughout the paper:

$$1 \leqslant i, j, k, l, s \leqslant m, \qquad 1 \leqslant \alpha \leqslant n.$$

It is also agreed that repeated lower and upper indices are summed over the respective ranges. It is assumed that all submanifolds and maps are sufficiently smooth, say, at least C^3. E denotes the Euclidean space of dimension $m + n$ ($\geqslant 3$). The origin of it is fixed and denoted by O. By M we denote a connected orientable submanifold of dimension m ($\geqslant 2$) immersed in E, and represented by the position vector field

$$\Phi = \Phi(u^1, \ldots, u^m),$$

where $\{u^i\}$ are the local coordinates on M. M is locally identified with $\Phi(M)$.

The following notations will also be adopted.

$\langle , \rangle$—the inner product in E;

$\Phi_i = \partial_i \Phi$ where $\partial_i = \partial/\partial u^i$, and in a similar fashion ξ_i, Φ_{ij}, ξ_{ij}, etc. for various sections and functions on M;

$T(M), N(M)$—the tangent and the normal bundles of M;

$T_x(M), N_x(M)$—the corresponding fibers at the point $x \in M$;

$G = (G_{ij})$—the metric on M induced from E via Φ;

Let ξ and η be two unit length sections of $N(M)$; ξ_x, η_x stand for $\xi(x), \eta(x)$.

$b(\xi) = (b_{ij}(\xi))$ denotes the second fundamental form of M with respect to ξ;

$g(\xi, \eta) = (g_{ij}(\xi, \eta)) = (\langle \xi_i, \eta_j \rangle)$—the third mixed fundamental form. We also write $g(\xi)$ for $g(\xi, \xi)$.

2. Let $\{N(\alpha)\}$ denote a local orthonormal frame field in $N(M)$. In the local adapted frame $\Phi_1, \Phi_2, \ldots, \Phi_m, N(1), \ldots, N(n)$ we have

(1) $$\xi_i = b_i^j(\xi)\Phi_j + \sum_\alpha < \xi_i, N(\alpha) > N(\alpha),$$

where $b_i^j(\xi) = -b_{il}(\xi)G^{lj}, (G^{lj}) = G^{-1}$. Also

(2) $$g(\xi, \eta) = b(\xi)G^{-1}b(\eta) + g^\perp(\xi, \eta),$$

where $g^\perp(\xi, \eta)$ denotes the matrix $\langle \xi_i^\perp, \eta_j^\perp \rangle$ with the symbol "$\perp$" standing for the normal component.

If ξ (or η) is parallel in $N(M)$ in the induced connection then $g^\perp(\xi, \eta) \equiv 0$.

Recall that ξ is said to be nondegenerate if $|b(\xi)| \neq 0$; here and in the following $| \cdot |$ denotes the operation of taking the determinant.

Suppose that ξ is nondegenerate and parallel (n.p.). Then it follows from (2) that $g(\xi)$ is positive definite and it gives a Riemannian metric on M. The vector fields $\xi_1, \ldots, \xi_m$ form a local frame in $T(M)$ and we have the Gauss equation

$$(3) \qquad \xi_{ij} = \Gamma_{ij}^k(\xi)\xi_k - \sum_\alpha g_{ij}(\xi, N(\alpha))N(\alpha),$$

where $\Gamma_{ij}^k(\xi)$ are the Christoffel symbols of the second kind with respect to $g(\xi)$. It will be convenient to rewrite (3) in the form

$$(4) \qquad \xi_{ij} = \Gamma_{ij}^k(\xi)\xi_k - g_{ij}(\xi)\xi + \xi'_{ij},$$

where ξ'_{ij} is the part of the normal component of ξ_{ij} orthogonal to ξ.

3. Let Σ denote a unit hypersphere in E centered at the origin O. Suppose that ξ is a section of $N(M)$. Translating it parallel to itself in E to the origin O we get the spherical mapping $\gamma_\xi: M \to \Sigma$. If ξ is a n.p. section of $N(M)$, then γ_ξ is an immersion. If ξ is parallel and γ_ξ is an immersion, then ξ is nondegenerate.

4. The principal radii of curvature with respect to a n.p. section ξ of $N(M)$ are denoted by $R_{\xi 1}, \ldots, R_{\xi m}$ and defined as the roots of determinantal equation

$$|b(\xi)g^{-1}(\xi) - RI| = 0,$$

where I is the identity matrix, and $g^{-1}(\xi)$ is the inverse to $g(\xi)$. Since $g(\xi)$ is positive definite, the quantities $\{R_{\xi i}\}$ are well defined. Moreover, in this case they don't vanish. The elementary symmetric function of order k in $R_{\xi i}$ is defined as

$$S_k(\xi) = \sum_{i_e \neq i_r} R_{\xi i_1} \cdots R_{\xi i_k},$$

and it is the coefficient at $(-R)^{m-k}$ of the polynomial equation

$$(-R)^m + S_1(\xi)(-R)^{m-1} + \cdots + S_m(\xi)$$

$$\equiv |b(\xi)g^{-1}(\xi) - RI| = 0.$$

5. Let M be a compact submanifold of E with boundary or without. When the boundary ∂M is not void we always assume that it is a smooth $(m - 1)$ dimensional manifold.

For the rest of the paper we will study submanifolds which admit a globally defined, smooth n.p. section ξ of $N(M)$. Obviously, the length of ξ is constant and we suppose it to be equal to one.

It will be convenient to denote by $\mathfrak{M}$ the class of compact submanifolds admitting such sections.

Note (added in proof). The local structure of a submanifold admitting a parallel normal vector field is described in a paper by Yu. G. Lumiste and A. V. Chakmazyan, Submanifolds with a parallel normal vector field, *Izv. Vysh. Uch. Zav., Matematika*, **18** (1974), no. 5, 148–157; Engl. transl.: *Soviet Math. (Iz. Vuz.)*, **18**, no. 5. 123–131.

2. Deformation fields

1. Let $M \in \mathfrak{M}$ and ξ some fixed n.p. section of $N(M)$. Let Z be a smooth section of $T(M) \oplus N(M)$, $Z : M \to E$. Consider a map $\Phi^t : M \to E$ where

$$\Phi^t(x) = \Phi(x) + tZ(x), \qquad x \in M, \quad t \in [0, \epsilon], \quad \epsilon > 0.$$

Φ^t is called a deformation and Z-the deformation field of Φ^t. Assume that ϵ is small enough so that Φ^t is a family of immersions.

Definition 2.1. Let $x \in M$, and let $\bar{x}$ be the image of x under Φ^t. We say that a deformation Φ^t preserves ξ if after a parallel translation of ξ_x in E to the corresponding point $\bar{x}$ of the deformed submanifold, ξ_x remains in the normal bundle of Φ^t for all $x \in M$ and $t \in (0, \epsilon]$.

This definition can be reworded in terms of the spherical map γ_ξ. Namely, since ξ is in the normal bundle of Φ^t for all $t \in [0, \epsilon]$, we can consider

$$b_{ij}^t(\xi) = b_{ij}(\xi) + t\langle Z_{ij}, \xi \rangle,$$

and, clearly, $|b^t(\xi)| \neq 0$ because ξ is nondegenerate. Therefore, the mappings $\gamma_\xi^t : \Phi^t(M) \to \Sigma$ define a family of immersions of M into Σ. But ξ remains parallel to itself in E throughout the deformation, and thus $\gamma_\xi^t(M) = \gamma_\xi(M)$ for all t.

We denote by $\mathfrak{F}(M, \xi)$ the space of smooth deformations of M which preserve a fixed n.p. section ξ of $N(M)$.

PROPOSITION 2.1. *Let $M \in \mathfrak{M}$ and ξ a n.p. section of $N(M)$. A deformation Φ^t belongs to $\mathfrak{F}(M, \xi)$ if and only if its deformation field Z can be represented in the form*

$$(5) \qquad Z = g^{ij}(\xi) f_i(\xi) \xi_j + \sum_\alpha \langle Z, N(\alpha) \rangle N(\alpha).$$

where $(g^{ij}(\xi)) = g^{-1}(\xi)$, and $f(\xi) = \langle Z, \xi \rangle$.

Proof. Though this proposition has been proved in [10], we present its proof here because of its importance and also for completeness of exposition.

Denote by $f(\alpha) = \langle Z, N(\alpha) \rangle$. Then

$$f(\xi) = \langle Z, \xi \rangle = \sum_\alpha f(\alpha) \langle \xi, N(\alpha) \rangle.$$

Suppose that Z is such that Φ^t preserves ξ. Then

$$\langle \Phi_i^t, \xi \rangle = \langle \Phi_i, \xi \rangle + t \langle Z_i, \xi \rangle = 0,$$

and, therefore, $f_i(\xi) = \langle Z_i, \xi \rangle + \langle Z, \xi_i \rangle = \langle Z, \xi_i \rangle$. Now, since ξ_1, ..., $\xi_m, N(1), \ldots, N(n)$ form a local frame, a standard manipulation gives (5).

Conversely, let Z be represented in the form (5). It is clear that the functions $f(\alpha)$ are smooth, since Z is smooth. We have to show that $\langle Z_k, \xi \rangle = 0$ for all k. In view of (3) and because

$$f_k(\xi) = \sum_\alpha \left[f_k(\alpha) \langle \xi, N(\alpha) \rangle + f(\alpha) \langle \xi, N_k(\alpha) \rangle \right],$$

we get

$$\langle Z_k, \xi \rangle = g^{ij}(\xi) f_i(\xi) \langle \xi_{jk}, \xi \rangle + f_k(\xi)$$

$$= -g^{ij}(\xi) f_i(\xi) g_{jk}(\xi) + f_k(\xi) = 0.$$

Here we have also used the fact that $g_{jk}(\xi) = -\langle \xi_{jk}, \xi \rangle$. The proposition is proved.

COROLLARY. *Let* $M \in \mathfrak{M}$ *and* ξ *a n.p. section of* $N(M)$. *Suppose that there exists a deformation* $\Phi^t \in \mathfrak{F}(M, \xi)$ *and let* Z *be the deformation field of* Φ^t. *We have the following*:

a) *if the normal component of* Z *vanishes identically on* M, *then* $Z \equiv 0$;

b) *if the tangential component of* Z *vanishes identically on* M, *then* $f(\xi) = \langle Z, \xi \rangle = const$ *on* M. *In particular, if* $Z = p\xi$, *where* p *is a smooth function on* M, *then the deformation* Φ^t *is actually an equidistant displacement of points of* M *along the section* ξ.

Note that a deformation preserving ξ preserves also the metric $g(\xi)$. We will often make use of this comment.

2. In [15] Yano introduced the concept of parallel variations. In our terminology the latter means that a deformation Φ^t is parallel if the tangent spaces to $\Phi^t(M)$ are parallel at corresponding points for all t. We will make use of this notion for submanifolds of the class $\mathfrak{M}$.

So, let $M \in \mathfrak{M}$, and ξ a n.p. section of $N(M)$. Denote by $\mathfrak{P}(M)$ the space of parallel deformations of M. It is clear that $\mathfrak{P}(M) \subset \mathfrak{F}(M, \xi)$. Because of this inclusion (5) is also valid for $\Phi^t \in \mathfrak{P}(M)$. In general, the two spaces are different, but they coincide in the case when M is a hypersurface in E with nonvanishing Gaussian curvature.

Yano has also given a necessary and sufficient condition for a deformation to be parallel. Let $\{\xi = N(1), N(2), \ldots, N(n)\}$ be a local orthonormal basis in $N(M)$. If ξ is a n.p. section of $N(M)$, then a deformation is parallel if and only if its deformation field Z satisfies the equations

$$(6) \qquad g^{ij}(\xi) f_i(\xi) g_{js}(\xi, N(\beta)) + \sum_{\alpha} f(\alpha) \langle N(\alpha), N_s(\beta) \rangle - f_s(\beta) = 0,$$

$$s = 1, \cdots, m, \quad \beta = 2, \cdots, n.$$

When $\beta = 1$, the corresponding equations are satisfied identically. Also, in the second term we have $\langle N(\beta), N_s(\beta) \rangle = 0$ for any β and s. In the case where Z is parallel in $N(M)$, it is a deformation field of a parallel deformation [15].

3. Let M be a hypersurface with nonvanishing Gaussian curvature. It is a simple consequence of Proposition 2.1 that any smooth function f defined on the spherical image of M generates a deformation which preserves the unit normal vector field ξ (and, therefore, parallel) on M. Namely, put

$$Z = g^{ij}(\xi)f_i\xi_j + f\xi.$$

For submanifolds of codimension > 1 which admit a n.p. section ξ of the normal bundle the situation is just as simple provided we want to construct deformations preserving ξ. For parallel deformations the situation is more complicated, since the functions defining Z must satisfy the equations (6). Writing out the standard integrability conditions one can give the necessary and sufficient conditions for local existence of Z, or global, if M is simple connected. Such conditions have been recently given by Mutō [7]. In his paper Mutō does not require existence of a n.p. section of $N(M)$, but assumes the Gauss map of $\Phi^t(M)$ into Grassmann manifold $G(m, n)$ to be regular for all t. He also observed that a nontrivial deformation (that is, different from parallel displacement of homothety) can be obtained by taking a product immersion of two manifolds each of which is subject to a homothety.

Generalizing slightly this example one can construct parallel deformations of submanifolds of class $\mathfrak{M}$. Let M_1 and M_2 be two strictly convex hypersurfaces, and Z_1, Z_2 the deformation fields of M_1 and M_2, respectively, defined as in the beginning of this section. Consider the natural imbedding of $M = M_1 \times M_2$ in Euclidean space of dimension $= \dim M_1 + \dim M_2 + 2$. Clearly, $\xi = (\xi_1, \xi_2)$ is a n.p. section of $N(M)$, and, therefore, $M \in \mathfrak{M}$. The deformation field $Z = (Z_1, Z_2)$ generates a parallel deformation of M.

3. Infinitesimal deformations.

1. Let $M \in \mathfrak{M}$, and ξ a n.p. section of $N(M)$. Denote by G^t the metric on $\Phi^t(M)$, where $\Phi^t \in \mathfrak{F}(M, \xi)$.

Definition 3.1. A deformation $\Phi^t \in \mathfrak{F}(M, \xi)$ is called an (infinitesimal) isometric deformation if $(\partial/\partial t)G^t|_{t=0} = 0$.

Throughout the paper ∇_{ij} denotes the operator of the second covariant derivative in the metric $g(\xi)$. For a smooth function f on M

$$\nabla_{ij}f = f_{ij} - \Gamma^k_{ij}(\xi)f_k.$$

PROPOSITION 3.1. *Let* $M \in \mathfrak{M}$ *and* ξ *a n.p. section of* $N(M)$. *Let* $\Phi^t \in \mathfrak{F}(M, \xi)$. *Denote by* $f(\xi) = \langle Z, \xi \rangle$ *and* $f(\alpha) = \langle Z, N(\alpha) \rangle$. *Then the deformation* Φ^t *is isometric if and only if*

$$(7) \qquad \tilde{b}_i^l(\xi) \left[\nabla_{lj} f(\xi) + \sum_\alpha g_{lj}(\xi, N(\alpha)) f(\alpha) \right]$$

$$+ \tilde{b}_j^l(\xi) \left[\nabla_{li} f(\xi) + \sum_\alpha g_{li}(\xi, N(\alpha)) f(\alpha) \right] = 0,$$

where $\tilde{b}_j^l(\xi) = b_{is}(\xi) g^{sl}(\xi)$.

The proof of this proposition is obtained by straightforward computations combined with the following

PROPOSITION 3.2. *Let* M, ξ, *and* Φ^t *be as in Proposition* 3.1. *Then*

$$(8) \qquad \langle Z_k, \xi_l \rangle = \langle Z_l, \xi_k \rangle = -\langle Z_{kl}, \xi \rangle$$

$$= \nabla_{kl} f(\xi) + \sum_\alpha g_{kl}(\xi, N(\alpha)) f(\alpha)$$

$$= \nabla_{kl} f(\xi) + g_{kl}(\xi) f(\xi) - \langle \xi'_{kl}, Z \rangle.$$

Proof. Use the fact $\langle Z_k, \xi \rangle = 0$ and formula (3).

2. Let $b^{ij}(\xi)$ denote the elements of the matrix $b^{-1}(\xi)$. Transvecting (7) with $b^{ij}(\xi)$ we obtain

$$(9) \qquad \Delta_2 f(\xi) + \sum_\alpha g^{jl}(\xi) g_{lj}(\xi, N(\alpha)) f(\alpha) = 0,$$

where Δ_2 is the Laplace operator on M with respect to the metric $g(\xi)$. For the second term in (9) we have

$$(10) \qquad \sum_\alpha g^{jl}(\xi) g_{lj}(\xi, N(\alpha)) f(\alpha) = \langle H(\xi), Z \rangle,$$

where $-H(\xi)$ is the m times mean curvature vector of the submanifold $\gamma_\xi(M)$ in E. Thus, (9) assumes the form

$$(11) \qquad D_\xi(Z) \equiv \Delta_2 f(\xi) + \langle H(\xi), Z \rangle = 0.$$

Since $H(\xi) = m\xi + H'(\xi)$, where $-H'(\xi)$ is the m times mean curvature vector of $\gamma_\xi(M)$ in Σ, we can write

(12) $D_\xi(Z) = \Delta_2 f(\xi) + m\xi + \langle H'(\xi), Z \rangle.$

The operator $D_\xi(Z)$ is an invariant differential operator on M, and obviously any deformation field Z satisfying (7) will also satisfy the equation $D_\xi(Z) = 0$.

3. Let M and ξ be as before, and $S_k(\xi)$ the k-th elementary symmetric function of the principal radii of curvature. If Φ^t preserves ξ, then ξ remains nondegenerate for all small t, and we can consider elementary symmetric functions of principal radii of curvature with respect to ξ for the immersions Φ^t. Let $S_k^t(\xi)$ be such a function of order k.

Definition 3.2. Let $\Phi^t \in \mathfrak{F}(M, \xi)$. We say that Φ^t is a deformation (infinitesimal) with respect to $S_k(\xi)$ if

(13) $\left. \dfrac{\partial}{\partial t} S_k^t(\xi) \right|_{t=0} = 0.$

We derive now a differential equation corresponding to (13). The function $S_k(\xi)$ is the sum of determinants obtained by taking combinations of k columns with $b_{ij}(\xi)$ and $m - k$ columns with $g_{ij}(\xi)$ in $|g(\xi)|^{-1}|b(\xi) - Rg(\xi)|$ (see section 4, §1). Denote by $S_k^{ij}(\xi)$ the sum of the cofactors of the elements $b_{ij}(\xi)$ over all determinants defining $S_k(\xi)$.

PROPOSITION 3.3. *Let* $M \in \mathfrak{M}$ *and* $\Phi^t \in \mathfrak{F}(M, \xi)$ *for some n.p. section* ξ. *Then* Φ^t *is a deformation with respect to* $S_k(\xi)$ *if and only if*

(14) $\dfrac{S_k^{ij}(\xi)}{|g(\xi)|} \left(\nabla_{ij} f(\xi) + \sum_\alpha g_{ij}(\xi, N(\alpha)) f(\alpha) \right) = 0.$

Proof. The second fundamental form with respect to ξ of the immersion Φ^t is given

$b_{ij}^t(\xi) = b_{ij}(\xi) + t \langle Z_{ij}, \xi \rangle.$

The function $S_k^t(\xi)$ is constructed from $|g(\xi)|^{-1}|b^t(\xi) - Rg(\xi)|$. Writing our explicitly the expansion of $S_k^t(\xi)$ in t we find that the coefficient at t

is equal to $|g(\xi)|^{-1}S_k^{ij}\langle Z_{ij},\xi\rangle$. Applying now the proposition 3.2 we obtain (14).

Using the last part of the formula (8) we can represent the left-hand side of (14) in the form

$$(15) \qquad L_{\xi k}(Z) \equiv P_{\xi k}(f) + (m - k + 1)S_{k-1}(\xi)f(\xi) + \langle H_k'(\xi), Z\rangle,$$

where

$$P_{\xi k}(f) = \frac{S_k^{ij}(\xi)}{|g(\xi)|}\nabla_{ij}f(\xi), \qquad S_0(\xi) = 1,$$

and

$$H_k'(\xi) = -\frac{S_k^{ij}(\xi)}{|g(\xi)|}\xi_{ij}'.$$

$H_k'(\xi)$ is a uniquely defined section of $N(M) \ominus \xi$ on M. If $k = 1$, then $-H_1'(\xi)$ is the m times mean curvature vector of $\gamma_\xi(M) \subset \Sigma$, the same as $-H'(\xi)$ in (12). For more details concerning $H_k'(\xi)$ see [9]; the case $k = 1$ is also discussed in [2], p. 203. If we combine the last two terms in (15) we get

$$(16) \qquad L_{\xi k}(Z) = P_{\xi k}(f) + \langle H_k(\xi), Z\rangle,$$

where

$$H_k(\xi) = \sum_\alpha \frac{S_k^{ij}(\xi)}{|g(\xi)|}g_{ij}(\xi, N(\alpha))N(\alpha),$$

which is again a uniquely defined section of the normal bundle $N(M)$. Again, $H_1(\xi) = H(\xi)$.

Note also that

$$(17) \qquad P_{\xi 1}(f) = \Delta_2 f,$$

and, thus, formally $L_{\xi 1}(Z) \equiv D_\xi(Z)$.

4. A deformation Φ^t is called trivial if there exists a parallel translation of E such that the deformation field Z is the restriction of this parallel translation to M.

Let c be a constant vector in E. If we project c on $T_x(M) \oplus N_x(M)$, $x \in M$, we obtain a section C defined over M. Consider a deformation $\Phi' = \Phi + tC$. Obviously, it preserves ξ. Because of (3) we have

$$\nabla_{ij}\langle C, \xi \rangle = -\sum_\alpha g_{ij}(\xi, N(\alpha))\langle C, N(\alpha)\rangle.$$

Substituting it in the equation (7) or (14), we see that those equations are identically satisfied by the deformation field C. Thus, geometrically trivial deformations generate analytically nontrivial solutions of the above equations. In many cases the goal is to show that no other solutions exist. In such cases M is called rigid with respect to the corresponding type of deformation.

4. Elliptic immersions.

1. The differential equations of infinitesimal deformations we can investigate in more or less satisfactory fashion only in case when they are elliptic. Except for the case of operators $L_{\xi_1}(Z)$ and $D_\xi(Z)$, the fact that ξ is nondegenerate, does not guarantee it.

In order to describe ellipticity of differential operators $L_{\xi_k}(Z)$ for $k > 1$ in geometric terms we introduce

Definition 4.1. Let $M \in \mathfrak{M}$ and ξ a n.p. section of $N(M)$. We say that M is immersed elliptically in E, and Φ is called elliptic, if there exists at least one point on M where the second fundamental form is definite for any tangential direction.

We denote by $\mathfrak{M}_e$ the subset of $\mathfrak{M}$ of elliptically immersed submanifolds.

We note immediately that if Φ is an elliptic immersion with respect to some section ξ, then $b(\xi)$ is definite everywhere on M; otherwise ξ would not be nondegenerate. In certain special cases, for example, when codimension is one, the ellipticity is actually an implication of compactness of M and the fact that $b(\xi)$ is nondegenerate. In fact when M is a compact hypersurface without boundary and its standard unit normal vector field is nondegenerate, the condition $M \in \mathfrak{M}_e$ is automatically satisfied; the hypersurface M in this case is strictly convex (see [6], p. 41).

Note that the concept of ellipticity, used in [5], is different from ours.

Without loss of generality we may assume (and we always will) that if Φ is an elliptic immersion, then $b(\xi)$ is positive definite. Note also that since M is assumed to be compact, $b(\xi)$ is uniformly bounded away from zero.

2. Here we summarize some basic properties of the differential operators corresponding to deformations with respect to $S_k(\xi)$.

PROPOSITION 4.1. *Suppose that $M \in \mathfrak{M}$ and ξ is a n.p. section of $N(M)$. Then the quadratic form $S_1^{ij}(\xi)\nu_i\nu_j$ is positive definite and uniformly bounded away from zero on M; here $\nu_1, \dots, \nu_m$ are arbitrary real numbers such that $\nu^2 = \sum_i \nu_i^2 \neq 0$. Let now k be an integer, $2 \leqslant k \leqslant m$, and assume that $M \in \mathfrak{M}_e$ with respect to ξ. Then the quadratic forms $S_k^{ij}(\xi)\nu_i\nu_j$ are positive definite and uniformly bounded away from zero on M. Thus, in either case the operator $P_{\xi k}(\ )$ is uniformly elliptic on M.*

Since the section ξ is parallel, $b(\xi)$ satisfies the classical Codazzi equations. We can use this fact to show that the operators $P_{\xi k}(\ )$ are formally self-adjoint, that is,

$$P_{\xi k}(\ ) = \frac{1}{\sqrt{|g(\xi)|}} \partial_i \left[\frac{S_k^{ij}(\xi)}{\sqrt{|g(\xi)|}} \partial_j \right].$$

An analog of Hopf's Lemma holds also for the operators $P_{\xi k}(\ )$.

PROPOSITION 4.2. *Let $M \in \mathfrak{M}_e$ for some n.p. section ξ of $N(M)$, and $\partial M = \varnothing$. Let f be a smooth function on M such that for some k, $2 \leqslant k \leqslant m$, $P_{\xi k}(f)$ does not change its sign on M. Then f must be a constant. The assertion remains true for $k = 1$ without the assumption that M is immersed elliptically. The proposition also holds in case when $\partial M \neq \varnothing$, provided the derivative of f in conormal direction vanishes on ∂M.*

In case where $\partial M = \varnothing$ Propositions 4.1 and 4.2 were given in a slightly different form in [9].

Finally we note that all properties of $L_{\xi 1}(Z)$ stated in Propositions 4.1 and 4.2 are also valid for $D_\xi(Z)$.

3. In this section we indicate a few examples of elliptic immersions.

3.1. A piece of strictly convex hypersurface in E is imbedded elliptically with respect to its normal vector field.

3.2. Any submanifold of a hypersphere in E is an elliptically immersed submanifold with respect to its position vector field.

3.3. The natural imbeddings in Euclidean space of products of strictly convex hypersurfaces, as they were described in section 3, §2, give examples of elliptically imbedded submanifolds. In particular, such an imbedding of a product of spheres into a hypersphere is an elliptic imbedding.

3.4. Other examples can be obtained by a local parallel deformation of an already given elliptically immersed submanifold.

5. Results on global rigidity.

1. There are essentially two different ways in which one can investigate deformations. On one hand we can ask the question of what are the special properties of the submanifold M that imply existence of only special types of deformations. For example, it is known that a closed convex hypersurface does not admit deformations with respect to $S_k(\xi)$ except for those which are parallel translations of M (see [13], p. 47). Another approach is to impose certain additional restrictions on the deformation field Z and observe the implications for both the submanifold and the deformations Φ^t. An example of this approach is the Corollary to Proposition 2.1. In most cases we combine, of course, both approaches together.

2. The operators $L_{\xi k}(Z)$ are geometrically more significant than $D_\xi(Z)$ because any solution of the equation $L_{\xi k}(Z) = 0$ generates a deformation of the submanifold. There is no such correspondence between deformations and solutions of the equation $D_\xi(Z) = 0$. However, the results of this section are concerned with situations where these equations have only solutions corresponding to trivial deformations. In such circumstances there is no need to distinguish between $D_\xi(Z)$ and $L_{\xi 1}(Z)$, since formally they coincide. For that reason the results below are stated only for deformations with respect to $S_k(\xi)$, but everything that holds for deformations with respect to $S_1(\xi)$ also holds for isometric deformations.

3. For a deformation field Z denote by $\overline{Z}$ the component of Z in $T(M) \oplus \xi$, and by $\overline{\overline{Z}}$ the tangential component.

PROPOSITION 5.1. *Let $M \in \mathfrak{M}$, $\partial M = \varnothing$, ξ a n.p. section of $N(M)$, and $\Phi^t \in \mathfrak{F}(M, \xi)$ a deformation of M with respect to $S_k(\xi)$ for some k. If*

$k > 1$, *assume that* $M \in \mathfrak{M}_e$. *If, in addition,* $\langle H_k(\xi), Z \rangle \leqslant 0$ *or* $\langle H_k(\xi),$ $Z \rangle \geqslant 0$ *on* M *then* $\overline{\overline{Z}} = c\xi$, *where* $c = const$; *that is, in the direction* ξ Φ^t *amounts to an equidistant displacement. If* $\partial M \neq \emptyset$ *and one of the above inequalities is satisfied in* $M \setminus \partial M$, *then the assertion is still true provided* $\overline{\overline{Z}} = 0$ *on* ∂M.

Proof. The equation for the deformation field Z in the form (16) is

$$L_{\xi k}(Z) = P_{\xi k}(f) + \langle H_k(\xi), Z \rangle = 0$$

on M (or $M \setminus \partial M$ if $\partial M \neq \emptyset$),

where $f = \langle Z, \xi \rangle$. Now the assertion follows from Proposition 4.2 and the Corollary to Proposition 2.1.

COROLLARY 5.1. *Let* M, ξ, *and* Φ^t *be as before and* $\partial M = \emptyset$. *If* $\partial M \neq \emptyset$ *assume that* $\overline{\overline{Z}} = 0$ *on* ∂M. *Then the component* $\overline{Z}$ *of* Z *is uniquely determined* (*up to an additive constant in* $\langle Z, \xi \rangle$) *by projections of* Z *on* $H_k(\xi)$. *In other words, two deformations having equal projections on* $H_k(\xi)$ *must have equal tangential components and equal projections on* ξ (*up to an additive constant*).

4. Here we will show that under certain additional conditions Z is determined by its component in $N(M) \ominus \xi$.

THEOREM 5.1. *Let* $M \in \mathfrak{M}$, $\partial M \neq \emptyset$, ξ *a n.p. section of* $N(M)$, *and* $\Phi^t \in \mathfrak{F}(M, \xi)$ *a deformation of* M *with respect to* $S_k(\xi)$ *for some* k. *When* $k > 1$ *assume that* $M \in \mathfrak{M}_e$. *Then the deformation field* Z *is uniquely determined by its component in* $N_x(M) \ominus \xi_x$, $x \in M \setminus \partial M$, *and its values along* ∂M.

Proof. Let $f = \langle Z, \xi \rangle$. Then the equation (14) of deformations with respect to $S_k(\xi)$ can be written with the use of (15) as

$$L_{\xi k}(Z) \equiv P_{\xi k}(f) + (m - k + 1)S_{k-1}(\xi)f + \langle H_k'(\xi), Z \rangle = 0$$

on $M \setminus \partial M$,

where $S_{k-1}(\xi) > 0$.

Let $\hat{Z}$ be a deformation field on M with the same properties as Z and such that its component in $N(M) \ominus \xi$ coincides with the corresponding component of Z. Put $\hat{f} = \langle \hat{Z}, \xi \rangle$. Then the function $\varphi = \hat{f} - f$

satisfies the equation

(18) $P_{\xi k}(\varphi) + (m - k + 1)S_{k-1}(\xi)\varphi = 0$ on $M \setminus \partial M$.

If the values of Z coincide with that of $\hat{Z}$ along ∂M, then by Proposition 2.1 we have

(19) $\varphi|_{\partial M} = 0$,

(20) $\text{grad } \varphi|_{\partial M} = 0$,

(21) $(Z' - \hat{Z}')|_{\partial M} = 0$,

where the gradient in (20) is taken in the metric $g(\xi)$, and " $'$ " in (21) denotes the corresponding components in $N(M) \ominus \xi$. The principal part of (18) is a uniformly elliptic self-adjoint operator. If the equation (18) has a smooth nonzero solution satisfying (19), then it means that this solution is an eigenfunction for (18), (19). There exist only a finite number of linearly independent eigenfunctions of (18), (19) and any other solution is a linear combination of those eigenfunctions. By a theorem of Courant on nodal domains the nodal set of φ divides M into a finite number of subdomains, such that in each of them the corresponding restriction of the solution preserves its sign inside it, and vanishes only on its relative boundary (see, for example, [3]). Pick any such domain whose boundary contains a portion of ∂M. Then there must exist a point x_0 on ∂M where the normal derivative of φ is different from zero. The normal direction here is understood as a direction in $T_{x_0}(M)$ orthogonal to $T_{x_0}(\partial M)$. But because of the condition (20) $\text{grad } \varphi = 0$ everywhere on ∂M. Thus, we arrive at a contradiction, and $\varphi \equiv 0$. The Theorem is proved.

COROLLARY 5.2. *Let M and ξ be as in Theorem 5.1. For any function $S_k(\xi)$, $k = 1, \cdots, m$, the submanifold M admits no deformations with respect to $S_k(\xi)$ preserving ξ and such that the deformation field Z lies in the space $T_x(M) \oplus \xi_x$, $x \in M$, and leaves the boundary ∂M fixed. (It is assumed that $M \in \mathfrak{M}_e$ in case $k > 1$.)*

COROLLARY 5.3. *Let M, ξ, and Φ^t be as before. Let C be a deformation field on M generated by a constant vector in E. Suppose the component Z' of Z lying in $N(M) \ominus \xi$ coincides with the corresponding* *component of C everywhere on $M \setminus \partial M$, and, in addition, $Z = C$ on ∂M. Then $Z = C$ everywhere on M.*

5. Our next result is concerned with parallel deformations.

THEOREM 5.2. *Let $M \in \mathfrak{M}$, $\partial M = \varnothing$, ξ a n.p. section of $N(M)$, and Φ', $\hat{\Phi}' \in \mathfrak{P}(M, \xi)$ two deformations of M with respect to $S_k(\xi)$ for some $k = 1, \cdots, m - 1$. If $k > 1$ assume that $M \in \mathfrak{M}_e$. Suppose the deformation fields Z and $\hat{Z}$ of Φ' and $\hat{\Phi}'$ correspondingly are such that their components Z' and $\hat{Z}'$ in $N(M) \ominus \xi$ coincide. Then $\mathbf{Z} = \hat{Z} - Z$ is a constant section of $T(M) \oplus \xi$, that is, $\mathbf{Z}_i = 0$. Moreover, if $\mathbf{Z}$ is not identically zero, then M with the metric $g(\xi)$ is isometrically diffeomorphic to a unit hypersphere σ in Euclidean space E^{m+1}, and*

$$\mathbf{Z} = g^{ij}(\xi)\varphi_i\xi_j + \varphi\xi,$$

where φ is a linear combination of spherical harmonics of the first order on σ.

Proof. Put $f = \langle Z, \xi \rangle$ and $\hat{f} = \langle \hat{Z}, \xi \rangle$. Then the function $\varphi = \hat{f} - f$ satisfies the equation

$$(22) \qquad \frac{S_k^{ij}(\xi)}{|g(\xi)|} \left(\nabla_{ij}\varphi + g_{ij}(\xi)\varphi \right) = 0.$$

Let q denote the tensor whose representation in local coordinates is $q_{ij} = \nabla_{ij}\varphi + g_{ij}(\xi)\varphi$. Then the expression $S_k^{ij}(\xi)q_{ij}$, up to a factor, is the mixed discriminant of the forms $b(\xi)$, $g(\xi)$, and q, and the first two are positive definite. Differential expressions of this form were studied in [11]. From the main theorem there it follows that if q is a Codazzi tensor (i.e., satisfies the classical Codazzi equations) with respect to $g(\xi)$, then (22) implies that $q \equiv 0$. Since Z and $\hat{Z}$ are parallel in $N(M)$, $\langle Z_i, \xi_j \rangle$ and $\langle \hat{Z}_i, \xi_j \rangle$ satisfy the Codazzi equations. From here, Proposition 3.2, and the fact that $Z' = \hat{Z}'$ we conclude that $q_{ij} = \langle Z_i - \hat{Z}_i, \xi_j \rangle$ is a Codazzi tensor, and therefore, the function φ satisfies on M the following system of equations

$$(23) \qquad \nabla_{ij}\varphi + g_{ij}(\xi)\varphi = 0.$$

Obviously $\varphi \neq \text{const}$, unless $\varphi \equiv 0$. Let $\varphi \neq 0$. Since $\varphi = \langle \hat{Z} - Z, \xi \rangle$, it follows from Proposition 2.1 that $\mathbf{Z} \equiv \hat{Z} - Z = g^{ij}(\xi)\varphi_i\xi_j + \varphi\xi$.

For the derivative of $\mathbf{Z}$ in E, we have $\mathbf{Z}_s = \nabla_s \mathbf{Z} + \nabla_s^\perp \mathbf{Z}$, where ∇_s is the derivative in $g(\xi)$, and $\nabla_s^\perp$ is the derivative in $N(M)$. The sections Z and $\hat{Z}$ are parallel in $N(M)$, hence $\nabla_s^\perp \mathbf{Z} = 0$. On the other hand, $\nabla_s \mathbf{Z} = g^{ij}(\xi)(\nabla_{si}\varphi)\xi_j + \varphi\xi_s = 0$ in view of (23). Thus, $\mathbf{Z} = \text{const}$. Moreover, it is known, Obata [8] (see also [1], p. 180), that existence of a solution to (23) implies that M is isometrically diffeomorphic to a unit hypersphere in Euclidean space E^{m+1}, and then any such solution is a linear combination of $m + 1$ linearly independent spherical harmonics of the first order. This completes the proof of the Theorem.

A "rigidity" result similar to Corollary 5.3 can be now stated for submanifolds without boundaries.

COROLLARY 5.4. *Let* $M \in \mathfrak{M}$, $\partial M = \emptyset$, *and* ξ *a n.p. section of* $N(M)$. *Let* $\Phi^t \in \mathfrak{P}(M, \xi)$ *be a deformation with respect to* $S_k(\xi)$ *for some* k, $k = 1, \cdots, m - 1$, *whose deformation field lies in* $T_x(M) \oplus \xi_x$, $x \in M$. *If* $k > 1$, *assume that* $M \in \mathfrak{M}_e$. *Then this deformation field is constant.*

COROLLARY 5.5. *Let the conditions of Corollary 5.4 be satisfied. Assume, in addition, that the genus of* M *is different from zero. Then the deformation field* Z *vanishes identically on* M.

6. When M is a submanifold of codimension 2, Theorem 5.2 can be improved in the following way.

THEOREM 5.3. *Suppose that* $M \in \mathfrak{M}$, $\partial M = \emptyset$, *and the codimension of* M *in* E *is equal to two. Let* ξ *be a n.p. section of* $N(M)$ *and* $\Phi^t \in \mathfrak{P}(M, \xi)$ *a deformation with respect to* $S_k(\xi)$ *for some* $k = 1$, $\cdots, m - 1$. *Assume that* $M \in \mathfrak{M}_e$, *if* $k > 1$. *Furthermore, suppose that the deformation field* Z *of* Φ^t *lies in* $T_x(M) \oplus \xi_x$, $x \in M$. *Then either* Z *vanishes identically on* M, *or* M *is contained in a hyperplane of* E, *and* Z *is a restriction to* M *of a parallel translation in that hyperplane.*

Proof. From the proof of Theorem 5.2 it follows that at most $Z = \mathbf{Z}$. If M lies in a hyperplane, then $\mathbf{Z} = \text{constant}$ vector. Assume now that M is not contained in any hyperplane of E. Then if the function $\varphi = \langle Z, \xi \rangle$ is not identically zero, it follows from the proof of Theorem 5.2 that M, equipped with the metric $g(\xi)$, is isometrically diffeomorphic to a unit hypersphere σ in Euclidean space E^{m+1}. The spherical image $\gamma_\xi(M)$ on the hypersphere Σ is also isometrically diffeomorphic to σ. On the other hand, it is known [12] that a unit hypersphere can be immersed isometrically in Σ only as a totally geodesic submanifold, that is, $\gamma_\xi(M)$ is a great hypersphere of Σ. Since ξ is nondegenerate, it can be shown [9] that the position vector field of M lies in the space spanned by $\{\xi_1, \ldots, \xi_m, \xi\}$, that is, M lies in a hyperplane of E. This contradiction proves the theorem.

THE UNIVERSITY OF IOWA

REFERENCES

[1] Berger, M., Gauduchon, P., Mazet, E.: Le Spectre d'une Variété Riemannienne, *Lecture Notes in Mathematics*, Berlin-Heidelberg-New York: Springer Verlag, 1971.

[2] Chen, B.-Y.: Geometry of Submanifolds, New York: Marcel Dekker, Inc., 1973.

[3] Cheng, S. Y.: Eigenfunctions and nodal sets, *Comment. Math. Helvetici* 51 (1976), 43–55.

[4] ———, Yau, S. T.: On the regularity of the solution of n-dimensional Minkowski problem, *Comm. Pure Appl. Math.* 29 (1976), 495–516.

[5] Kaneda, E., Tanaka, N.: Rigidity for isometric imbeddings, *J. of Math. of Kyoto Univ.* 18, no. 1 (1978), 1–70.

[6] Kobayashi, S., Nomizu, K.: Foundations of Differential Geometry, vol. 2, New York, London, Sydney, Interscience, 1969.

[7] Mutō, Y.: Deformability of a submanifold in a Euclidean space whose image by the Gauss map is fixed, *Proceedings of the A.M.S.* 76, no. 1 (1979), 140–144.

[8] Obata, M.: Certain conditions of a Riemannian manifold to be isometric with a sphere, *J. Math. Soc. Japan* 14 (1962), 333–340.

[9] Oliker, V. I.: Submanifolds with nondegenerate parallel normal vector fields, *Pacific J. Math.* 82, no. 1 (1979), 58–71.

[10] ———: Infinitesimal deformations preserving parallel normal vector fields, *Lecture Notes in Mathematics*, v. 792, Berlin-Heidelberg-New York: Springer-Verlag, 1979, 383–405.

[11] ———: An application of A. D. Aleksandrov's inequality to the problem of characterization of spheres, Proceedings of the Colloquium on Global Differential Geometry/Global Analysis, Berlin, 1979, *Lecture Notes in Mathematics*, v. 838, Berlin-Heidelberg-New York: Springer-Verlag, 1981, 275–288.

[12] O'Neill, B., Stiel, E.: Isometric immersions of constant curvature manifolds, *Michigan Math. J.* 10 (1963), 335–339.

[13] Pogorelov, A. V.: Multidimensional Minkowski Problem (translated from Russian), New York: John Wiley, 1978.

[14] Tenenblat, K.: On infinitesimal isometric deformations, *Proceedings of the A.M.S.* 75, no. 2 (1979), 269–275.

[15] Yano, K.: Infinitesimal variations of submanifolds, *Kodai Math. J.* 1 (1978), 30–44.

ON A CRITERION OF INSTABILITY FOR NONAUTONOMOUS ORDINARY DIFFERENTIAL EQUATIONS

By Nelson Onuchic and Luiz Carlos Pavlu

---

1. Introduction. The main objective of this work is to extend the criterion of instability given by Onuchic in [5] to a certain class of nonautonomous ordinary differential equations.

In order to obtain the result above mentioned we basically use the invariance properties of ω-limit sets of solutions, bounded in the future, of differential equations.

We will consider the system of ordinary differential equations of the form:

(PE) $\dot{x} = P(t,x) + S(t,x) + Q(t,x)$

where we assume that the following hypotheses are satisfied:

$P(t,x)$ is defined and continuous on $R \times D$ where D is a region, $D \subset R^n$.

Here R^n denotes a normed, real n-dimensional vector space with any convenient norm $|.|$.

(a) For each fixed $x \in D$, $P(t,x)$ is bounded in $t \in R$.

(b) For each compact subset $K \subset D$, $P(t,x)$ is uniformly continuous on the set $R \times K$.

(H_1) $Q(t,x)$ is continuous for $t \geq 0$ and $x \in D$.

Furthermore, for every compact subset B of D and every continuous function $x(t) \in B$ defined on $[0, \infty)$ we have

$$\int_s^{s+t} Q(\tau, x(\tau)) \, d\tau \to 0 \qquad \text{as } s \to \infty$$

uniformly for t on $[0, 1]$.

(H_2) $S(t,x)$ is continuous on $[0, \infty) \times D$. Let A be a closed fixed subset of D. Assume that $S(t,x)$ satisfies the following property with respect to the set A:

For each $\epsilon > 0$ and each compact subset K of D there correspond $\delta = \delta(\epsilon, K) > 0$ and $T_0 = T_0(\epsilon, K)$ so that $t \geqslant T_0$, $x \in K$ and $\text{dist}(x, A) < \delta$ imply $|S(t, x)| < \epsilon$.

Under these assumptions the system (PE) may be considered as a perturbation of the system

(E) $\dot{x} = P(t, x)$

Markus [3] and Opial [9] obtain results on the asymptotic behavior of perturbed systems of autonomous systems by relating the method of Liapunov with the concept of an invariant set.

La Salle in [2] generalizes the concept of invariant set to periodic systems.

Miller in [4] generalizes this concept to perturbed systems of uniformly almost periodic systems.

Onuchic in [6] applies Miller's results. However he considers different hypotheses about the perturbations.

Rodrigues in [10] extended Miller's results to a class of systems of the form (PE), where P, Q and S satisfy the hypotheses mentioned above, by using an idea suggested by Sell in [11].

In 2 we will present the necessary results about invariance and in 3 we will present our main result: a criterion of instability for nonautonomous differential equations.

Finally, in 4, we will make an application of interest.

2. Invariance for Ordinary Differential Equations. The following conventions will be used here:

A function $x(t)$ defined in $I = [0, \infty)$ with values in D is called bounded in D if there is a compact subset $B \subset D$ such that $x(t) \in B$ for all $t \in I$.

Let $x(t)$ be a continuous function defined in the future, that is, for all $t \geqslant$ some t_0.

A point p of R^n is said to be an ω-limit point of $x(t)$ if there exists a sequence $\{t_m\}$, $t_m \to \infty$ as $m \to \infty$, such that $x(t_m) \to p$ as $m \to \infty$.

The set of all ω-limit points of $x(t)$ is denoted by Ω, $(\Omega(x(\cdot)))$, and is called the ω-limit set of $x(t)$.

If $x(t)$ is bounded in the future, that is, $x(t)$ is bounded in some interval $[a, \infty)$, $a > -\infty$, it is easily seen that Ω is a nonempty, connected and compact set with $x(t) \to \Omega$ as $t \to \infty$, that is, dist$(x(t), \Omega) \to 0$ as $t \to \infty$.

In order to generalize the concept of an invariant set to systems of the form (E) where $P(t, x)$ satisfies (a) and (b), the following results are necessary:

If $P(t, x)$ satisfies (a) and (b), then $P(t, x)$ is bounded on sets $R \times K$, for each compact subset $K \subset D$.

Let $E = \mathcal{C}(R \times D, R^n)$ be the set of all continuous functions from $R \times D$ into R^n with the compact open topology.

If $P \in E$ we have:

$$P : (t, x) \in R \times D \to P(t, x) \in R^n$$

Let $\tau \in R$. We define $P_\tau \in E$ by $P_\tau(t, x) = P(t + \tau, x)$.

If $P \in E$, let $\mathcal{P} = \{P_\tau \in E \mid \tau \in R\} \subset E$ and let us state the conditions:

(i) For every $(t_0, x_0) \in R \times D$, the set $\{P_\tau(t_0, x_0) \mid \tau \in R\}$ is relatively compact set in R^n.

(ii) $\mathcal{P}$ is equicontinuous in $R \times D$.

PROPOSITION. *The conditions* (a) *and* (b) *are respectively equivalent to* (i) *and* (ii).

COROLLARY 1. *If* $P \in E$ *satisfies* (a) *and* (b), $\mathcal{P}$ *is relatively compact in* E.

COROLLARY 2. *Let* $P \in E$ *with the hypotheses* (a) *and* (b).

Given any real sequence $\{\tau_m\}$, *there is a subsequence* $\{\tau_{m_j}\}$ *and a function* $P^* \in E$, *such that* $P_{\tau_{m_j}} \to P^*$, *on* E, *as* $j \to \infty$, *that is,* $P(t + \tau_{m_j}, x) \to P^*(t, x)$ *uniformly on* $[-l, l] \times M$, *for every* $l > 0$, $l \in R$ *and for every compact subset* $M \subset D$.

Furthermore, P^* *satisfies* (a) *and* (b).

Let $P \in E$ *and* $\mathcal{P} = \{P_\tau \in E \mid \tau \in R\}$ *and let* $\overline{\mathcal{P}}$ *be the closure of* $\mathcal{P}$ *in* E.

We define $\hat{\Omega}(P) = \{P^* \in E \mid \exists \text{ seq}\{t_m\}, \ t_m \to \infty$ *as* $m \to \infty$ *and* $P_{t_m} \to P^*$, *in* E, *as* $m \to \infty\}$.

It is easily seen that $\hat{\Omega}(P) \subset \overline{\mathcal{P}}$.

THEOREM 1. *Let* (a), (b), (H_1) *and* (H_2) *be satisfied. Let* $x(t)$ *be a solution of* (PE) *which is defined and bounded in the future, in* D. *Let* $x(t) \to A$ *as* $t \to \infty$. *Then for each point* $z \in \Omega(x(\cdot))$ *there exists a* *function* $P^* \in \hat{\Omega}(P)$, *a sequence* $t_m \to \infty$ *as* $m \to \infty$ *and a function* $y(t)$ *such that*:

1) $y(0) = z$, $\dot{y}(t) = P^*(t, y(t))$ $\forall t \in R$.

2) $x(t + t_m) \to y(t)$ *as* $m \to \infty$ *uniformly on compact subsets of* R.

3) $P_{t_m} \to P^*$, *in* E, *as* $m \to \infty$.

Remarks. The above Proposition, its Corollaries and Theorem 1 were obtained by Rodrigues in [10] by carrying out an idea suggested by Sell in [11].

Theorem 1 generalizes Miller's Theorem 1 in [4]. In fact, if $P(t, x)$ is almost periodic in t, uniformly in x, then $P(t, x)$ satisfies (a) and (b). The next theorem is another form of Theorem 1 and will give us a natural way of defining semi-invariance with respect to system (E).

THEOREM 2. *Let* (a), (b), (H_1) *and* (H_2) *be satisfied. Let* $x(t)$, *bounded in the future in* D, *be a solution of* (PE). *Then the set* $\Omega(x(\cdot))$ *is nonempty, and for each* $z \in \Omega(x(\cdot))$ *there exists a function* $P^* \in \hat{\Omega}(P)$, *a sequence* $t_m \to \infty$ *as* $m \to \infty$ *and a function* $y(t)$ *such that*:

α) $\dot{y}(t) = P^*(t, y(t))$ $\forall t \in R$ *and* $y(0) = z$.

β) $x(t + t_m) \to y(t)$ *uniformly in compact subset of* R, *as* $m \to \infty$, *and* $y(t) \in \Omega(x(\cdot))$ $\forall t \in R$.

γ) $P_{t_m} \to P^*$, *in* E, *as* $m \to \infty$.

δ) $x(t) \to \Omega(x(\cdot))$, *as* $t \to \infty$.

The conclusion of Theorem 2 suggests a convenient definition of semi-invariance with respect to system (E).

This definition is given as follows:

If M is a subset of D, then M is called semi-invariant with respect to the system

(E) $\dot{y} = P(t, y)$

if and only if for each $z \in M$ there exists a function $P^* \in \hat{\Omega}(P)$ and a solution $y(t)$ of the system

(E*) $\dot{y} = P^*(t, y)$

with $y(0) = z$, such that $y(t)$ exists and remains in M for $-\infty < t < \infty$.

COROLLARY 3. *If $x(t)$ is a solution of* (PE) *bounded in the future in* D, *then* $\Omega(x(\cdot))$ *is semi-invariant with respect to the system* (E).

The proof follows from Theorem 2.

Remark. When $P(t, y) = P(y)$, the system (E) is an autonomous system; we see that $\hat{\Omega}(P) = \{P\}$ and hence the definition of semi-invariance given for autonomous systems is a particular case of the above definition because we necessarily have $P^* = P$.

COROLLARY 4. *Consider the system* (E) *and* (PE) *with* $S(t, x) \equiv 0$. *Let* (a), (b) *and* (H$_1$) *be satisfied.*

If $x(t)$ *is a solution of* (PE), *bounded in the future, then* $\Omega(x(\cdot))$ *is a semi-invariant set with respect to the system* (E).

COROLLARY 5. *Consider the system* (E) *and let* (a) *and* (b) *be satisfied.*

If $y(t)$ *is a solution of* (E) *bounded in the future, then* $\Omega(y(\cdot))$ *is a semi-invariant set with respect to the system* (E).

3. Sufficient Conditions for Instability. Consider the differential systems:

(E) $\dot{y} = P(t, y)$

(E*) $\dot{y} = P^*(t, y)$

(EP$_1$) $\dot{x} = P(t, x) + Q(t, x)$

where $P : R \times \Gamma_H \to R^n$ is continuous, $\Gamma_H = \{x \in R^n \mid |x| < H, 0 < H \leqslant \infty\}$, and satisfies:

(a) For each fixed x in Γ_H, $P(t, x)$ is bounded in $t \in R$.

(b) For each compact subset $K \subset \Gamma_H$, $P(t, x)$ is uniformly continuous on the set $R \times K$, and furthermore $Q(t, x)$ satisfies:

(H$_1$) $Q : I \times \Gamma_H \to R^n$ is continuous and for every compact subset B of Γ_H and every continuous function $x(t) \in B$ defined on I it follows that

$$\int_s^{s+t} Q(\tau, x(\tau)) \, d\tau \to 0 \qquad \text{as } s \to \infty$$

uniformly for t on $[0, 1]$.

THEOREM 3. *Let* $P(t, 0) \equiv 0$ *and* $G(t, 0) \equiv 0$ *and suppose that* (a), (b) *and* (H$_1$) *are satisfied with respect to the system* (EP$_1$).

Suppose that there is a function $V = V(x)$ *continuous on* Γ_H *and there is an open set* U *of* Γ_H *such that the following conditions are satisfied:*

1) $V(x) > 0$ *on* U *and* $V(x) = 0$ *on* ∂U *where* ∂U *denotes the boundary of* U *with respect to* Γ_H.

2) $0 \in \partial U$.

3) $\dot{V}_{(EP_1)}(x) = W(t, x) \geqslant 0$ *on* $I \times U$.

4) *There does not exist* $y^*(t)$, *solution of* (E*), *bounded in* U *for* $-\infty < t < \infty$, *such that* $V(y^*(t)) = k$ *(constant) for all* $t \in R$.

Then, for each $x_0 \in U$, $t_0 \in I$ *and each solution* $x(t, t_0, x_0)$ *of* (EP_1), *there is a sequence* $\{t_m\}$, *such that* $|x(t_m, t_0, x_0)| \to H$ *as* $m \to \infty$, *and hence, by* (2), *the solution* $x(t) \equiv 0$ *of* (EP_1) *is unstable.*

Proof.   Assume the contrary.

Then there exist $(t_0, x_0) \in I \times U$, H_1, $0 < H_1 < H$, and a solution $\bar{x}(t, t_0, x_0)$ of (EP_1), such that $|\bar{x}(t, t_0, x_0)| \leqslant H_1$ for all $t \geqslant t_0$.

We claim that $\bar{x}(t, t_0, x_0) \in U$ for all $t \geqslant t_0$.

Indeed, if this were not the case, there would exist $\bar{t} > t_0$ such that $\bar{x}(\bar{t}, t_0, x_0) \in \partial U$ and $\bar{x}(t, t_0, x_0) \in U$ for $t_0 \leqslant t < \bar{t}$.

But this would imply $V(\bar{x}(t, t_0, x_0)) > 0$ for $t_0 \leqslant t < \bar{t}$ and $V(\bar{x}(\bar{t}, t_0, x_0)) = 0$ which is impossible since by (3) $V(\bar{x}(t, t_0, x_0))$ must be nondecreasing on $[t_0, \bar{t})$. Then $\bar{x}(t, t_0, x_0) \in U$ for $t_0 \leqslant t < \infty$ and we have that $\Omega(\bar{x}(\cdot))$ is a nonempty, connected and compact set.

Since $\bar{x}(t, t_0, x_0) \in U$ and $V(\bar{x}(t, t_0, x_0)) \geqslant V(x_0) > 0$ for all $t \geqslant t_0$, it follows that $\Omega(\bar{x}(\cdot)) \subset U$.

Since $V(\bar{x}(t, t_0, x_0))$ is nondecreasing and $\Omega(\bar{x}(\cdot))$ is nonempty it follows that $V(\bar{x}(t, t_0, x_0)) \to \xi$, $\xi \in R$, as $t \to \infty$.

Consequently $V(x)$ is a constant function on $\Omega(\bar{x}(\cdot))$.

From Corollary 4 $\Omega(\bar{x}(\cdot))$ is a semi-invariant set; hence, from the definition of semi-invariant set, given $y_0 \in \Omega(\bar{x}(\cdot))$, there exist $P^* \in \hat{\Omega}(P)$ and a solution $y^*(t)$ of the system (E*) $\dot{y} = P^*(t, y)$ with $y^*(0) = y_0$, such that $y^*(t)$ exists and remains in $\Omega(\bar{x}(\cdot))$ for $-\infty < t < \infty$.

Then $V(y^*(t)) = \xi \; \forall t \in R$ and, taking into account hypothesis (4) we have a contradiction.

The proof is complete.

Remarks.   The criterion of Instability presented here is more general than the ones due to Onuchic in [5] or Hale in [1], when their results are applied to ordinary differential equations.

The Theorem of Cetaev for autonomous systems is a particular case of Theorem 3.

Indeed, consider the autonomous system

(5) $\dot{x} = f(x)$

where f is continuous in Γ_H.

Let $f(0) = 0$. Suppose that there is $V(x)$ continuous on Γ_H and that there is an open set U of Γ_H such that the following conditions are satisfied:

(i) $V(x) > 0$ on U and $V(x) = 0$ on ∂U.
(ii) $0 \in \partial U$.
(iii) $\dot{V}_{(5)}(x) > 0$ on U.

Then the solution $x = 0$ of (5) is unstable.

The criterion presented here in Theorem 3 can be extended, to a certain class of delay-differential equations.

4. Application.   We will need the following preliminary

LEMMA.   *Let $f(x)$ be continuous on $(-\rho_0, \rho_0)$, $0 < \rho_0 \leqslant \infty$ and suppose that $xf(x) \geqslant 0$ for all x, $|x| < \rho_0$.*

Let $x(t)$ be a nonconstant solution of

(6) $\ddot{x} = f(x)$

defined for all $t \in R$.

Then, there does not exist ρ, $0 < \rho < \rho_0$, such that $|x(t)| \leqslant \rho$ for every $t \in R$.

The proof is done by using elementary techniques.

THEOREM 4.   *Consider the second order differential equation*

(7) $\ddot{x} = h(t, x, \dot{x})\dot{x} + f(x) + g(t, x, \dot{x})$

or equivalently:

(7′) $\begin{cases} \dot{x} = y \\ \dot{y} = h(t, x, y)y + f(x) + g(t, x, y) \end{cases}$

and the following set of assumptions on the function in (7′).

Let $\beta_{\rho_0}: \{(x, y) \in R^2 | |x| + |y| < \rho\}, 0 < \rho_0 \leqslant \infty$ and
(h_1) $f(x)$ is a continuous function and $xf(x) \geqslant 0$ for all $x \in (-\rho_0, \rho_0)$.

(h_2) $h(t, x, y)$ *is a continuous nonnegative function on* $R \times \beta_{\rho_0}$ *and satisfies*:

a) $h(t, x, y)$ *is bounded in* $t \in R$ *for each fixed* $(x, y) \in \beta_{\rho_0}$.

b) $h(t, x, y)$ *is uniformly continuous on* $R \times B$ *where* B *is any compact subset of* β_{ρ_0}.

(h_3) $g(t, x, y)$ *is continuous and* $yg(t, x, y) \geqslant 0$ *for all* $t \geqslant 0$, (x, y) $\in \beta_{\rho_0}$. *Furthermore for every compact subset* B *of* β_{ρ_0} *and all continuous functions* $(x(t), y(t))$ *defined in* $[0, \infty)$ *with values in* B, *we have that*

$$\int_s^{s+t} g(\tau, x(\tau), y(\tau)) \, d\tau \to 0 \qquad as \ s \to \infty$$

uniformly for t *on* $[0, 1]$.

Then the solution $(x(t), y(t)) = (0, 0)$ of $(7')$ is unstable and there exists an open subset U of β_{ρ_0} with $(0, 0) \in \partial U$ where no solution of $(7')$ that starts in $U \times \{t_0\}$, for any $t_0 \in [0, \infty)$, is bounded in the future in β_{ρ_0}.

Proof. First we take $(7')$ in the form

$$(EP_1) \qquad \begin{bmatrix} \dot{x} \\ \dot{y} \end{bmatrix} = \begin{bmatrix} y \\ h(t, x, y)y + f(x) \end{bmatrix} + \begin{bmatrix} 0 \\ g(t, x, y) \end{bmatrix}$$

It is easily seen that if

$$P = \begin{bmatrix} y \\ h(t, x, y)y + f(x) \end{bmatrix}$$

$P^* \in \hat{\Omega}(P)$ is of the form

$$P^* = \begin{bmatrix} y \\ h^*(t, x, y)y + f(x) \end{bmatrix}$$

where $h^* \in \hat{\Omega}(h)$.

Then (E) and (E*) are respectively:

$$(8) \qquad \begin{cases} \dot{x} = y \\ \dot{y} = h(t, x, y)y + f(x) \end{cases}$$

and

$$(8^*) \qquad \begin{cases} \dot{x} = y \\ \dot{y} = h^*(t, x, y)y + f(x) \end{cases}$$

We define

$$V(x, y) = \frac{y^2}{2} - \int_0^x f(s)\,ds \qquad \text{and}$$

$$U = \{(x, y) \in \beta_{\rho_0} \mid V(x, y) > 0\}$$

Then $V \in C^1$ on β_{ρ_0} and the hypotheses of Theorem 3 hold. Indeed, it is easily seen that (1) and (2) are true.

$$\dot{V}_{(7)}(x, y) = y\dot{y} - f(x) = y\big[h(t, x, y)y + f(x) + g(t, x, y)\big]$$

$$= y^2 h(t, x, y) + yg(t, x, y) \geqslant 0$$

on $[0, \infty) \times U$ and then (3) is true.

Finally, suppose for the moment that (4) is not true, that is, suppose that there exists a solution $(x^*(t), y^*(t))$ of (8*) bounded in U with $V(x^*(t), y^*(t)) = k$ (constant), $\forall t \in R$. Then $\Omega(x^*(\cdot), y^*(\cdot)) = \Omega^*$ is a nonempty, connected and compact set contained in U.

Further, $V(x, y) = k$, $\forall (x, y) \in \Omega^*$.

From Corollary 5 Ω^* is semi-invariant with respect to (8*), that is, given $(x_0, y_0) \in \Omega^*$ there exists a function $h^{**} \in \hat{\Omega}(h^*)$ and a solution $(x^{**}(t), y^{**}(t))$ of the system

$$(8^{**}) \qquad \begin{cases} \dot{x} = y \\ \dot{y} = h^{**}(t, x, y)y + f(x) \end{cases}$$

with $(x^{**}(0), y^{**}(0)) = (x_0, y_0)$ such that $(x^{**}(t), y^{**}(t))$ exists and remains in Ω^* for all $t \in R$.

Since $V(x, y) = k$ on Ω^*, we have

$$\dot{V}_{(8^{**})}(x, y) = y^2 h^{**}(t, x, y) = 0$$

Since $y^2 > 0$ on Ω^*, we have $h^{**}(t, x, y) = 0$ for all $t \in R$. Then:

$$\begin{cases} \dot{x}^{**}(t) = y^{**}(t) \\ \dot{y}^{**}(t) = h^{**}(t, x^{**}(t), y^{**}(t))y^{**}(t) + f(x^{**}(t)) = f(x^{**}(t)) \end{cases}$$

that is, $x^{**}(t)$ is a solution of $\ddot{x} = f(x)$ which is bounded in β_{ρ_0} and, taking into account the Lemma, we have a contradiction.

The above considerations show that we can apply Theorem 3, thus obtaining the desired result.

Remark. A recent paper on stability and instability related to the work under consideration, is found in [7].

UNIVERSIDADE DE SAO PAULO
FUNDAÇÃO UNIVERSIDADE FEDERAL DE SÃO CARLOS

REFERENCES

[1] Hale, J. K., Sufficient conditions for stability and instability of autonomous functional differential equations. *J. Diff. Equations* **1** (1965), 452–482.

[2] La Salle, J. P., An Invariance Principle in the Theory of Stability, *Differential Equations and Dynamical Systems*, A.P. (1967), 277–286.

[3] Markus, L., Asymptotic autonomous differential systems, *Contributions to the Theory of Nonlinear Oscillations* 3 Princeton U.P. (1956), 17–29.

[4] Miller, R. K., Asymptotic behavior of solutions of nonlinear differential equations, *Trans. A.M.S.* 115-3, (1965), 400–416.

[5] Onuchic, N., On a criterion of instability for differential equations with time delay. Periodic orbits, stability and resonances. *GEO Giacaglia* (1970), 339–342.

[6] Onuchic, N., Invariance properties in the theory of ordinary differential equations with applications to stability problems. *SIAM J. Control* **9** (1971), 97–104.

[7] Onuchic, N., Invariance properties for ordinary differential equations: stability and instability. *Nonlinear Analysis: Theory, Methods and Applications* **2**, 1 (1978), 69–76.

[8] Onuchic, N., Invariance and stability for ordinary differential equations, *J. Math. Analysis and Applications* **63**, 1 (1978), 9–18.

[9] Opial, Z., Sur la dependance de solutions d'une systéme de equations differentielles de leurs seconds membres—Applications aux systémes presque autonomes. *Ann. Polon. Math.* **8** (1960), 75–89.

[10] Rodrigues, H. M., Invariança para Sistemas não Autônomos de Equações Diferenciais com Retardamento e Aplicações. Master's Dissertation, ICMSC-USP (1970), São Carlos, S.P. Brazil.

[11] Sell, G. R., Nonautonomous differential equations and topological dynamics I—The basic theory. *Trans. A.M.S.* **127**, 2 (1967), 241–262.

THE TOTAL CURVATURE OF ALGEBRAIC SURFACES

By Robert Osserman

---

1. Statement of results. It seems reasonable to expect that an algebraic surface in euclidean space, even though non-compact, would have finite total curvature. One would in fact expect the value of the total curvature to be bounded in terms of the algebraic degree. However, there do not seem to be many explicit results to that effect. One such theorem was proved some years ago by the author [13]:

THEOREM 0. *Let $P(x, y)$ be a polynomial of degree d. Let S be the graph $z = P(x, y)$ and let K denote the Gauss curvature of S. Then*

$$(1) \qquad \int\int_S |K|\, dA \leqslant 2\pi(d - 1)^2,$$

and

$$(2) \qquad \left|\int\int_S K\, dA\right| \leqslant 2\pi(d - 1).$$

Furthermore, the bound in (2) is sharp, with equality for all harmonic polynomials. The inequality in (1) is strict whenever $d > 2$, but the bound is best possible in the sense that for every $d > 2$, no smaller quantity on the right of (1) will work for all polynomials of degree d.

In the present paper we first prove a number of results directly generalizing Theorem 0, and then derive a somewhat different type of result in the complex case. Specifically, we prove the following:

THEOREM 1. *Let S be a surface in $\mathbb{R}^n$ given by a polynomial map*

$$(3) \qquad x_1 = P_1(\xi, \eta), \ldots, x_n = P_n(\xi, \eta)$$

and let K denote the Gauss curvature of S. If the maximum degree of the polynomials P_j is d, then

$$(4) \qquad \int\int_{S^*} |K|\, dA \leqslant 2\pi(d - 1)^2,$$

where S^* is the (open) set of regular points of S. (If S is immersed, $S^* = S$.)

COROLLARY. *Let S be a complete surface immersed in $\mathbb{R}^n$ by a polynomial map* (3). *Then S is conformally equivalent to the plane.*

This corollary generalizes the theorem of Heinz Huber [11] asserting the same conclusion for the graph of a polynomial in $\mathbb{R}^3$. Huber's proof makes use of length estimates of the sort first applied to conformal type by Ahlfors [1] and later made rigorous by Hartman [7]. In our case the corollary follows immediately from Theorem 1 by virtue of a theorem of Blanc and Fiala [3] (see also A. Huber [10]) relating conformal type to total curvature.

Another consequence of Theorem 1 is that for polynomial surfaces of the form (3), the integral $\iint K \, dA$ is well defined. The next result gives a bound for that quantity that in certain cases is a strong improvement of (4).

THEOREM 2. *In the notation of Theorem 1, if we assume that the map* (3) *is an immersion, then*

$$(5) \qquad \int\!\!\int_S K \, dA \geqslant -2\pi(d - 1).$$

COROLLARY. *If, furthermore, $K \leqslant 0$ on S, then*

$$(6) \qquad \int\!\!\int_S |K| \, dA = \left| \int\!\!\int_S K \, dA \right| \leqslant 2\pi(d - 1).$$

Note that if in Theorem 2, S is complete, then the Cohn-Vossen inequality gives

$$(7) \qquad \int\!\!\int_S K \, dA \leqslant 2\pi,$$

complementing (5). Thus for complete immersed polynomial surfaces the right-hand inequality in (6) is also valid.

We may note that equality holds in (5) and (6) in the case of minimal surfaces of the form (3) where ξ, η are isothermal parameters; i.e., the polynomials P_j satisfy

$$(8) \qquad \sum_{j=1}^{n} \left(\frac{\partial P_j}{\partial \xi} \right)^2 = \sum_{j=1}^{n} \left(\frac{\partial P_j}{\partial \eta} \right)^2, \quad \sum_{j=1}^{n} \frac{\partial P_j}{\partial \xi} \frac{\partial P_j}{\partial \eta} = 0.$$

In fact, one has in that case a kind of converse to Theorem 1: *every complete simply-connected minimal surface in $\mathbb{R}^n$ with finite total curvature may be represented by a polynomial map* (3) *satisfying also* (8). (See Hoffman-Osserman [8], Theorem 6.3.)

Our next result is a generalization of Theorem 1 to higher-dimensional polynomial maps.

THEOREM 3. *Let M be defined by a polynomial map of $\mathbb{R}^m \to \mathbb{R}^n$:*

$$(9) \qquad x_1 = P_1(\xi_1, \ldots, \xi_m), \ldots, x_n = P_n(\xi_1, \ldots, \xi_m).$$

Denote by d the maximum degree of the P_j, and by $K\,dA$ the generalized Gauss-Bonnet integrand on M. Then

$$(10) \qquad \int_{M^*} |K|\, dA \leq \frac{c_m}{2}(d-1)^m$$

where c_m denotes the surface area of the unit m-sphere, and M^* is the set of regular points of M.

Thus, in the case $m = 2$, this reduces precisely to Theorem 1.

Our final result concerns the complex case.

THEOREM 4. *Let S be a surface in $\mathbb{R}^4$ defined by an equation*

$$(11) \qquad P(z, w) = 0$$

in $\mathbb{C}^2$, where P is a polynomial of degree d. Then the Gauss curvature K of S satisfies

$$(12) \qquad \int\int_S |K|\, dA = -\int\int_S K\, dA \leq 2\pi d(d-1).$$

This theorem answers a question posed by Smale, that arose in connection with his work on computational complexity (see [14], section 4).

I should like to thank Arnold Kas for some useful comments.

2. Proof of the theorems.

Proof of Theorem 1. We make use of the fact that for an arbitrary surface S in $\mathbb{R}^n$ the total absolute curvature, $\int\int_S |K|\, dA$, is majorized by the (suitably normalized) total absolute curvature $\tau(S)$ in the sense of Chern-Lashof [4,5]. (See Hoffman-Osserman [9], Theorem 4.) Specifically, at any regular point p of S, and for any unit normal ν to S at p, let $B(\nu)$ be the second fundamental form of S in the direction ν. Then the Gauss curvature K of S at p satisfies

(13) $$K(p) = \frac{n-2}{c_{n-3}} \int_{S^{n-3}} \det B(\nu)\, d\omega_\nu$$

where S^{n-3} denotes the unit sphere in the normal space to S at p, $d\omega_\nu$ is the area element of S^{n-3} and c_{n-3} is the total area. On the other hand, Chern-Lashof [5] denote

(14) $$K^*(p) = \frac{1}{c_{n-1}} \int_{S^{n-3}} |\det B(\nu)|\, d\omega_\nu$$

and set

(15) $$\tau(S) = \int_S K^*\, dA.$$

Thus we have

$$\int_S |K|\, dA \leqslant (n-2) \frac{c_{n-1}}{c_{n-3}} \tau(S).$$

But a standard recursion relation for the c_k shows that the quantity $(n-2)c_{n-1}/c_{n-3}$ is independent of n, and is equal to 2π. Hence

(16) $$\int_S |K|\, dA \leqslant 2\pi\tau(S).$$

Theorem 1 will therefore follow from the stronger result

(17) $$\tau(S^*) \leqslant (d-1)^2.$$

To prove (17), we recall that if

$$g : US^* \to S^{n-1}$$

is the map of the bundle US^* of unit spheres in the normal spaces at regular points of S into the unit sphere $S^{n-1} \subset \mathbb{R}^n$ defined by

$$g : (p, \nu) \mapsto \nu,$$

then $c_{n-1}\tau(S^*)$ is just the total area, counting multiplicities, of the image. By Sard's Theorem, the measure of the image set is equal to the measure of regular values in the image. To prove (17) it is therefore sufficient to show that for each regular value v of the map g, $g^{-1}(v)$ contains at most $(d-1)^2$ points. Consider then a fixed unit vector v. It is the image under g of a point (p, v) if and only if v is orthogonal to the tangent plane to S at p; i.e.

(18) $$\sum_{j=1}^{n} v_j \frac{\partial P_j}{\partial \xi} = 0, \qquad \sum_{j=1}^{n} v_j \frac{\partial P_j}{\partial \eta} = 0.$$

The solutions of (18) are the common zeros of a pair of polynomials of degree at most $d-1$. The condition that v is a regular value of g is equivalent to $\det(B(v)) \neq 0$, or that the function

$$u(\xi, \eta) = v \cdot x = \sum v_j p_j(\xi, \eta)$$

has regular Hessian, which in turn means that the zero sets of the polynomials in (18) have only simple intersections. By Bézout's Theorem there can be at most $(d-1)^2$ intersections, and this proves Theorem 1.

Remark. The fact that the same constant 2π occurs in Theorem 1 as in Theorem 0 is initially surprising. In Theorem 0, the surfaces considered are graphs, and their Gauss maps are contained in a hemisphere. Since the area of the hemisphere is 2π and each (regular) point is covered at most $(d-1)^2$ times, the meaning of inequality (1) is clear. On the other hand, for parametric surfaces in $\mathbb{R}^3$, the image under the Gauss map can cover the entire sphere, and one would expect the corresponding constant to be 4π. On closer analysis, however, one sees that for polynomials of degree at most d, a given *tangent plane* occurs at most $(d-1)^2$ times, and the normals at those points are distributed between the two possible values v and $-v$ corresponding to the given plane. Thus, in computing the area of the Gauss image, it is the area 2π of the projective plane that is the appropriate multiple of $(d-1)^2$.

Proof of Theorem 2. Let S_r be the part of S defined by

$$x = P(\xi, \eta), \qquad \xi^2 + \eta^2 < r^2$$

and let Γ_r be the boundary of S_r. Then the Gauss-Bonnet formula gives

(19) $$\iint_{S_r} K \, dA = 2\pi - \int_{\Gamma_r} k_g \, ds.$$

Hence

(20) $$\iint_{S_r} K \, dA \geqslant 2\pi - \int_{\Gamma_r} k \, ds$$

since the geodesic curvature k_g is bounded above the curvature k of Γ_r. We use Milnor's expression [12] for the total curvature as 2π times the average number of relative maxima on $x \cdot v$ on Γ_r, averaged over all unit vectors v. For a given v, let μ denote the number of relative maxima of $x \cdot v$ on $\xi^2 + \eta^2 = r^2$. Then there must also be μ relative minima, hence 2μ critical points. Each of those critical points must be a common solution of the two equations

(21) $$\xi^2 + \eta^2 = r^2$$

$$\eta P_\xi - \xi P_\eta = 0$$

where

$$P(\xi, \eta) = \sum v_j P_j(\xi, \eta) = v \cdot x$$

is a polynomial of degree at most d. By Bézout's Theorem, there are at most $2d$ solutions. Hence $\mu \leqslant d$. Since that holds for each direction v, the same is true for the average. Thus

(22) $$\int_{\Gamma_r} k \, ds \leqslant 2\pi d$$

and combining with (20) yields (5).

Remark. One might conjecture that a polynomial map (3) that was an immersion would be a proper map and the image surface automatically complete. However, G. Brumfiel has shown how to construct examples disproving such a conjecture. For instance, the surface

$$x_1 = h + \xi, \qquad x_2 = h + \xi^2, \qquad x_3 = \eta h + \xi,$$

$$x_4 = \eta h + \xi^3, \qquad h = \xi\eta - 1,$$

is immersed, but the image of the hyperbola $h = 0$ is a curve terminating at the origin. When $n = 2$, the conjecture appears to be still unresolved.

Proof of Theorem 3.   The argument is exactly as in the proof of Theorem 1. We use the expression

$$(23) \qquad K = \frac{1}{2} \frac{c_m}{c_{n-1}} \int_{S^{n-m-1}} \det(B(v)) \, d\omega_v$$

for the Gauss-Bonnet integrand that is the one used by Allendoerfer [2] and Fenchel [6] in their original proofs of the generalized Gauss-Bonnet Theorem:

$$(24) \qquad \int_M K \, dA = \frac{c_m}{2} \chi(M)$$

for a compact manifold M of dimension m immersed in $\mathbb{R}^n$. Then

$$(25) \qquad \int_{M^*} |K| \, dA = \frac{1}{2} \frac{c_m}{c_{n-1}} \int_{M^*} \left| \int_{S^{n-m-1}} \det(B(v)) \, d\omega_v \right| dA$$

$$\leqslant \frac{1}{2} \frac{c_m}{c_{n-1}} \int_{M^*} \int_{S^{n-m-1}} |\det(B \cdot v)| \, d\omega_v \, dA$$

$$= \frac{c_m}{2} \tau(M^*)$$

where again $\tau(M^*)$ is the total absolute curvature of M^* in the sense of Chern-Lashof and represents the average number of sheets over the unit sphere S^{n-1} of the image of the unit-sphere normal bundle over M. But exactly as in the proof of Theorem 1, there are at most $(d - 1)^m$ sheets over regular values, by Bézout's Theorem, and the singular values have measure zero. Hence

$$(26) \qquad \tau(M^*) \leqslant (n - 1)^d$$

and using (25) gives (10).

Remark.   It would be interesting to generalize Theorem 2 also to higher dimensions. It seems likely that the argument used for Theorem 2 could be extended using an appropriate form of the generalized Gauss-Bonnet Theorem for manifolds with boundary.

Proof of Theorem 4.   We may as well assume that the polynomial $P(z,w)$ is irreducible, since we may apply (12) to each irreducible component and add. The surface S defined by equation (11) is a generalized minimal surface in $\mathbb{R}^4$. The following facts are then known. (For details, see Hoffman-Osserman [8].)

1. The image of S under the generalized Gauss map g lies on a projective line L in $\mathbb{C}P^3$.

2. The total curvature of S is equal in absolute value to the area of $g(S)$, counting multiplicities, where the area is computed relative to the suitably normalized Fubini-Study metric on $\mathbb{C}P^3$.

3. The area of L in the given metric is 2π.

Thus, inequality (12) follows if we show that each point of L has at most $d(d-1)$ pre-images under g. But the tangent planes to S are complex lines of the form

(27) $P_z(a,b)(z-a) + P_w(a,b)(w-b) = 0,$

where

(28) $P(a,b) = 0.$

For given complex numbers A, B, the line

$Az + Bw = C$

is represented as a tangent plane (27) for some C if and only if the numbers a, b are simultaneous solutions of the equations (28) and

(29) $BP_z(a,b) - AP_w(a,b) = 0.$

But (28) and (29) are polynomial equations of degree d and $d-1$ respectively, and since P is irreducible, there are at most $d(d-1)$ common solutions, by virtue of Bézout's Theorem. This completes the proof of Theorem 4.

Remark.   As indicated by Smale [14], Griffiths has given another proof of Theorem 4.

STANFORD UNIVERSITY

REFERENCES

[1] L. V. Ahlfors, Sur le type d'une surface de Riemann, *C. R. Acad. Sci.* Paris **201** (1935), 30–32.

[2] C. B. Allendoerfer, The Euler number of a Riemann manifold, *Amer. J. Math.* **62** (1940), 243–248.

[3] C. Blanc and F. Fiala, Le type d'une surface et sa courbure totale, *Comment. Math. Helv.* **14** (1941–42), 230–233.

[4] S.-S. Chern and R. K. Lashof, On the total curvature of immersed manifolds, *Amer. J. Math.* **79** (1957), 306–318.

[5] S.-S. Chern and R. K. Lashof, On the total curvature of immersed manifolds II, *Mich. Math. J.* **5** (1958), 5–12.

[6] W. Fenchel, On total curvatures of Riemannian manifolds: I, *J. London Math. Soc.* **15** (1940), 15–22.

[7] P. Hartman, Geodesic parallel coordinates in the large, *Amer. J. Math.* **86** (1964), 705–725.

[8] D. Hoffman and R. Osserman, The geometry of the generalized Gauss map, *Amer. Math. Soc. Memoir* **236** (1980), 1–105.

[9] D. Hoffman and R. Osserman, The area of the generalized Gaussian image and the stability of minimal surfaces in S^n and $\mathbb{R}^n$.

[10] A. Huber, On subharmonic functions and differential geometry in the large, *Comment. Math. Helv.* **32** (1957), 13–72.

[11] H. Huber, Uber den konformen Typus von Flächen im euklidischen Raum, *Math. Annalen* **146** (1962), 180–188.

[12] J. W. Milnor, On the total curvature of knots, *Annals. of Math.* **52** (1950), 248–257.

[13] R. Osserman, Some geometric properties of polynomial surfaces, *Comment. Math. Helv.* **37** (1963), 214–220.

[14] S. Smale, The fundamental theorem of algebra and complexity theory, *Bull. Amer. Math. Soc. N.S.* **4** (1981), 1–36.

ON A DELAY-DIFFERENTIAL SYSTEM

By G. Pecelli*

---

1. Introduction. In this note we present some existence, bounded-ness and bifurcation results for the delay-differential system

$$\dot{U}(t) = \gamma U(t)(1 - U(t)/K) - x(t)U(t)F(U(t))$$

(1.1)

$$\dot{x}(t) = -Dx(t) + m\int_{-1}^{0}\int_{-1}^{0} x(t + s)U(t + \sigma)\,dA(s,\sigma).$$

Systems of this type have arisen in the study of prey-predator interactions [4, 5, 7, 8], in particular as models of ecosystems in which stable oscillations may take place.

In the sequel we assume the reader to be familiar with the necessary portions of the theory of Functional-Differential Equations [3] and, for the last section, with [2]. Proofs will be only briefly indicated, since the author expects that details and other related material will appear elsewhere, being too lengthy and technical for these Proceedings.

2. Some general results. We consider the initial value problem given by (1.1) with initial conditions $x(s) = \varphi(s)$, $U(s) = \psi(s)$ for $-1 \leqslant s \leqslant 0$, where $\varphi(s)$, $\psi(s)$ are continuous real valued functions. We also assume that γ, K, D and m are positive constants; that $A(s,\sigma)$ generates a Stieltjes measure on the unit square with $\int_{-1}^{0}\int_{-1}^{0} dA(s,\sigma) = 1$ and $\int_{-1}^{0}\int_{-1}^{0}\varphi(s)\psi(\sigma)\,dA(s,\sigma) \geqslant 0$ for all continuous nonnegative φ and ψ; and that F is smooth, nonnegative on $[0, \infty)$, positive on $[0, K]$, $F(D/m) = m$ (with no loss of generality: rescale x) and $F'(D/m) \leqslant 0$. With these assumptions one can prove:

THEOREM 2.1. *For every pair $(\psi(s), \varphi(s))$ of positive, real valued, continuous functions on $[-1, 0]$, there exist unique functions $U(t), x(t)$, defined on $[-1, \infty)$, positive, bounded and continuous there, differentiable on $[0, \infty)$, which are solutions of (1.1) on $[0, \infty)$ with the initial data*

---

* The author gratefully acknowledges the hospitality of S.U.N.Y.-Albany during the preparation of this research, the access to MACSYMA provided by the M.I.T. Mathlab facilities and ARPAnet.

$U(s) = \psi(s), x(s) = \varphi(s)$ on $[-1, 0]$. Furthermore $U(t) < K$ for all large t.

This result is exactly what one would expect. The proof involves standard existence and uniqueness arguments, comparisons with the differential equation $\dot{z} = \gamma z(1 - z/K)$ and Gronwall's inequality for Functional-Differential Equations.

When the effect of the memory functional is *concentrated exclusively* at one of the corners of the unit square, we can also prove:

THEOREM 2.2. *If* $D > mK$, *then* $x(t) \to 0$ *as* $t \to +\infty$.

The theorem can be proven in a straightforward manner in the case when (1.1) exhibits no delay in the variable $x(t)$, and using Liapounov theory [3, Ch. 5, Theorem 2.1] when it does.

The two theorems can be interpreted as providing conditions for a reasonable evolution of a predator (x)-prey(U) system, the second theorem giving conditions under which the predator is not efficient enough to make a living in the chosen environment.

3. Critical points, linearized analysis and eigenvalue crossings. The system (1.1) has three critical points of interest, the origin $(\psi, \varphi) = (0, 0)$ and $(\psi, \varphi) = (K, 0)$ which are both on the boundary of the cone of positive functions, and the point $(\psi, \varphi) = (D/m, \gamma(1 - D/mK))$ which is interior to the cone if $D < mK$. The origin is clearly unstable; Theorem 2.2 and some easy considerations provide cases when $(K, 0)$ is globally asymptotically stable from the interior of the cone of positive initial data. It is the third critical point to which we now turn our attention.

Since the transcendental equations for the spectrum are otherwise too complicated, we assume

$$\int_{-1}^{0} \int_{-1}^{0} x(t + s) U(t + \sigma) \, dA(s, \sigma) = x(t) U(t - 1),$$

which leaves us with generalizations of the models examined in [5] and, without delay, in [4]. We first replace U by $U + D/m$ and x by $x + \gamma(1 - D/mK)$, expand, replace D/m by $1/\alpha$ and K by K/α (this turns out to have convenient effects on the computations) to obtain a system whose characteristic equation [3. Ch. 7, Lemma 2.1] is

$$(3.1) \qquad K\lambda^2 + \gamma\big[1 + (K - 1)F'(1/\alpha)/\alpha\big]\lambda + D\gamma(K - 1)e^{-\lambda} = 0,$$

where $K > 1$.

THEOREM 3.1. *For the equation* (3.1) *with* $D\gamma < 4\pi^2$, *there exists* $K_0 > 1$ *such that*:

(i) *For* $1 < K < K_0$, (3.1) *has all its zeros in the open left half plane*;

(ii) *For* $K = K_0$, (3.1) *has exactly two zeros of algebraic (and hence geometric) multiplicity one on the imaginary axis, and all its remaining zeros in the open left half plane*;

(iii) *The derivative with respect to K of the real parts of the two zeros of* (ii) *at* $K = K_0$ *is not zero*;

(iv) *For* $K > K_0$, *near* K_0, (3.1) *has exactly two zeros of multiplicity one in the open right half plane, and all its remaining zeros in the open left half plane*;

(v) *If* $D\gamma < 9\pi^2/4$, *then the nearness restriction in* (iv) *can be removed*.

This theorem provides the main step towards an application of the Hopf bifurcation. Except for (iii), the proof is very similar to that in [5, Lemma 3.1], requiring only more care and more cases. The restriction $D\gamma < 4\pi^2$ is possibly nonessential and insures that the zeros of two functions intertwine in an easily verifiable manner. (iii) involves some fairly straightforward comparisons and implicit function arguments.

To complete our analysis, and thus show the onset of stable oscillations as K increases through K_0, we need to verify some last hypothesis (see [1, 2, 6]). Chow and Mallet-Paret [2] show that the method of averaging is, at least in theory, applicable, provide an algorithm and some examples. Because of the computational difficulties involved, the author turned to MACSYMA (the symbolic manipulation language at M.I.T.) as an untiring and careful amanuensis. The next section reports on some of the results obtained.

4. Some results. The computation of the bifurcation constant (as given in [2]) as a function of the parameters of (1.1) was carried out using MACSYMA. This is a very sophisticated symbolic manipulator with *exact* rational arithmetic. To obtain an expression of reasonable size it was first assumed that $F'(D/m) = F''(D/m) = F'''(D/m) = 0$. This leads to substantial simplification and to a generalization of [5]. The bifurcation constant can then be computed explicitly:

$$\mathcal{K}(\gamma, D, m, K_0, \eta) = \left(\frac{m}{D}\right)^2 \tilde{\mathcal{K}}(\gamma, D, K_0, \eta),$$

where η is the positive imaginary part of the crossing eigenvalues and $\tilde{\mathcal{K}}$ is a rational function of its arguments. Since $\tilde{\mathcal{K}}$ remains too complicated for an exact analysis, it was assumed that $D = 1$, and ten distinct functions were computed:

$$\mathfrak{K}_i = \tilde{\mathfrak{K}}(i/10, 1, K_0, \eta), \qquad i = 1, \ldots, 10.$$

It bears repeating that up to this point all computations are *exact*. At this point K_0 and η were computed from (3.1) via an iteration scheme, the values inserted into the $\mathfrak{K}_i$'s to obtain

$$\mathfrak{K}_1 = -.00915 \quad \mathfrak{K}_2 = -.01740$$
$$\mathfrak{K}_3 = -.02527 \quad \mathfrak{K}_4 = -.03271$$
$$\mathfrak{K}_5 = -.03982 \quad \mathfrak{K}_6 = -.04665$$
$$\mathfrak{K}_7 = -.05326 \quad \mathfrak{K}_8 = -.05971$$
$$\mathfrak{K}_9 = -.06603 \quad \mathfrak{K}_{10} = -.07228$$

Using the results in [2], we have thus shown:

If $F'(D/m) = F''(D/m) = F'''(D/m) = 0$, $D = 1$ and $\gamma = i/10$, $i = 1, \ldots, 10$, then, as K increases past K_0 (now independent of α), (1.1) develops a *stable periodic orbit*, bifurcating from the critical point $(1/\alpha, \gamma(1 - 1/K))$, for any value of $m > 0$.

This should be contrasted to the result in [5], where the existence of a periodic orbit was established for a large enough retardation, but no conclusion could be reached as to its stability.

For the reader's amusement we include

$$\mathfrak{K}_{10} = -\big((45K^{10} - 172K^9 + 203K^8 - 43K^7 - 112K^6$$

$$+ 158K^5 - 86K^4 - 38K^3 + 36K^2 + 8K + 16)\eta^2$$

$$- 15K^{10} + 79K^9 - 66K^8 - 220K^7 + 495K^6$$

$$- 350K^5 + 42K^4 + 46K^3 - 20K^2 + 24K - 16\big)$$

$$\div \big((900K^{10} - 3540K^9 + 5024K^8 - 2384K^7 - 1136K^6$$

$$+ 1680K^5 - 432K^4 - 208K^3 - 64K^2 + 64K + 128)\eta^2$$

$$- 240K^9 + 1300K^8 - 2660K^7 + 2512K^6 - 1008K^5$$

$$+ 272K^4 - 432K^3 + 192K^2 + 192K - 128\big).$$

A more detailed analysis of $\tilde{\mathcal{H}}(\gamma, D, K_0, \eta)$ is planned for the future.

HUNTER COLLEGE, C.U.N.Y.

REFERENCES

---

[1] Chaffee, N., The bifurcation of one or more closed orbits from an equilibrium point of an autonomous differential equation, *Journal of Differential Equations* **4** (1968), 661–179.

[2] Chow, S.-N., and J. Mallet-Paret, Integral averaging and bifurcation, *Journal of Differential Equations* **26** (1977), 112–159.

[3] Hale, J., *Theory of Functional Differential Equations*, Springer Verlag, New York, 1977.

[4] Hsu, S.B., S.P. Hubbell and P. Waltman, Competing Predators, *S.I.A.M. Journal of Applied Mathematics* **35** (1978), 617–625.

[5] Leung, A., Periodic solutions for a prey-predator differential delay equation, *Journal of Differential Equations* **26** (1977), 391–403.

[6] Marsden, J.E. and M.F. McCracken, The Hopf bifurcation and its applications, Springer Verlag, New York, 1976.

[7] Ros, G.G., A difference-differential model in population dynamics, *Journal of Theoretical Biology* **37** (1972), 477–492.

[8] Wangersky, P., and W. Cunningham, Time lag in prey predator population models, *Ecology* **38** (1957), 136–139.

SUSPENDING SUBSHIFTS*

By CHARLES C. PUGH and MICHAEL SHUB**

---

1. Introduction. Every full shift on k-symbols is conjugate to a basic set of an Axiom AS diffeomorphism of S^2, Smale [12], but the same is false for a subshift of finite type (ssft), as shown by an example of Franks [5]. The corresponding question on a general compact 2-manifold M^2, instead of S^2, remains open. Partial results may be found in Franks [5] and Batterson [1].[1]

Here we show that the situations for flows is quite different. For Axiom A flows one can pass from basic sets to sub-basic sets on the same manifold, Theorem 6 below. In particular for ssft's we have

THEOREM 1. *The suspension of any basic subshift of finite type (ie a ssft with dense periodic orbits and a dense orbit) is conjugate to a basic set of some Axiom AS flow on S^3 — or any other $M^m, m \geqslant 3$.*

Theorem 6 is proved in §4. We are indebted to R. Mañé for simplifying our proof of Theorem 6. Sections 2 and 3 set the notations we use, define the terms, and put the problem in context.

Remarks. 1) The Axiom AS flow we produce in Theorem 1 has singularities. It is not known if they can be eliminated on S^3.

2) Basic ssft's are identical to those which arise from irreducible matrices. See [4, 11] and §2.

2. Cascades. Before proving Theorem 1, we explain the general problem to which it relates. Background references are Smale [12] and Shub [11]. Let $f: M \to M$ be a diffeomorphism, M compact. Its *cascade* is $\{f^n : n \in \mathbb{Z}\}$. Let Ω be the nonwandering set of f. Then f obeys *Axiom A* if

    (Aa) Ω is a hyperbolic set of f

    (Ab) The periodic points are dense in Ω.

---

*Supported by NSF Grant MCS 77-17907

**Supported by NSF Grant MCS 78-02721

[1] P. Blanchard and J. Franks have produced subshifts of finite type which cannot occur on any compact M^2.

If, in addition, all invariant manifolds of Ω meet transversally then f is said to obey Axiom AS and it follows that f is structurally stable. Smale shows that Axiom A implies a finite disjoint decomposition

$$\Omega = \Omega_1 \cup \cdots \cup \Omega_k$$

where the Ω_j are the *basic sets* of Ω. Each Ω_j is compact, f-invariant, contains an f-orbit which is dense in Ω_j, and $f|\Omega_j$ is expansive. "Expansive" means that for some $\delta > 0$ and some metric d, if $d(f^n x, f^n y) \leqslant \delta$ for all $n \in \mathbb{Z}$ then $x = y$.

Embedding Problem. Given a compact metrizable Λ, a homeomorphism $g : \Lambda \to \Lambda$, and a manifold M, when are there an Axiom A (or AS) diffeomorphism $f : M \to M$ and an embedding $i : \Lambda \hookrightarrow M$ onto a basic set of f making

$$
\begin{array}{ccc}
M & \xrightarrow{\;f\;} & M \\
\uparrow i & & \uparrow i \\
\Lambda & \xrightarrow{\;g\;} & \Lambda
\end{array}
$$

commute? Obviously, necessary conditions are

(1) g is expansive, g has a dense orbit, the periodic points of g are dense in Λ, and (Λ, g) embeds in some diffeomorphism of M.

A well behaved sort of (Λ, g) is furnished by symbolic dynamics. Let $2 \leqslant k < \infty$ be fixed and let $\sum = \sum^k$ denote the bi-infinite sequences of letters $0, 1, \ldots, k - 1$, i.e.

$$\sum = \{0, 1, \ldots, k - 1\}^{\mathbb{Z}}$$

The general element of $\sum$ is written $\underline{a} = (\ldots a_{-1} \cdot a_0 a_1 a_2 \ldots)$ where each a_i is a letter from 0 to $k - 1$. Note the convenient decimal point. Put the product topology on $\sum$. It is compact, zero dimensional, and metrizable. Let $\sigma : \sum \to \sum$ be the map:

$$\underline{a} = (\ldots a_{-1} \cdot a_0 a_1 a_2 \ldots) \mapsto (\ldots a_{-1} a_0 \cdot a_1 a_2 \ldots) = \sigma(\underline{a})$$

which shifts the decimal point to the right, i.e., σ shifts the entries of $\underline{a}$ to the left. σ is an homeomorphism called the *full shift on k symbols*. In [12], Smale shows

THEOREM 2. *Each full shift is conjugate to a basic set of some Axiom AS diffeomorphism on S^2, or any other m-manifold $m \geqslant 2$.*

Definition. Let $\sigma : \sum \to \sum$ be the full shift on k symbols. If $\sum' \subset \sum$ is σ-invariant then $\sigma \mid \sum'$ is a *subshift*. It is of *finite type* if for some fixed N there is a list $L \subset \{0, 1, \ldots, k - 1\}^N$ such that

$$\underline{a} \in \sum' \text{ iff each } N\text{-string of } \underline{a} \text{ appears in the list } L.$$

An N-string of $\underline{a}$ is just N consecutive entries of $\underline{a}$. All ssft's are compact. This definition of ssft is conjugacy-invariant and is equivalent to the transition matrix definition in [11]. See [4, p. 121].

Definition. A ssft is *basic* if it has a dense orbit and a dense set of periodic points.

In [14] R.F. Williams shows

THEOREM 3. *Each basic ssft is conjugate to a basic set of some Axion AS diffeomorphism on S^3, or any other m-manifold $m \geqslant 3$.*

Thus the Embedding Problem for zero-dimensional sets Λ is fairly well understood. Let $g : \Lambda \to \Lambda$ be a homeomorphism of the compact metrizable space Λ and ask: when does it embed onto a basic set? The necessary condition of expansiveness in (1) implies that (Λ, g) is conjugate to some subshift $\sum' \subset \sum$, provided Λ has dimension zero [10]. By [4, p. 244], for a subshift to be conjugate to a basic set it must be ssft. Thus

(2) A zero-dimensional (Λ, g) is conjugate to a basic set iff it is a basic ssft.

Here, what is left to know is which subshifts embed as basic sets in which 2-manifolds.

Basic sets of higher dimension are grossly messier than shifts, so (2) is far from a full solution to the Embedding Problem for cascades. See the example of Guckenheimer [6, 2] for a basic set that is topologically bizarre.

3. Flows. In a straight forward manner, the notion of Axiom AS generalizes from cascades to flows. The Embedding Problem is the same.

Definition. Let $g : \Lambda \to \Lambda$ be a homeomorphism. The identification
space $\mathrm{Susp}(\Lambda, g)$ is the set

$$[0, 1] \times \Lambda /_{(1, x) \sim (0, \, gx)}$$

The *suspension* of g is the flow ϕ on $\mathrm{Susp}(\Lambda, g)$ whose trajectory through
$(0, x) \in 0 \times \Lambda$ is

$$\phi_t(0, x) = (t, x) \qquad 0 \leqslant t \leqslant 1$$

The time-one map of ϕ caries $0 \times \Lambda$ onto itself by $g : (0, x) \mapsto (0, gx)$.

Let (S, ϕ) be the suspension of the full k-shift. We have

THEOREM 4. (S, ϕ) *embeds onto a basic set in any* $M^m, m \geqslant 3$.

Proof. See [12] as for Theorem 2.

A result of Bowen essentially solves the embedding problem for 1
dimensional basic sets.

THEOREM 5 [3]. *A one dimensional basic set for a flow is conjugate
to the suspension of a subshift of finite type.*

Theorem 3 would allow us to embed a suspension of a basic ssft in
any M^4. Here we accomplish the embedding in any M^3.

4. Embedding Sub-objects. Let $\sum$ be the full shift on k symbols,
let $\sum'$ be a basic ssft in $\sum$, and let S', S be the suspension of $\sum'$, $\sum$. By
Theorem 3, embed S as a basic set of some Axiom A flow ϕ on (any)
M^3. Call Λ the image of S' in M^3.

THEOREM 6. *There is an Axiom A no cycle flow* ψ *on* M^3 *such that*
Λ *is a basic set of* $\psi, \phi|\Lambda = \psi|\Lambda$, *and* $\Omega(\psi) - \Lambda$ *is finite.*

In fact, our proof of Theorem 6 is more general and gives

THEOREM 7. *Let* Λ *be any locally maximal compact invariant set for
the flow* ϕ *on* M. *Then there is a flow* ψ *on* M *such that*
   (i) $\psi \equiv \phi$ *on a neighborhood of* Λ
   (ii) $\Omega(\psi) \subset \Lambda \cup P$ *where* P *consists of finitely many hyperbolic fixed
points of* ψ, $p_1, p_2, \ldots, p_n$
   (iii) *There are no cycles among* $p_1, \ldots, p_n, \Lambda$
   (iv) *Moreover, if* Λ *is hyperbolic for* ϕ, *has a dense orbit and dense set
of periodic points then* ψ *is Axiom A and one of its basic sets is* Λ.

Note that (iii) and (iv) include Theorem 6. We need the following lemma, referred to in [9], and for which we give a direct proof below, avoiding the subtleties of Wilson [15].

LEMMA 1.   *Let Λ be a compact invariant set for the C^r flow ϕ on M and suppose U is a neighborhood of Λ in which Λ is the maximal ϕ-invariant subset. Then there exists a C^∞ Lyapunov function $\lambda : U \to \mathbb{R}$ for ϕ at Λ.*

This means

$$\lambda \,|\, \Lambda \equiv 0 \equiv D\lambda \,|\, \Lambda$$

$$\dot{\lambda}(x) \overset{\text{def}}{=} \frac{d}{dt} \lambda(\phi_t x)\Big|_{t=0} > 0 \qquad x \in U - \Lambda.$$

Remark.   The authors cannot agree on the sign of $\dot{\lambda}(x)$ in the definition of Lyapunov function. In this section we have taken it positive.

Proof of Theorem 7. Λ is the maximal ϕ-invariant set in some neighborhood $U \supset \Lambda$. Let $\lambda : U \to \mathbb{R}$ be the Lyapunov function supplied by Lemma 1. Since λ is nonsingular off Λ, it extends smoothly to all of M

$$\lambda : M \to \mathbb{R}$$

such that, except for Λ, λ has only nondegenerate critical points: $\text{sing}(\lambda) = \Lambda \cup P$ when $P = \{ p_1, \ldots p_n \}$ and $P \cap U = \emptyset$

Fix a smooth bump function $\beta : M \to [0, 1]$ which has support in U and is identically equal to 1 on a neighborhood of Λ. Set

$$Y = \beta X + (1 - \beta)\text{grad}\,\lambda$$

where $X = \dot{\phi}$. Let ψ be the Y-flow; we claim that ψ verifies Theorem 7. (i) is clear.

Consider $\langle \beta X, \text{grad}\,\lambda \rangle$ where $\langle , \rangle$ is the inner product defining the gradient. If $\beta(x) = 0$ or $x \in \Lambda$ then this quantity equals 0. If $\beta(x) \neq 0$ and $x \notin \Lambda$ then $x \in U$, $\beta(x) > 0$, $\text{grad}\,\lambda(x) \neq 0$, and by Lemma 1, λ increases along the X-trajectory at x, so

$$\langle \beta(x)X(x), \text{grad}\,\lambda(x) \rangle > 0$$

Thus

$$\langle \beta X, \operatorname{grad} \lambda \rangle \geqslant 0$$

everywhere on M. Adding $(1 - \beta)$ grad λ to βX only improves the inequality and we get

$$\langle Y(x), \operatorname{grad} \lambda(x) \rangle \geqslant 0 \qquad x \in M$$

with equality only at $\operatorname{sing}(\lambda) = \Lambda \cup P$. This proves (iii) and

$$\Omega(\psi) \subset \operatorname{Sing}(\lambda)$$

which gives (ii). For λ is continuous and $\lambda(\psi_t x)$ increases (strictly) monotonically with t if $x \in M - \operatorname{sing}(\lambda)$, so all $x \in M - \operatorname{sing}(\lambda)$ wander under the flow ψ.

(iv) Since $\beta \equiv 1$ on a neighborhood of Λ, $D\phi_t|\Lambda = D\psi_t|\Lambda$ for all t and Λ is a hyperbolic set for ψ. Since the periodic orbits are dense in Λ, $\Lambda \subset \Omega(\psi)$. Thus, by (ii), ψ is Axiom A and Λ is basic set.   Q.E.D.

Remark.   The function λ may be chosen so that $\lambda(p_j)$ are distinct from one another and from $0 = \lambda(\Lambda)$, $j = 1, \ldots, n$. Thus, there is a $c > 0$ such that for $M_c = \lambda^{-1}[-c, c]$

(v) $M_c \cap \Omega(\psi) \subset \Lambda$

(vi) Y is transverse to $\partial M_c = \lambda^{-1}(-c) \cup \lambda^{-1}(c)$

(vii) On $M - M_c$, Y is a gradient-like Morse smale flow.

Moreover by the Kupka-Smale Theorem, all the stable and unstable manifolds of $p_1, \ldots, p_n$ may be assumed transverse: $W^u(p_i) \pitchfork W^s(p_j)$, $1 \leqslant i, j \leqslant n$. To get Theorem 1 we must deal with $W^u(\Lambda), W^s(\Lambda)$ as well. This we do in §5.

Now we return to our construction of Lyapunov functions.

Proof of Lemma 1.   If $x \in U$ then the orbit of x exits U in at least one direction of time or else $x \in \Lambda$. If $\phi_t x \in U$ for all $t \geqslant 0$ then $\omega(x)$ is a ϕ-invariant subset of U, which is contained in Λ by maximality of Λ. Similarly for $t \leqslant 0$. Thus, each $x \in U$ has an orbit O_x of precisely one of the four types

(a) $O_x \subset \Lambda$

(b) O_x exits U in forward and reverse time

(c) O_x exits U in forward time and has $\alpha(x) \subset \Lambda$

(d) O_x exits U in reverse time and has $\omega(x) \subset \Lambda$.

The points obeying (c) form the local unstable set for Λ, W^u, and those obeying (d) form the local stable set W^s. Note that all orbits off Λ cross ∂U in at least one direction of time, and that by proper choice of U, $(W^u \cap \partial U)$, $(W^s \cap \partial U)$ are compact disjoint subsets. Denote by B the set of points with orbits of type (b); B is open in U. Let $x \in B$. Choose $x_0 = \phi_{t_0}x \in M - U$ for some $t_0 < 0$, choose a smooth compact $(m - 1) - $ disc D at x_0 transverse to ϕ, suppose $\phi_\tau x_0 \in M - U$ for some $\tau > -t_0$, and consider the flowbox

$$F = \{\phi_t y : y \in D \text{ and } 0 \leqslant t \leqslant \tau\}$$

Choose D small enough so that D and $\phi_\tau D$ lie outside U, see Figure 1. This is a flowbox of type (b) around x.

Let $x \in W^u$. Choose $x_1 = \phi_{t_1}x \in M - U$ for some $t_1 > 0$, choose a smooth compact $(m - 1)$disc D at x_1 transverse to ϕ, choose $\tau > t_1$, and consider the flowbox

$$F = \{\phi_t y : y \in D \text{ and } -\tau \leqslant t \leqslant 0\}$$

F is a flowbox whose forward endface is D and reverse endface is $\phi_{-\tau}(D)$. Choose D small enough so that $D \subset M - U$. This is a flowbox of type (c) around x.

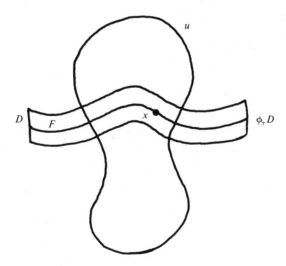

Figure 1.   A flowbox of type (b).

Let $x \in W^s$ and construct a flowbox of type (d) around x symmetric to one of type (c).

On each type of flowbox there is a C^∞ function $\lambda_F: F \to \mathbb{R}$ such that

$$\dot{\lambda}_F > 0 \quad \text{on Interior}(F)$$

(3) $\lambda_F \equiv 0 \quad \text{on} \quad \partial F \cap U$

$$\max|\lambda_F(x)| \leqslant 1 \qquad \max \dot{\lambda}_F(x) \leqslant 1.$$

For consider the three graphs of λ_F over the straightened-out flowbox in Figure 2. Set $\lambda_F \equiv 0$ on $U - F$, so $\lambda_F: F \cup U \to [-1,1]$. The flowbox chart is C^r, so λ_F is C^r. [If we want more differentiability we can C^r-approximate λ_F by a C^∞ map $\tilde{\lambda}_F: F \cup U \to [-1,1]$ such that (3) holds for $\tilde{\lambda}_F$ also. This requires a little care near ∂F.]

Cover $U - \Lambda$ by the interiors of countably many such flowboxes $F_1, F_2, \ldots$ and set

(4) $$\lambda(x) = \sum_{n=1}^{\infty} \epsilon_n \lambda_{F_n}(x) \qquad x \in U$$

where $0 < \epsilon_n < 1/2^n$. Observe that λ is continuous, $\dot{\lambda}$ is continuous, $\dot{\lambda} > 0$ for all $x \in U - \Lambda$, and $\lambda|\Lambda \equiv 0 \equiv D\lambda|\Lambda$. To make $\lambda \in C^r$ we choose $\epsilon_n \to 0$ rapidly.

(5) $$\epsilon_n|\lambda_{F_n}|_r < \frac{1}{2^n} \qquad n = 1, 2, \ldots$$

where $|\ |_r$ denotes the C^r size of a function. Then λ is C^r. [To get $\lambda \in C^\infty$, even if ϕ is only C^1, we replace λ_{F_n} with $\tilde{\lambda}_{F_n}$ in (4) and instead

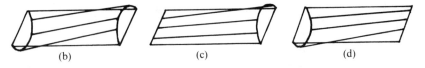

(b)   (c)   (d)

Figure 2.   Lyapunov functions on flowboxes.

of (5), we choose $\epsilon_n \to 0$ so rapidly that

$$\epsilon_n |\tilde{\lambda}_{F_n}|_r < \frac{1}{2^n} \qquad n \geqslant r$$

for each $r = 1, 2, 3, \ldots$. Then λ is C^∞.] Q.E.D.

We also state the one-sided result

LEMMA 2. *Suppose A is a compact forward invariant set for ϕ which has a neighborhood U in M such that*

(6) $$O_-(x) \subset U \Rightarrow O_-(x) \subset A \qquad x \in U$$

where O_- denotes the reverse ϕ-orbit. Then ϕ has a Lyapunov function $\lambda : U \to \mathbb{R}$, and $\lambda < 0$ on $U - A$.

Remark. (6) implies A attracts in U.

Proof. Each $x \in U - A$ has $O_-(x) \subset U$. Forward invariance of A then says $O_-(x) \cap A = \varnothing$, so we get a flowbox F around each $x \in U - A$ of type (d) above. Cover $U - A$ with the interiors of $F_1, F_2, F_3, \ldots$ and set $\lambda = \sum \epsilon_n \lambda_{F_n}$ as above. Q.E.D.

5. Proof of Theorem 1. We suppose that M has dimension 3 and return to the construction of the flow ψ in Theorem 7 and the remark after its proof. We must make the stable manifolds of the fixed points transverse to the unstable manifolds of the orbits in Λ. There are infinitely many of the latter so we must go beyond the Kupka-Smale Theorem. Each orbit O in Λ has $W^u(O)$ and $W^s(O)$ of dimension 2. Since the flow is transverse to $\lambda^{-1}(c)$ so are the $W^u(O)$. Thus $W^u(\Lambda) \cap \lambda^{-1}(c)$ is "laminated" [7, §7] by the individual 1-dimensional curves $W^u(O) \cap \lambda^{-1}(c)$ where O ranges over all ψ-orbits in Λ. Since dim $\Lambda = 1$, $W^u(\Lambda) \cap \lambda^{-1}(c)$ has empty interior.

Suppose q is a fixed point of ψ.

If dim $W^s(q) = 1$ then $W^s(q) \cap \lambda^{-1}(c)$ is at most two points, q_1 and q_2 and we perturb ψ near q to make q_1, q_2 lie off $W^u(\Lambda) \cap \lambda^{-1}(c)$.

If dim $W^s(q) = 3$ then q is a sink and $W^s(q)$ is transverse to all $W^u(O)$ for any orbit O.

If dim $W^s(q) = 2$ and $W^s(q) \cap \lambda^{-1}(c) \neq \varnothing$, note first that $q \in \lambda^{-1}(c, \infty)$. As the flow is Morse-Smale on $\lambda^{-1}(c, \infty)$, we see that $W^s(q) \cap \lambda^{-1}(c)$ is a one-dimensional manifold which can accumulate only at points of $W^s(q') \cap \lambda^{-1}(c)$ where dim $W^s(q') = 1$. These points already lie off the compact set $W^u(\Lambda) \cap \lambda^{-1}(c)$. Thus, generally, this 1-manifold can be tangent to the one dimensional lamination $W^u(q) \cap \lambda^{-1}(c)$ in isolated points, so we can push these tangencies off $W^u(\Lambda) \cap \lambda^{-1}(c)$ since $W^u(\Lambda) \cap \lambda^{-1}(c)$ has no interior in $\lambda^{-1}(c)$. See Figure 3.

In each case, we get transversality, so working one at a time on the fixed points of ψ we make ψ satisfy Axiom AS. Do the same for unstable manifolds of fixed points, $W^s(\Lambda)$ and $\lambda^{-1}(-c)$. This completes the proof in dimension 3.

Next, suppose $m = \dim(M) \geqslant 4$ and that we have embedded our suspended subshift Λ as a basic set of an Axiom AS flow ψ on S^{m-1}. We assume ψ has some point sinks by induction.

Let B^m be the ball of radius 2 in $\mathbb{R}^m$ and extend ψ from $S^{m-1} \subset B^m$ to a flow ψ on B^m such that

S^{m-1} is an attractor, ∂B^m is a repellor, the origin is a point source, $\psi|\partial B^m$ is a north-pole south-pole flow.

Let q_0, q_1 be the source and sink of $\psi|\partial B^m$. Clearly ψ is Axiom A. The only transversality in question for Axiom AS is $W^u(q_1) \cap W^s(\Omega_i)$ where Ω_i are the basic sets of $\psi|S^{m-1}$. Since $W^u(q_1)$ has dimension 1, we push it into the basin of attraction of one of the point sinks of $\psi|S^{m-1}$. This gives Axiom AS on B^m and leaves Λ as a basic set.

In M^m, glue a copy of ψ on B^m onto any m-ball B and extend ψ to M-B as a Morse-Smale gradient flow. The resulting flow obeys Axiom AS on M.   Q.E.D.

Clearly, we wanted to prove Theorem 6 with Axiom AS instead of Axiom A. Forcing transversality between a single stable manifold and a

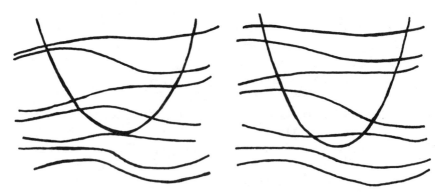

Figure 3.   Removing tangency.

lamination is no easy task when both have high dimension. Consider Smale's hook [13] and Newhouse's hooked horseshoe [8]. In our case, we are willing to make large perturbations to get transversality, so we retain hope for an Axiom AS Theorem 6.

UNIVERSITY OF CALIFORNIA, BERKELEY
QUEENS COLLEGE, C.U.N.Y.

REFERENCES

[1] S. Batterson, Constructing Smale diffeomorphisms on compact surfaces, *Trans. A.M. S.* **256** (79), 237–245.
[2] L. Block, Diffeomorphisms obtained from endomorphisms, *Trans. A.M.S.* **214** (75), 403–413.
[3] R. Bowen, One dimensional hyperbolic sets for flows, *J.D.E.* **12** (72), 173–179.
[4] M. Denker, C. Grillenberger, K. Sigmund, *Ergodic Theory on Compact Spaces*, Lecture Notes in Math no. 527. Springer-Verlag, N.Y., 1976.
[5] J. Franks, Homology and Dynamical Systems, preprint, 5–24, 5–25.
[6] J. Guckenheimer, Endomorphisms of the Riemann sphere, *Proc. Symp. Pure Math*, A.M.S. **14** (68), 95–124.
[7] M. Hirsch, C. Pugh, and M. Shub, *Invariant manifolds*. Springer-Verlag, N.Y., 1977.
[8] S. Newhouse, Non density of Axiom $A(a)$ on S^2, *Proc. Symp. Pure Math*, A.M.S., **14** (68), 191–202.
[9] C. Pugh and M. Shub, The Ω-Stability Theorem for flows, *Inventiones Math* **11** (70), 150–158.
[10] W. Reddy, Lifting expansive homeomorphisms to symbolic flows, *Math. Syst. Theory* **2** (68), 91–92.
[11] M. Shub, Stabilité globale des systèmes dynamiques, Asterisque, 56, Paris, 1978.
[12] S. Smale, Differential dynamical systems, *Bull. A.M.S.* **73** (67), 747–817.
[13] ———, Structurally stable systems are not dense, *Amer. J. Math.* **88** (66), 491–496.
[14] R. F. Willaims, Classification of Subshifts of finite type, *Annals of Math* **98** (73), 120–153; Errata, *ibid*, **99** (74), 380–381.
[15] F. W. Wilson, Smoothing derivatives of functions and applications. *Trans. A.M.S.* **139** (69), 413–428.

ABSOLUTE CONTINUITY OF POLAR FACTORS OF HYPONORMAL OPERATORS*

By C. R. PUTNAM

---

1. Let T be a bounded operator on a separable Hilbert space H with spectrum $\sigma(T)$ and point spectrum $\sigma_p(T)$. Suppose that T has a polar factorization $T = UP$ where U is unitary and P is nonnegative, thus

$$(1.1) \qquad T = UP, \; U \text{ unitary and } P = (T^*T)^{1/2}.$$

If $0 \notin \sigma(T)$ such a factorization always exists and is unique (Wintner [9]). Using a generalization due to von Neumann ([4], p. 307), Hartman [2] showed that (1.1) holds if and only if 0 has identical multiplicities in $\sigma_p(T)$ and $\sigma_p(T^*)$.

An operator T is hyponormal if

$$(1.2) \qquad T^*T - TT^* = D \geqq 0,$$

and completely hyponormal if, in addition, T has no normal part. If $T_z = T - z$ then $T^*T - TT^* = T_z^*T_z - T_zT_z^*$ and it is clear from (1.2) that $\sigma_p(T) \subset \sigma_p(T^*)$ whenever T is hyponormal and that $\sigma_p(T)$ is empty whenever T is completely hyponormal. Further, it follows from the earlier discussion that if T is hyponormal and if $0 \notin \sigma_p(T^*)$ then T has a (unique) polar factorization (1.1). Consequently, if T is completely hyponormal, (1.1) holds if and only if $0 \notin \sigma_p(T^*)$.

If A is a selfadjoint operator with spectral resolution $A = \int t\, dE_t$ then the set, $H_a(A)$, of vectors f in H for which $\|E_t f\|^2$ is an absolutely continuous function of t, is a reducing subspace of A; A is said to be absolutely continuous if $H_a(A) = H$. (See [1], p. 104, [3], p. 516, [5], p. 19.) Similar concepts can be defined for a unitary operator. In case T is completely hyponormal it is known that $\mathrm{Re}(T) = (1/2)(T + T^*)$ and $\mathrm{Im}(T) = (1/2i)(T - T^*)$ are absolutely continuous ([5], 42–43). Moreover, if also $0 \notin \sigma_p(T^*)$, so that T has a polar factorization (1.1), then U

---

*This work was supported by a National Science Foundation Research Grant.

is absolutely continuous. This was noted in [7] (Lemma 4) and is a consequence of Th. 2.3.1 or [5], p. 21.

Henceforth, let T be completely hyponormal and have the factorization (1.1). It was shown in [6] (Th. 3, p. 422) that if $z \neq 0$ and $z = |z|e^{it}$ and if $z \in \sigma(T)$ then $e^{it} \in \sigma(U)$. Further ([6], Th. 5, p. 423), if $e^{it} \in \sigma(U)$, then there exist $z_n = |z_n|e^{it_n} \neq 0$, $z_n \in \sigma(T)$, such that $t_n \to t$ as $n \to \infty$. Also ([6], Th. 6, p. 424), if there exists a wedge

(1.3) $W = \{z : z = re^{it}, r > 0, a < t < b\}$

satisfying

(1.4) $\sigma(T) \cap W$ is empty,

then

(1.5) $P^2 = T^*T \left(\text{equivalently, } P = (T^*T)^{1/2}\right)$

is absolutely continuous.

(As noted above, the absolute continuity of U in the statement of [6], Th. 6, p. 424, is assured even without the wedge hypothesis (1.4).)

It may be noted that if $\sigma(T)$ contains an open annulus centered at the origin then (1.5) need not hold. In fact, let H be the usual bilateral sequence l^2 space and let $A = (a_{ij})$ and $B = (b_{ij})$ be doubly infinite diagonal matrices defined by $a_{ij} = \delta_{ij}\lambda_j$ and $b_{ij} = \delta_{ij}\lambda_{j-1}$, where $\{\lambda_n\}$, $n = 0, \pm 1, \pm 2, \ldots$, is any bounded, strictly increasing sequence of real numbers. For definiteness, suppose that $\lim_{n \to -\infty}\lambda_n = 1$ and $\lim_{n \to \infty}\lambda_n = 2$. Then $A = UBU^*$ where U is the unitary bilateral shift, $A - B = D \geq 0$, and $0 \notin \sigma_p(D)$; see, e.g., [5], p. 22. If $T = B^{1/2}U^*$ then $T^*T - TT^* = D$ and it is clear that T is completely hyponormal. Since $T^*T = A$ has a pure point spectrum then (1.5) fails to hold. In addition, $\sigma(T) = \{z : 1 \leq |z| \leq 2^{1/2}\}$, as can be seen, for instance, from Th. 1 and Th. 9 of [6].

It may be noted that the wedge hypothesis (1.4) is used to ensure that the spectrum of the unitary U in (1.1) is not the entire unit circle. Clearly, if (1.4) is assumed, there is no loss of generality in supposing that $1 \notin \sigma(U)$, so that U is the Cayley transform of a bounded self-adjoint operator A, thus

(1.6) $\quad U = (A - i)(A + i)^{-1}, \qquad$ equivalently,

$$A = i(I + U)(I - U)^{-1}.$$

Further, (1.2) becomes

(1.7) $\quad P^2 - UP^2U^* = D \geqq 0,$

and hence

(1.8) $\quad AP^2 - P^2A = \tfrac{1}{2}i(A + i)D(A - i).$

Since A and P are bounded, the absolute continuity of P, as well as of A, follows from known results; see [5], p. 21.

The preceding argument breaks down, however, if $\sigma(U)$ is the entire unit circle. Although, as noted above, U is still absolutely continuous, the operator A is now unbounded and, as the foregoing example shows, the absolute continuity of P cannot be inferred.

For positive constants c and d, with $0 < d < 1$, let $C(c,d)$ be the open set defined by

(1.9) $\quad C(c,d) = \{z = x + iy : x > 0, -cx^{2-d} < y < cx^{2-d}\},$

so that $C(c,d)$ is a "cusp wedge" symmetric with respect to the positive real axis and with vertex at the origin. There will be proved the following

THEOREM. *Let T be completely hyponormal and have the polar factorization* (1.1). *Suppose that for $c > 0$ and $0 < d < 1$,*

(1.10) $\quad 0 \in \sigma(T)$ *and* $\sigma(T) \cap C(c,d)$ *is empty,*

where $C(c,d)$ is defined in (1.9). *Then the operator P satisfies* (1.5).

Remarks. Although, under the hypothesis (1.10), U may have as its spectrum the entire unit circle (so that the corresponding Cayley transform A is then unbounded), nevertheless, P is absolutely continuous in this case. It will remain undecided, however, whether this latter assertion remains true if, in the hypothesis (1.10), the set $C(c,d)$ is replaced by, say, the half-line $(0, \infty)$, or even if $d = 0$ is allowed in the definition of $C(c, d)$ in (1.9). It was noted above that if $\sigma(T)$ contains an open annulus centered at the origin then (1.5) need not hold. The question naturally arises as to whether the *absence* of such an annulus in the spectrum of a completely hyponormal T is by itself sufficient to imply the absolute continuity of P or the least, for instance, that $H_a(P) \neq 0$. (The hypothesis (1.10) is, of course, a much stronger hypothesis.) Although we cannot answer this question, nevertheless, it is true that the absence of such an annulus does imply that the spectrum of T^*T has positive linear Lebesgue measure (even if T fails to have a polar factorization (1.1), so that 0 may belong to $\sigma_p(T^*)$). See [6], Th. 9, p. 426.

2. Proof of Theorem. As was noted above, U is absolutely continuous, so that, in particular, the selfadjoint (possible unbounded) operator A can be defined as in (1.6). If $\{G_t\}$ denotes the spectral family of U then

$$(2.1) \qquad U = \int_0^{2\pi} e^{it}\, dG_t \quad \text{and} \quad A = i \int_0^{2\pi} (1 + e^{it})(1 - e^{it})^{-1}\, dG_t.$$

It will be shown first that

$$(2.2) \qquad AP^2 \text{ is bounded.}$$

Let f be an arbitrary vector in H and consider the integral

$$(2.3) \qquad \int_0^{2\pi} |(1 + e^{it})(1 - e^{it})^{-1}|^2\, dp(t), \qquad \text{where} \quad p(t) = \|G_t P^2 F\|^2.$$

In order to estabilsh (2.2) it is sufficient to show that for a fixed small $h > 0$, each of the integrals $\int_0^h t^{-2}\, dp(t)$ and $\int_{2\pi-h}^{2\pi-0}(t - 2\pi)^{-2}\, dp(t)$ is bounded by const $\|f\|^2$, where "const" depends on h but not on f. It is sufficient to establish the first relation, that is,

$$(2.4) \qquad \int_0^h t^{-2}\, dp(t) \leqq \text{const}\, \|f\|^2,$$

since, in view of the symmetry of the cusp wedge (1.9), a similar argument can be applied to the other integral.

Let $\Delta = (a,b) \subset [0, 2\pi]$. A multiplication by $G(\Delta)$ on the left and right of each term of (1.7) yields

(2.5) $G(\Delta)P^2G(\Delta) - UG(\Delta)P^2G(\Delta)U^* = G(\Delta)DG(\Delta) \geqq 0.$

Hence, if $T(\Delta) = (G(\Delta)U)[G(\Delta)P^2G(\Delta)]^{1/2}$, then $T(\Delta)$ is hyponormal on $G(\Delta)H$. It was shown in [7] (§3, formula line (3.3)) that

(2.6) $\sigma(T(\Delta)) \subset \sigma(T) \cap W_\Delta^-,$

where $W = W_\Delta$ is the wedge of (1.3). If, for $0 \leqq t < 2\pi$, $m(t) = \max\{r : r \geqq 0$ and $re^{it} \in \sigma(T)\}$, it follows from (2.6) that $\Delta = (a, b \subset [0, 2\pi]$,

(2.7) $\|G(\Delta)P^2G(\Delta)\|^{1/2} = \|T(\Delta)\| = \sup\{m(t) : a < t < b\}.$

(Note that $G([a,b]) = G((a,b))$, in view of the absolute continuity of U.)

It is seen from the definition of $p(t)$ in (2.3) that $p(t) \leqq \|G_tP\|^2\|Pf\|^2 \leqq \|G_tP^2G_t\|\|P\|^2\|f\|^2$. Since $G_t = G((0,t))$, it follows from (2.7) that $p(t) \leqq [\sup\{m^2(s) : 0 < s < t\}]\|P\|^2\|f\|^2$. Also, it is clear that if $t = \text{Arc}\tan(y/x)$ with $y = cx^{2-d}$ (see (1.9)), then, as $t \to 0$, or, equivalently, as $x \to 0$, one has $p(t) \leqq (x^2 + y^2)\|P\|^2\|f\|^2 \leqq \text{const}~(x^2)\|f\|^2$. But, as $t \to 0$, $t \sim y/x = cx^{1-d}$ and so $p(t) \leqq \text{const}~\|f\|^2(t^{2+e})$ for some positive constant e. An integration by parts of the integral in (2.4) yields $\int_0^h t^{-2}dp(t) = t^{-2}p(t)|_{0+}^h + 2\int_{0+}^h t^{-3}p(t)dt \leqq \text{const}~\|f\|^2$, that is, (2.4). Hence, as already noted, (2.2) is established.

Next, choose $\Delta = (a, b)$ with $0 < a < b < 2\pi$. Then, by (2.5) and an argument similar to that used in deriving (1.8) from (1.7), $(AG(\Delta))(G(\Delta)P^2G(\Delta)) - (G(\Delta)P^2G(\Delta))AG(\Delta) = \frac{1}{2}iG(\Delta)(A + i)D(A - i)G(\Delta)$. Note that $AG(\Delta)$ is a bounded operator. In particular, if $f = G(\Delta)f$, then $(AP^2f, f) - (P^2Af, f) = \frac{1}{2}i\|D^{1/2}(A - i)f\|^2$. Since $AP^2 = (P^2A)^*$ (cf. [8], p. 29) then

(2.8) $(AP^2f, f) - (f, AP^2f) = \frac{1}{2}i\|D^{1/2}(A - i)f\|^2,$

where $f = G(\Delta)f$.

Since U is absolutely continuous, hence, in particular, $1 \notin \sigma(U)$, it is clear that the set of vectors $K = \{G(\Delta)f : \Delta = (a, b),~0 < a < b < 2\pi\}$ is dense in H. It follows from (2.2) and (2.8) that $D^{1/2}(A - i) \subset M$, where M is bounded. Thus $M^* \subset (D^{1/2}(A - i))^* = (A + i)D^{1/2}$, and hence $(A + i)D^{1/2} = M^*$ is bounded. Next, let P^2 have the spectral resolution $P^2 = \int_0^\infty t\, dE_t (= \int_0^{\|P\|^2} t\, dE_t)$, and let $\delta = (a, b)$, where $0 < a < b$. Clearly, $AE(\delta) = (AP^2)(P^{-2}E(\delta))$ is bounded. If $f = E(\delta)f$ and $s = \frac{1}{2}(a + b)$, then the left side, q, of (2.8) becomes

$$q = \left(A \int_\delta (t - s)\, dE_t f, E(\delta)f \right) - \left(E(\delta)f, A \int_\delta (t - s)\, dE_t f \right)$$

$$= \left(AP^2 \int_\delta t^{-1}(t - s)\, dE_t f, E(\delta)f \right)$$

$$- \left(E(\delta)f, AP^2 \int_\delta t^{-1}(t - s)\, dE_t f \right).$$

Consequently,

$$(2.9) \qquad |q| \leq 2\|AP^2\| \tfrac{1}{2} a^{-1}|\delta| \, \|E(\delta)f\|^2 = ka^{-1}|\delta| \, \|E(\delta)f\|^2,$$

where k is a constant independent of f and δ.

For the above fixed δ and vector f, choose vectors f_n in K satisfying $f_n \to E(\delta)f$ (strongly) as $n \to \infty$. It follows from (2.8) and (2.9) that $\frac{1}{2}\|Mf_n\|^2 \leq ka^{-1}|\delta| \, \|E(\delta)f_n\|^2$ and hence that $\frac{1}{2}\|ME(\delta)f\|^2 \leq ka^{-1}|\delta| \times \|E(\delta)f\|^2$. In particular, $\|ME(\delta)\| \leq 2ka^{-1}|\delta|$, and hence, if β is any Borel subset of $[0, \|P\|^2]$ not containing 0 in its closure, $\|ME(\beta)\| \leq 2k \int_\beta t^{-1}\, dt$. In particular, if $\beta = Z$ has measure zero then

$$(2.10) \qquad ME(Z) = 0.$$

Since $0 \notin \sigma_p(T)$ then $0 \notin \sigma_p(P^2)$, and so (2.10) holds for *every* Borel set, Z, of Lebesgue measure 0 on $[0, \|P\|^2]$.

Since $P^2A \subset (AP^2)^*$, the latter operator being bounded (by (2.2)), then, (2.8),

$$(2.11) \qquad AP^2 - P^2A = \tfrac{1}{2}iM^*M \qquad \text{holds on} \quad D_A,$$

where D_A is the domain of A. That is, $AP^2f - P^2Af = \frac{1}{2}iM^*Mf$ for f in D_A. Let Z' be any Borel subset of $[0, \|P\|^2]$ of Lebesgue measure zero and not containing 0 in its closure. Then $R_{E(Z')} \subset D_A$ and, in fact, $AE(Z') = (AP^2)(P^{-2}E(Z'))$ is bounded.

In view of (2.10) with Z replaced by Z' it follows from (2.11) that

$$(2.12) \qquad AP^2E(Z') = P^2AE(Z').$$

Multiplication on the left by P^2 yields $P^2 A P^2 E(Z') = P^4 A E(Z')$, while a multiplication on the right gives $A P^4 E(Z') = P^2 A P^2 E(Z')$, so that $A P^4 E(Z') = P^4 A E(Z')$. More generally, one obtains $A P^{2n} E(Z') = P^{2n} A E(Z')$ for $n = 0, 1, 2, \ldots$, and, since $P \geqq 0$, also $A P^n E(Z') = P^n A E(Z')$ for $n = 0, 1, 2, \ldots$. Hence, on choosing a sequence of polynomials on $[0, \|P\|^2]$ converging strongly to $E(Z')$, one obtains $A E(Z') = E(Z') A E(Z')$.

Thus, $R_{E(Z')}$ reduces A and hence also U. Since $R_{E(Z')}$ clearly reduces P^2 (hence P) then it reduces the operator $T = UP$. In view of (2.12), A and P^2, when restricted to $R_{E(Z')}$, commute; hence, so also do U and P^2, as also U and P. Thus, $T | R_{E(Z')}$ is normal and, in view of the supposed complete hyponormality of T, $E(Z') = 0$. Since $0 \notin \sigma_p(P^2)$, it is clear that $E(Z) = 0$ for any Borel subset, Z, of $[0, \|P\|^2]$ having Lebesgue measure zero and the proof of the Theorem is now complete.

PURDUE UNIVERSITY

REFERENCES

---

[1] P. R. Halmos, Introduction to Hilbert space and the theory of spectral multiplicity, Chelsea Pub. Co., N.Y., 1951.

[2] P. Hartman, On the essential spectra of symmetric operators in Hilbert space, *Amer. Jour. Math.* **75** (1953), 229–240.

[3] T. Kato, Perturbation theory for linear operators, *Die Grundlehren der Math. Wiss.* **132**, Springer-Verlag, N.Y., 1966.

[4] J. von Neumann, Über adjungierte Funktionaloperatoren, *Ann. of Math.* **33** (1932), 294–310.

[5] C. R. Putnam, Commutation properties of Hilbert space operators and related topics, *Ergebnisse der Math.* **36** Springer-Verlag, N.Y., 1967.

[6] C. R. Putnam, Spectra of polar factors of hyponormal operators, *Trans. Amer. Math. Soc.* **188** (1974), 419–428.

[7] C. R. Putnam, A polar area inequality for hyponormal spectra, *Jour. Operator Theory* **4** (1980), 191–200.

[8] B. Sz.-Nagy, Spektraldarstellung linearer Transformationen des Hilbertschen Raumes, *Ergebnisse der Math.* **39**, Springer-Verlag, 1967.

[9] A. Wintner, On non-singular bounded matrices, *Amer. Jour. Math.* **54** (1932), 145–149.

SUBHARMONIC SOLUTIONS OF A FORCED WAVE
EQUATION*

By Paul H. Rabinowitz

Introduction. In a recent paper [1], we established the existence of subharmonic solutions of forced Hamiltonian systems of ordinary differential equations. The goal of this note is to show that subharmonics also occur for a class of semilinear wave equations.

To be more precise, let $z(t) = (z_1(t), \ldots, z_{2n}(t))$, $H : \mathbb{R}^{2n} \to \mathbb{R}$, and consider the Hamiltonian system of ordinary differential equations:

$$(0.1) \qquad \frac{dz}{dt} = \mathcal{J} H_z(t,z), \qquad \mathcal{J} = \begin{pmatrix} 0 & -I \\ I & 0 \end{pmatrix}$$

where I denotes the identity matrix in $\mathbb{R}^n$. Suppose $H(t,0) = 0$, $H(t,z) \geqslant 0$, and H is T periodic in t. It was shown in [1] that if H satisfies appropriate additional conditions near $z = 0$ and $z = \infty$, then (0.1) possesses an infinite number of distinct subharmonic solutions, i.e. for each $k \in \mathbb{N}$, (0.1) has a solution $z_k(t)$ of period kT and infinitely many of the functions z_k are distinct. For single second order equations of the form

$$(0.2) \qquad v'' + g(t,v) = 0$$

with g T-periodic in t, more delicate such results were obtained earlier under related hypotheses by Jacobowitz [2]. Further work on this question was carried out by Hartman [3] who weakened the hypotheses of [2] and improved the conclusions.

We will show how analogues of some of the results of [1] can be obtained for a family of forced semilinear wave equations. Thus consider

$$(0.3) \qquad \begin{cases} u_{tt} - u_{xx} + f(x,t,u) = 0 & 0 < x < l \\ u(0,t) = 0 = u(l,t) \end{cases}$$

---

*This research was sponsored in part by the Office of Naval Research under Contract No. N00014-76-C-0300. Reproduction in whole or in part is permitted for any purpose of the United States Government.

where f is T periodic in t. It was shown in [4] that (0.3) possesses a nontrivial classical T periodic solution provided that $T \in l\mathbb{Q}$, i.e. T is a rational multiple of l, and f satisfies appropriate conditions. Recently a slightly stronger result has been obtained by Brezis, Coron, and Nirenberg [5]. In the following section we will prove that the hypotheses required in [4] for the above existence theorem imply that (0.3) also has subharmonic solutions: for all $k \in \mathbb{N}$, (0.3) possesses a kT periodic solution u_k and infinitely many of these functions are distinct. The proof relies on an amalgam of ideas from [1] and [4].

1. The existence theorem. Suppose $f: [0, l] \times \mathbb{R}^2 \to \mathbb{R}$ and satisfies
(f_1) $f(x, t, 0) = 0$, $f_r(x, t, r) > 0$ for $0 \neq r$ near 0, and $f(x, t, r)$ is strictly monotonically increasing in r for all $r \in \mathbb{R}$.
(f_2) $f(x, t, r) = o(|r|)$ at $r = 0$
(f_3) There are constants $\mu > 2$ and $\bar{r} > 0$ such that $0 < \mu F(x, t, r) \equiv \int_0^r f(x, t, s) \, ds \leq rf(x, t, r)$ for $|r| \geq \bar{r}$
(f_4) There is a constant $T > 0$ such that $f(x, t + T, r) = f(x, t, r)$ for all x, t, r.
Note that (f_3) implies that

$$(1.1) \qquad F(x, t, r) \geq a_1 |r|^\mu - a_2$$

for some constants $a_1 > 0$, $a_2 \geq 0$ and for all $r \in \mathbb{R}$, i.e. F grows at a more rapid rate than quadratic at $r = \infty$.
We will prove the following theorem:

THEOREM 1.2. *Let* $f \in C^2([0, l] \times \mathbb{R}^2, \mathbb{R})$ *and satisfy* (f_1)–(f_4). *If* $T \in l\mathbb{Q}$, *then for all* $k \in \mathbb{N}$, *the problem*

$$(1.3) \qquad \begin{cases} u_{tt} - u_{xx} + f(x, t, u) = 0 & 0 < x < l \\ u(0, t) = 0 = u(l, t) \end{cases}$$

possesses a nonconstant kT *periodic solution* $u_k \in C^2$. *Moreover infinitely many of the functions* u_k *are distinct.*

Before giving the proof of Theorem 1.2, several remarks are in order. Since $T \in l\mathbb{Q}$ implies that $kT \in l\mathbb{Q}$ for all $k \in \mathbb{N}$, the first assertion of the theorem is a special case of Theorem 4.1 and Corollary 4.14 of [4]. However, since we do not know kT is a minimal period of u_k, the functions u_k may all represent the same T periodic function or possibly a finite number of distinct periodic functions. Thus what is new and of interest here is that in fact infinitely many of the functions u_k must be distinct.

To establish this result we will show that on the one hand, if only finitely many of the functions u_k were distinct, a corresponding variational formulation of (1.3) would have an unbounded subsequence of critical values, c_{k_j}, with corresponding critical points representing reparametrizations of the same function. The growth of the c_{k_j}'s will be like k_j^2. On the other hand it turns out that c_k grows at most linearly in k, a contradiction.

To make this statement, which contains variants of ideas in [1], more precise, a closer inspection must be made of the existence mechanism of [4]. For convenience we take $l = \pi$ and $T = 2\pi$. Fixing $k \in \mathbb{N}$, we seek a solution of (1.3) which is $2\pi k$ periodic in t. It is convenient to rescale time $t = k\tau$ so that the period becomes 2π and (1.3) transforms to

$$(1.4) \quad \begin{cases} u_{\tau\tau} - k^2(u_{xx} - f(x, k\tau, u)) = 0 & 0 < x < \pi \\ u(0, \tau) = 0 = u(\pi, \tau); u(x, \tau + 2\pi) = u(x, \tau) \end{cases}$$

The solution of (1.4) is obtained via an approximation argument. Three approximations are made. First observe that the wave operator part of (1.4), $u_{\tau\tau} - k^2 u_{xx}$ has an infinite dimensional null space, N, in the class of functions satisfying the periodicity and boundary conditions, namely

$$N = \operatorname{span}\{\sin jx \sin kj\tau, \sin jx \cos kj\tau \mid j \in \mathbb{N}\}$$

To provide some compactness for the problem in N, we perturb the wave operator by adding a term $-\beta v_{\tau\tau}$ to it where $\beta > 0$ and v denotes the L^2 orthogonal projection of u into N. Secondly the unrestricted rate of growth of $f(x, t, r)$ at $|r| = \infty$ creates technical problems which we bypass by suitably truncating f, i.e., we replace f by $f_K(x, t, r)$ where f_K coincides with f for $|r| \leq K$, satisfies (f_1)–(f_4) with μ replaced by a new constant $\bar{\mu} = \min(4, \mu)$ in (f_3). Moreover f_K grows like r^3 at ∞. (See Eq. (5.22) of [4]). Thus we replace (1.4) by

$$(1.5) \quad \begin{cases} u_{\tau\tau} - \beta v_{\tau\tau} - k^2(u_{xx} - f_K(x, k\tau, u)) = 0, & 0 < x < \pi \\ u(0, \tau) = 0 = u(\pi, \tau); u(x, \tau + 2\pi) = u(x, \tau) \end{cases}$$

Formally (1.5) can be cast as a variational problem, namely that of finding critical points of

$$(1.6) \qquad I(u; k, \beta, K) = \int_0^{2\pi} \int_0^\pi \left[\frac{1}{2} u_\tau^2 - \frac{\beta}{2} v_\tau^2 \right.$$

$$\left. - k^2 \left(\frac{1}{2} u_x^2 + F_K(x, k\tau, u) \right) \right] dx \, dt$$

where F_K is the primitive of f_K. Our final approximation is to pose this variational problem in the finite dimensional space

$$E_m = \text{span}\{\sin jx \sin n\tau, \sin jx \cos n\tau | 0 \leqslant j, n \leqslant m\}.$$

A critical point of $I|_{E_m}$ will be a solution of the L^2 orthogonal projection of (1.5) onto E_m.

A series of lemmas in [4] use (f_1)–(f_4) and the form of I to establish the existence of a nontrivial critical point u_{mk} of $I|_{E_m}$ as well as an estimate on the corresponding critical value c_{mk} of the form

$$(1.7) \qquad 0 < c_{mk} = I(u_{mk}; k, \beta, K) \leqslant M_k$$

where M_k is a constant independent of β, K, and m. Further arguments in [4] allow successively letting $m \to \infty$ and $\beta \to 0$ to get a solution u_k of

$$(1.8) \qquad \begin{cases} u_{\tau\tau} - k^2(u_{xx} - f_K(x, k\tau, u)) = 0 & 0 < x < \pi \\ u(0, \tau) = 0 = u(\pi, \tau); u(x, \tau + 2\pi) = u(x, \tau) \end{cases}$$

with $c_k \equiv I(u_k, k, 0, K) \leqslant M_k$. Moreover for $K = K(k)$ sufficiently large, $\|u_k\|_{L^\infty} \leqslant K$ so $f_K(x, k\tau, u_k) = f(x, k\tau, u_k)$ and u_k satisfies (1.4). Lastly a separate argument shows $c_k > 0$ so $u_k \not\equiv 0$ via (f_1) and the form of I.

Returning to the question of how many of the functions u_k are distinct, we will first study the dependence of M_k on k. To do so requires a closer look at how the bound M_k is determined. Lemma 1.13 of [4] provides a minimax characterization of $I(u_{mk}; k, \beta, K)$ which in turn yields the bound M_k. Let

$$W_{mk} = \text{span}\{\sin jx \sin n\tau, \sin jx \cos n\tau \mid 0 \leqslant j, n$$

$$\leqslant m \text{ and } n^2 \leqslant j^2 k^2\},$$

$$\varphi_k = \alpha_k \sin x \sin(k+1)\tau$$

and α_k is chosen so that $\|\varphi_k\|_{L^2} = 1$. Set $V_{mk} = W_{mk} \oplus \text{span}\{\varphi_k\}$. It was shown in [4] that

$$(1.9) \qquad 0 < c_{mk} \leqslant \max_{u \in V_{mk}} I(u; k, \beta, K)$$

(Note that $I \to -\infty$ as $\|u\|_{L^2} \to \infty$ via (f_3) so we have a max rather than a sup in (1.9)). Let $z = z_{mk}$ denote the point in V_{mk} at which the max is attained. We can write

$$(1.10) \qquad z = \|z\|_{L^2}(\gamma\xi + \delta\varphi_k)$$

where $\xi \in W_{mk}$ with $\|\xi\|_{L^2} = 1$ and $\gamma^2 + \delta^2 = 1$. Substituting (1.10) into (1.9) and using the form of I yields

$$(1.11) \qquad k^2 \int_0^{2\pi} \int_0^{\pi} F_K(x, k\tau, z)\, dx\, d\tau \leqslant \frac{1}{2} \int_0^{2\pi} \int_0^{\pi} (z_\tau^2 - k^2 z_x^2)\, dx\, d\tau$$

$$\leqslant \frac{\delta^2}{2} \|z\|_{L^2}^2 \int_0^{2\pi} \int_0^{\pi} (\varphi_{k_\tau}^2 - k^2 \varphi_{k_x}^2)\, dx\, d\tau$$

$$\leqslant \overline{M} \|z\|_{L^2}^2 k$$

where $\overline{M}$ is independent of k and m (as well as β and K). Since F_K satisfies (1.1) with a constant $\bar{\mu}$ independent of K, (1.11) shows that

$$(1.12) \qquad k\big(a_1\|z\|_{L^\mu}^{\bar{\mu}} - a_3\big) \leqslant \overline{M}\|z\|_{L^2}^2.$$

By the Hölder inequality we find that

$$(1.13) \qquad k\big(a_4\|z\|_{L^2}^{\bar{\mu}} - a_3\big) \leqslant \overline{M}\|z\|_{L^2}^2$$

which implies that

(1.14) $\|z\|_{L^2} \le \overline{M}_1$

with $\overline{M}_1$ independent of m, k, β, K. Returning to (1.9) and using (1.14) yields

(1.15) $c_{mk} = I(u_{mk}; k, \beta, K) \le \overline{M}_2 k$

with $\overline{M}_2$ independent of m, k, β, K. It follows that c_k satisfies the same estimate:

(1.16) $c_k = I(u_k; k, 0, K) \le \overline{M}_2 k$

To complete the proof of Theorem 1.2, we will show that (1.16) is violated if more than finitely many solutions u_k correspond to the same function in the original t variables. To present the idea in its simplest setting, suppose first that all of the functions $u_k(x, \tau)$ are reparameterizations of $u_1(x, t)$. Then $u_k(x, \tau) = u_1(x, k\tau) = u_1(x, t) \equiv u(x, t)$. For $K = K(k)$ sufficiently large we have

$$(1.17) \quad c_k = \int_0^{2\pi} \int_0^\pi \left[\frac{1}{2} u_{k\tau}^2 - k^2 \left(\frac{u_{kx}^2}{2} + F(x, k\tau, u_k) \right) \right] dx \, d\tau$$

$$= k \int_0^{2\pi k} \int_0^\pi \left[\frac{1}{2} (u_t^2 - u_x^2) - F(x, t, u) \right] dx \, dt$$

$$= k^2 \int_0^{2\pi} \int_0^\pi \left[\frac{1}{2} (u_t^2 - u_x^2) - F(x, t, u) \right] dx \, dt$$

$$= k^2 c_1$$

since u is 2π periodic in t. The positivity of c_1 and (1.17) show that c_k tends to inifinity like k^2 contrary to the bound (1.16). This argument shows (1.3) has at least one $2\pi k$ periodic solution distinct from $u_1(x, t)$.

For the general case we argue similarly. Suppose two solutions $u_j(x, \tau)$ and $u_k(x, \tau)$ correspond to the same function of (x, t), i.e. $u_j(x, \tau) = u_j(x, t/j) \equiv v(x, t) \equiv u_k(x, t/k)$. Thus $u_j(x, \tau) = v(x, j\tau)$ and $u_k(x, \tau) = v(x, k\tau)$. Since $v(x, t)$ is both $2\pi j$ and $2\pi k$ periodic in t, there are $j_1, k_1, \sigma \in \mathbb{N}$ such that $j = \sigma j_1, k = \sigma k_1$ and v is $2\pi\sigma$ periodic in t. (We can take σ to be the greatest common divisor of j and k). Arguing as in (1.17) yields

$$(1.18) \quad c_k = k \int_0^{2\pi k} \int_0^\pi \left[\frac{1}{2} (v_t^2 - v_x^2) - F(x,t,v) \right] dx \, dt$$

$$= \frac{k^2}{\sigma} \int_0^{2\pi\sigma} \int_0^\pi \left[\frac{1}{2} (v_t^2 - v_x^2) - F(x,t,v) \right] dx \, dt$$

$$\equiv \frac{k^2}{\sigma} A$$

and

$$(1.19) \quad c_j = \frac{j^2}{\sigma} A$$

Thus if there is a sequence u_{k_i} of solutions of (1.4) corresponding to the same function v, by (1.18)–(1.19) we have

$$(1.20) \quad c_{k_i} = \frac{k_i^2}{\sigma} A$$

where $\sigma \in \mathbb{N}$ is the greatest common divisor of $\{k_i\}$. Hence $c_{k_i} \to \infty$ like k_i^2 contrary to (1.16) and the proof of Theorem 1.2 is complete.

Remark 1.21. Note that if $F(x,t,r)$ and F_K satisfy

$$F, F_K \geqslant a_1 |r|^\nu$$

for some $\nu > 2$, it follows from (1.11) that

$$\|z\|_{L^2} \leqslant a_5 k^{-1/(\nu-2)}$$

and therefore

$$c_k \leqslant a_6 k^{1-(2/\nu-2)} = a_6 k^{\nu-4/\nu-2}$$

Thus if $\nu < 4, c_k \to 0$ as $k \to \infty$. Further restrictions on F (as in [1]) imply $u_k \to 0$ as $k \to \infty$.

Remark 1.22. Existence of infinitely many distinct subharmonic solutions was also established in [1] for a family of subquadratic Hamiltonian systems, i.e. Hamiltonian systems where H grows less rapidly than quadratically as $|z| \to \infty$. There are several existence theorems for periodic solutions of semilinear wave equations in which the primitive of the forcing term is subquadratic [6–10]. We believe the conclusions of this paper carry over to the subquadratic case via the arguments used here and in [1].

UNIVERSITY OF WISCONSIN

REFERENCES

[1] Rabinowitz, P. H., On subharmonic solutions of Hamiltonian systems, *Comm. Pure Appl. Math.* **33** (1980), 609–633.

[2] Jacobowitz, H., Periodic solutions of $x'' + f(t, x) = 0$ via the Poincaré-Birkhoff Theorem, *J. Diff. Eq.* **20** (1976), 37–53.

[3] Hartman, P., On boundary value problems for superlinear second order differential equations, *J. Diff. Eq.* **26** (1977), 37–53.

[4] Rabinowitz, P. H., Free vibrations for a semilinear wave equation, *Comm. Pure Appl. Math.* **31** (1978), 31–68.

[5] Brezis, H., J. M. Coron, and L. Nirenberg, Free vibrations for a nonlinear wave equation and a theorem of P. Rabinowitz, to appear *Comm. Pure Appl. Math.*

[6] Rabinowitz, P. H., Some minimax theorems and applications to nonlinear partial differential equations, *Nonlinear Analysis: A collection of papers in honor of Erich H. Rothe*, edited by L. Cesari, R. Kannan, and H.F. Weinberger, Academic Press (1978), 161–177.

[7] Coron, J. M., Résolution de l'equation $Au + Bu = f$ ou A est linéaire et B derive d'un potential convexe, to appear *Ann. Fac. Sc. Toulouse.*

[8] Brezis, H. and L. Nirenberg, Forced vibrations for a nonlinear wave equation, *Comm. Pure Appl. Math.* **31** (1978), 1–30.

[9] Bahri, A. and H. Brezis, Periodic solutions of a nonlinear wave equation, *Proc. Roy. Soc. Edinburgh* **85A** (1980), 313, 320.

[10] Brezis, H. and J. M. Coron, Periodic solutions of nonlinear wave equations and Hamiltonian systems, Preprint.

ON THE HELMHOLTZ EQUATION

By Richard Sacksteder

---

1. Introduction. It seems appropriate to discuss the Helmholtz (or reduced wave) equation

$$(1) \qquad \Delta u + \omega^2 u = 0 \qquad \left(\Delta = \sum_{i=1}^{3} \partial^2/(\partial x_i)^2, \, \omega \neq 0 \text{ real} \right)$$

here, since my first introduction to its theory came from papers of Hartman ([7] and [8]). Most of the results presented can be generalized to Euclidean spaces of arbitrary dimension, but the discussion here will be limited to the three dimensional case because our motivation (like that of Helmholtz himself [9], [10]) stems from attempts to understand the acoustics of musical instruments.

Equation (1) comes from seeking sinusoidal solutions of the wave equation

$$(2) \qquad c^2 \Delta u = \partial^2 u/\partial t^2$$

(of frequency $f = c\omega/2\pi$, $c = $ velocity of wave propagation) that are of the form

$$(3) \qquad \varphi(x,t) = u(x)e^{-i\omega ct}.$$

There are good reasons for considering solutions of (1) that are complex valued even though one may utimately be interested only in real valued solutions of (2):

1. A shift in the time transforms a solution φ of (2) into another, but if φ is as in (3) this obviously corresponds to multiplication of u by a complex number of absolute value 1.

2. A boundary condition at ∞ must be imposed to assure that no energy is being extracted from outer space as a result of the wave motion. This condition can be formulated elegantly for complex solu-

293

tions as

(4) $\qquad \dfrac{\partial u}{\partial r} - i\omega u = o(r^{-1}), \qquad (r = |x|), \qquad$ (uniformly along rays),

which is known as the Sommerfeld radiation condition.

Our concern here is with questions related to the classical boundary value problems for (1), especially in exterior domains. The existence theorems for these problems were proved long ago [11], [14], [15], [20], [25]; however, it seems worthwhile to give a simple and unified treatment of them here. In Section 2 the formal and analytic properties of various operators connected with the solution of (1) are developed and these results are applied to the classical boundary value problems in Section 3. Energy radiation and the Sommerfeld condition are discussed in Section 4. Spectral properties of the operators are investigated in Section 5 and some applications to acoustics are taken up in the last section.

Our approach to the Helmholtz equation emphasizes the correspondence the values of the normal derivative u_N and the values of u itself assumed on the boundary for solutions u of (1). This emphasis was originally motivated by the applications that were of interest, but the treatment also serves to unify, systematise, and extend many results that are scattered through the literature.

2. Properties of related operators. Here Σ will denote a compact two-dimensional surface in E^3 of class C^2. The unbounded component of $E^3 - \Sigma$ will be called Ω_+ and the union of the remaining components will be called Ω_-. It is not necessary to assume that Σ is connected; however, it will be assumed (without essential loss of generality) that Σ has no components that are "concentric", that is, it will be assumed that the complement of each component of Ω_- is connected. Spaces of complex-valued functions defined on Σ will be designated as follows: C = continuous functions, H = Hölder continuous functions, and H' = functions with Hölder continuous first derivatives. Clearly $C \supset H \supset H'$.

For any y on Σ and any $x \neq y$ in E^3 we define the following kernels for integral operators:

$$Q(x, y) = e^{i\omega r/4\pi r}, \qquad \text{where } r = |x - y|,$$

$K(x, y) = \partial/\partial N_y Q(x, y)$, where N_y is the unit outward normal to Σ at y, $K'(x, y) = K(y, x)$, the (unconjugated) transpose of K.

These kernels define integral operators that map C into H. The same letter will be used for the operator and the corresponding kernel so that, for example,

$$Kh(x) = \int_\Sigma K(x, y)h(y)\,dy \qquad \text{for } x \text{ on } \Sigma,$$

and where dy denotes the volume element on Σ. Another operator of interest is defined (see Lemma 2.2) for elements of h of H' by

$$Ph(x) = \partial/\partial N_x \int_\Sigma K(x, y)h(y)\,dy.$$

If we define for x in E^3

$$v(x) = \int_\Sigma Q(x, y)g(y)\,dy \qquad \text{and} \qquad w(x) = \int_\Sigma K(x, y)h(y)\,dy,$$

both v and w satisfy (1) and (4) in $E^3 - \Sigma$, and are real analytic there. In addition, v, but not w, is continuous in all of E^3. The well-known jump relations assert that for g and h in C and x on Σ.

(5) $$v_{N_\pm}(x) = \lim \partial v/\partial N(x') = \left(K' \mp \tfrac{1}{2}I \right) g(x)$$

as $x' \in \Omega_\pm$ approaches x,

(6) $$w_\pm(x) = \lim w(x') = \left(K \pm \tfrac{1}{2}I \right) h(x)$$

as $x' \in \Omega_\pm$ approaches x,

(cf. [3], [11], [14]). Here I denotes the identity map. The following lemmas are proved in the same way as the corresponding ones in potential theory, good sources for which are [5] and [22].

LEMMA 2.1. K, K', and Q are compact operators which map C into H and H into H'. Their eigenfunctions are elements of H'.

LEMMA 2.2. The operator P is well defined on H' and maps H' into C. Moreover P is formally symmetric, that is, if g and h are in H', $\int_\Sigma Ph(x)g(x)\,dx = \int_\Sigma h(x)Pg(x)\,dx$.

The relations between the operators are described in the following theorem.

THEOREM 2.1. *The operators defined above satisfy the following identities*:

(7) $KQ = QK^t$ (*maps C into H'*)

(8) $PK = K'P$ (*maps H' into H*)

(9) $PQ = (K^t)^2 - \frac{1}{4}I$ (*maps C into C*)

(10) $QP = K^2 - \frac{1}{4}I$ (*maps H' into H'*).

Moreover, $Qg = 0$ *if and only if* $K'g = \frac{1}{2}g$ *and* $Ph = 0$ *if and only if* $Kh = -\frac{1}{2}h$. *The invariant subspace corresponding to an eigenvalue of* K^t *different from* $\frac{1}{2}$ *is mapped by* Q *isomorphically onto the corresponding subspace for K. The invariant subspace corresponding to an eigenvalue of K different from* $-\frac{1}{2}$ *is mapped isomorphically by P onto the corresponding subspace for* K'.

The relation (9) shows that P makes sense on the range of Q as well as on H' (cf. lemma 2.2); therefore, if E denotes the smallest subspace of C containing H' and the range of Q, E is a natural domain for P. (P can still be interpreted as a two-sided normal derivative. cf. [5 p. 297].)

The formal relations (7)–(10) can be conveniently summarized by defining a 2×2 matrix of operators

$$T = \begin{pmatrix} -K^t, & P \\ -Q, & K \end{pmatrix}$$

operating on the column vectors in $D = \{(g, h)^t : g \in C, h \in E\}$.

COROLLARY 2.1. *T maps D into D and*

(11) $T^2 = \frac{1}{4}I$, *hence* $(\frac{1}{2}I \pm T)^2 = \frac{1}{2}I \pm T$.

T has eigenvalues $\pm\frac{1}{2}$ *and* $(\frac{1}{2}I \pm T)$ *act as the projections from D onto the spaces of eigenvectors, which together span D.*

The assertions (11) follow formally from (7)-(10), and the final statement follows easily from (11).

To prove the theorem, first note that if x_+ is in Ω_+, $Q(x_+, y)$ satisfies (1) for y in Ω_-. Green's identity therefore gives for any solution v of (1) in Ω_-,

$$\int_\Sigma K(x_+, y)v(y)\,dy = \int_\Sigma Q(x_+, y)v_{N_-}(y)\,dy.$$

and if v is as above, the continuity of v and (5) give

(12) $$\int_\Sigma K(x_+, y)Qg(y)\,dy = \int_\Sigma Q(x_+, y)\left(K' + \tfrac{1}{2}I\right)g(y)\,dy.$$

Letting x_+ approach a point on Σ and using (6) give $(K + \tfrac{1}{2}I)Qg = Q(K' + \tfrac{1}{2}I)g$, which essentially (7). The assertion about the range in (7) follows from Lemma 2.1. Taking the normal derivative at a point x of Σ on the right side of (12) gives by (5), $(K' - \tfrac{1}{2}I)(K' + \tfrac{1}{2}I)g(x)$. Therefore the normal derivative on the left is also defined and we get (9).

A similar argument with the function w defined above replacing v and using (6) and Lemma 2.2 gives

$$\int_\Sigma K(x_+, y)\left(K - \tfrac{1}{2}I\right)h(y)\,dy = \int_\Sigma Q(x_+, y)Ph(y)\,dy$$

for h in H'. Now letting x_+ approach Σ gives (10) and taking the normal derivative at a point of Σ gives (8) in view of (5) and Lemma 2.2, since $(K - \tfrac{1}{2}I)h$ is in H'. Since $v_{N_+} = 0$ is equivalent to $K'g = \tfrac{1}{2}g$ by (5) and since by the Kupradse-Rellich uniqueness theorem [12], [18], [21], $v_{N_+} \equiv 0$ if and only if $v \equiv Qg \equiv 0$ on Σ, the assertion about the nullspace of Q is proved. Similarly (using w) one sees that $Ph = 0$ if and only if $Kh = -\tfrac{1}{2}h$. The remaining assertions follow immediately from (7)–(10).

Finally, we note that if $\overline{T}$ denotes the operator that is obtained by taking complex conjugates of the operators used to define T (or equivalently by replacing ω by $-\omega$), $\overline{T}$ satisfies the relations (11). Moreover arguments like those used to derive (7)–(10) show that

(13) $$\overline{T}T = \tfrac{1}{2}(\overline{T} - T) + \tfrac{1}{4}I \quad \text{and} \quad T\overline{T} = \tfrac{1}{2}(T - \overline{T}) + \tfrac{1}{4}I.$$

3. Boundary value problems. The eigenspaces of T described in Corollary 2.1 can be interpreted as boundary data corresponding to solutions of (1). Let the eigenspace belonging to $\tfrac{1}{2}$ be denoted by D_+

and that belonging to $-\frac{1}{2}$ by D_-. Then if $F_+ = (g_+, h_+)'$ is in D_+,

(14) $\left(\frac{1}{2}I + K'\right)g_+ - Ph_+ = 0,$ and

(15) $Qg_+ + \left(\frac{1}{2}I - K\right)h_+ = 0.$

Defining v_+ and w_+ as v and w were defined above with g_+ and h_+ playing the roles of g and h, we find that if $u_+ = w_+ - v_+, u_+$ satisfies (1) and (4) in Ω_+. Moreover, by (5) and (6) $g_+(x) = \partial u_+/\partial N_+(x)$ and $h_+(x) = u_+(x)$ for x on Σ, that is F_+ represents the boundary data for u_+. Similarly, if F_- is in D_-, it represents the boundary data for a solution $u_-(x) = w_-(x) - v_-(x)$ defined in Ω_-. These observations can be summarized as follows:

THEOREM 3.1.   *Let F be an element of D. Then $F = F_+ + F_-$ where $F_+ = (\frac{1}{2}I + T)F$ is in D_+ and $F_- = (\frac{1}{2}I - T)F$ is in D_-. There is a unique solution u_+ of (1) in Ω_+ satisfying (4) and such that $\partial u_+/\partial N_+(x) = g_+(x)$ and $u_+(x) = h_+(x)$ on Σ. There is also a solution $u_-(x)$ of (1) in Ω_- such that $u_-(x) = h_-(x)$ and $\partial u_-/\partial N_-(x) = g_-(x)$ for x on Σ.*

Solving boundary value problems for (1) in Ω_+ subject to (4) or in Ω_- therefore reduces to finding appropriate elements of D_+ or D_-. For example, consider the Neumann problem for (1) subject to (4) in Ω_+. Then an element g_+ of C is given and the problem is to find an h_+ in E such that $F_+ = (g_+, h_+)'$ is in D_+, or equivalently that (14) and (15) are satisfied. To see how the existence of h_+ follows from Fredholm theory first note the condition for the solvability of (15) for h_+ is that $(Qg_+, f) = \int_\Sigma (Qg_+)(x)f(x)dx = 0$ for every f such that $\frac{1}{2}f = K'f$. But the symmetry of Q and (7) give $(Qg_+, f) = 2(Qg_+, K'f) = 2(g_+, KQf)$ and by Theorem 2.1, $Qf = 0$. To see that (14) can also be satisfied, let h_0 denote some solution of (15). (If $\frac{1}{2}I - K'$ has a non-trivial kernel the solution is not unique.) It is necessary to find $q = h_+ - h_0$ such that $Kq = \frac{1}{2}q$ and (14) is satisfied, that is

(16) $Pq = \left(\frac{1}{2}I + K'\right)g_+ - Ph_0.$

Note that since h_0 is a solution of (15), Lemma 2.1 shows that h_0 is in H and that Kh_0 is in H'. Then (15) shows that h_0 is in E, hence Ph_0 is defined. Therefore the right side of (16) makes sense. By Theorem 2.1, P maps the eigenspace of K belonging to $\frac{1}{2}$ isomorphically onto the corresponding eigenspace for K'. The right side of (16) is in the latter eigenspace by (9), (8), and the fact that h_0 satisfies (15). This solves the Neumann problem for Ω_+.

The Neumann problem for Ω_- is solved similarly, but if $K + \frac{1}{2}I$ has a non-trivial kernel it is necessary to assume that $(Qg_-, f) = 0$ for every f such that $-\frac{1}{2}f = K'f$. By (7) and the symmetry of Q this condition is equivalent to $(g_-, f') = 0$ for every f' such that $-\frac{1}{2}f' = Kf'$. The solution is only unique modulo elements of the kernel of $K + \frac{1}{2}I$.

The same method can be used to solve the Dirichlet problem provided it is assumed that the boundary values are given by an element of E. However, this undesirable smoothness condition can be removed by using some of the results from Section 5 below (cf. [11], [15], [20], [25]).

Returning now to the Neumann problem for Ω_+, define $H_\omega(g) = \omega h$, where $h = h_+$ is the solution of (14) and (15) corresponding to the boundary data $g = g_+$.

LEMMA 3.1. *H_ω is a compact operator on C mapping C into E.*

Proof. H_ω is well-defined by the uniqueness of the solution to the Neumann problem in Ω_+. The compactness of H_ω is obvious if $\frac{1}{2}$ is not an eigenvalue of K, because then $H_\omega = \omega(K - \frac{1}{2}I)^{-1}Q$ is the composition of a compact operator with a bounded operator. In general, $(\frac{1}{2}I - K)^{-1}$ can be defined on a space of finite co-dimension to give a unique solution $h_0 = (K - \frac{1}{2}I)^{-1}Qg_+$ of (15). Then the correspondence between $g = g_+$ and the right side of (16) is compact since the right side is contained in a finite dimensional subspace of C. The restriction of P^{-1} to this subspace makes sense and is an isomorphism, hence H_ω is compact.

The third boundary value problem for Ω_+ is defined as follows: Suppose that two elements p and q of C are given with $\text{Im}\,\omega p \geq 0$. Then the third boundary value problem is to find a solution u of (1) and (4) in Ω_+ that satisfies $\partial u/\partial N_+(x) + p(x)u(x) = q(x)$ for x on Σ, or equivalently to find an element g of C such that

$$(17) \qquad \left(I + p(x)\omega^{-1}H_\omega\right)g(x) = q(x) \qquad \text{for } x \text{ on } \Sigma.$$

By Lemma 3.1, H_ω hence pH_ω is compact. The existence of the solution then follows from the Fredholm alternative provided that it can be shown that the case $q = 0$ of (17) does not have a non-trivil solution. If g were such a solution, the integral

$$i\int_{\Sigma} \bar{g}(x)H_{\omega}g(x)\,dx = -i\int_{\Sigma} \bar{p}(x)\omega^{-1}|H_{\omega}g(x)|^2\,dx$$

would have non-positive real part because of Im $\omega p \geq 0$. But it is known (see Section 4) that any non-trivial solution u of (1) and (4) satisfies

$$\text{Real } i\omega \int_{\Sigma} u(x)\partial\bar{u}/\partial N_+(x)\,dx > 0.$$

This shows that (17) is always solvable for g.

4. The energy form. Let u be a solution of (1) and let φ be the corresponding solution of (2) in Ω_+ defined by (3). Then $U(x,t) = $ real $\varphi(x,t)$ is also a solution of (2). In accoustical applications the normal derivative U_{N_+} represents the infinitesimal displacement of points of Σ and $-U_t$ is a positive constant multiple of the deviation of the pressure from its steady state value. Therefore the energy put into the sound wave during a period of length $T = |2\pi/\omega c|$ is a positive constant multiple of the integral

$$-T^{-1}\int_{\Sigma}\int_0^T U_{N_+}(x,t)\,U_t(x,t)\,dx\,dt.$$

A simple computation shows that this is a positive constant multiple of

$$(19) \qquad E(u) = i\omega \int_{\Sigma}\left(u(x)\bar{u}_{N_+}(x) - u_{N_+}(x)\bar{u}(x)\right)dx.$$

These considerations motivate the introduction of the Hermitian form $< F_1, F_2 >$ defined for elements $F_i = (g_i, h_i)^t$ of D by

$$(20) \qquad \langle F_1, F_2 \rangle = i\omega \int_{\Sigma}\left(h_1(x)\bar{g}_2(x) - g_1(x)\bar{h}_2(x)\right)dx.$$

Note that if $F = (u_{N_+}, u)^t$, $< F, F >$ is just $E(u)$. The same considerations can be applied to solutions u of (1) in Ω_- but in that case $E(u)$ always vanishes as follows by applying Green's theorem to u and $\bar{u}$. More generally, if Σ' is homologous to Σ and u is a solution of (1) in the volume between these surfaces, (19) is unchanged by replacing Σ by Σ'.

In particular if Σ' is a sphere of radius R centered at the origin, letting $R \to \infty$ leads to (cf. [3], [13], [14], [26]):

LEMMA 4.1. *If u satisfies* (1) *and* (4) *in* $\Omega_+, E(u) \geqq 0$, *with equality only if* $u \equiv 0$. *Equivalently, if* $< F_1, F_2 >$ *is restricted to the boundary data for such functions u, that is, to elements of D_+, it becomes a positive definite Hermitian form.*

Functions u satisfying (1) in Ω_+ and the incoming wave condition (4') $\partial u/\partial r + i\omega u = o(r^{-1})$ $(r = |x|$, uniformly along rays) are characterized by having boundary data F in $\overline{D}_+ = \{F \in D : \overline{T}F = \frac{1}{2}F\}$. This assertion is obvious in view of what has been shown, since $\overline{T}$ is just T with ω replaced by $-\omega$ and (4') is just (4) with the same change. It is also clear that the restriction of $< F_1, F_2 >$ to such boundary data is negative definite. The remainder of this section is devoted to the problem of characterizing the boundary data F in D such that

$$(21) \qquad F = F_1 + F_2 = \left(\tfrac{1}{2}I + T\right)F_1 + \left(\tfrac{1}{2}I + \overline{T}\right)F_2$$

$$\text{for } F_1 \text{ in } D_+ \text{ and } F_2 \text{ in } \overline{D}_+.$$

This is equivalent to characterizing the data F corresponding to solutions $u = u_1 + u_2$ in Ω_+ such that u_1 satisfies (4) and u_2 satisfies (4'), or briefly u is the sum of an incoming and an outgoing wave.

It follows from results of Hartman and Wilcox [8] (sections 5 and 6) that solutions u do not in general admit such decompositions even if they are defined in all of E^3. In view of Theorem 3.1, it is sufficient in order to characterize the F's satisfying (21) to limit consideration to those F's in D_-. First note that $(\overline{T} - T)D_- = 0$. For if $F = (\frac{1}{2}I - T) \cdot G$, it follows from (13) that $(\overline{T} - T)F = 0$. Conversely if $(\overline{T} - T)F = 0$, $(\frac{1}{2}I + \overline{T})F = (\frac{1}{2}I + T)F$ is in $D_+ \cap \overline{D}_+$, hence Lemma 4.1 and the remarks below it show that $(\frac{1}{2}I + T)F = 0$, that is, F is in D_-. This proves the first assertion of:

THEOREM 4.2. $D_- = \text{Ker}(\overline{T} - T)$. *An element F in D_- satisfies* (21) *if and only if F is in Range* $(\overline{T} - T) \subset \text{Ker}(\overline{T} - T)$.
To complete the proof note that (11) and (13) imply that $(\overline{T} - T)^2 = 0$. If (21) is satisfied for F in D_-, applying $(\frac{1}{2}I - T)$ to both sides gives $F = (\overline{T} - T)F_2$. Conversely if $F = (\overline{T} - T)G$, $F = -(\frac{1}{2}I + T)G + (\frac{1}{2}I + \overline{T})G = F_1 + F_2$.

5. Spectral properties. The following lemma is easily verified by direct computation, using Theorem 2.1 and noting that Lemma 2.1 shows that eigenfunctions of K and K' must be in H'.

LEMMA 5.1. *Let $g \neq 0$ satisfy $K'g = \lambda g$. Then $F_+ = ((\lambda - \frac{1}{2}I)g,$ $Qg)'$ is an element of D_+ and $F_- = ((\lambda + \frac{1}{2}I)g, Qg)'$ is an element of D_-. F_+ is non-trivial if and only if $\lambda \neq \frac{1}{2}$ and F_- is always non-trivial. If $h \neq 0$ satisfies $Kh = \lambda h, G_+ = (Ph, (\lambda + \frac{1}{2}I)h)'$ is an element of D_+, and $G_- = (Ph, (\lambda - \frac{1}{2}I)h)'$ is an element of D_-. G_+ is non-trivial if and only if $\lambda \neq -\frac{1}{2}$ and G_- is always non-trivial.*

COROLLARY 5.1. $< F_-, F_- > = < G_-, G_- > = 0.$ $< F_+, F_+ >$ > 0 *unless* $\lambda = \frac{1}{2}$ *and* $< G_+, G_+ >> 0$ *unless* $\lambda = -\frac{1}{2}$.

COROLLARY 5.2. *(cf. [11], [14]). The only possible real eigenvalues of K and K' are $\pm \frac{1}{2}$.*

Corollary 5.2 follows from Lemma 5.1, because if λ is real $(\lambda + \frac{1}{2})$ $< F_+, F_+ > = (\lambda - \frac{1}{2}) < F_-, F_- > = 0$. But the term on the left can only be zero if $\lambda = \pm \frac{1}{2}$.

COROLLARY 5.3. *(cf. [14]. If $K'g = \frac{1}{2}g, K'\bar{g} = \frac{1}{2}\bar{g}$. If $Kh = -\frac{1}{2}h,$ $K\bar{h} = -\frac{1}{2}\bar{h}$.*

Proof. $F_- = (g, 0)'$ is in $D_- = \bar{D}_-$. Therefore $\bar{T}F_- = -\frac{1}{2}F_-$ and $T\bar{F}_- = -\frac{1}{2}\bar{F}_-$, but the first component of the latter is $K'\bar{g} = \frac{1}{2}\bar{g}$. The other half is similar.

The following lemma is also verified by direct computation.

LEMMA 5.2. *Assume the notation of Lemma 5.1 and suppose that $K'g' = \lambda g' + g$, or $Kh' = \lambda h' + h$. Then $F'_+ = ((\lambda - \frac{1}{2})g' + g, Qg')'$ is in D_+ and $F' = ((\lambda + \frac{1}{2})g' + g, Qg')'$ is in D_-, or $G'_+ = (Ph', (\lambda + \frac{1}{2})$ $h' + h)'$ is in D_+ and $G'_- = (Ph', (\lambda - \frac{1}{2})h' + h)'$ is in D_-. All are non-trivial.*

COROLLARY 5.4. *(cf. [14]) If λ is real the elements g' and h' of Lemma 5.2 cannot exist. That is, the real eigenvalues of K and K' are simple.*

Proof. If, for example, $\lambda = -\frac{1}{2}$ were not a simple eigenvalue of $K', F'_- = (g, Qg')'$ and $\bar{F}_- = (0, \overline{Qg})'$ would be in D_-. Applying

Green's formula to the solutions u' and u of (1) with boundary data given by F'_- and $\overline{F}_-$, gives $0 = \int_\Sigma (u' u_{N_-} - u u'_{N_-}) \, dx = \int_\Sigma g \, \overline{Qg} \, dx$. But $F_+ = (-g, Qg)'$ is in D_+, hence $< F_+, F_+ > = 2 \text{ real } i \int_\Sigma g \, \overline{Qg} \, dx > 0$, which contradicts the above equality. Similarly $\lambda = \frac{1}{2}$ must be a simple eigenvalue of K, hence the simplicity of all real eigenvalues of K and K' follows from the last part of Theorem 2.1 and Corollary 5.2.

Turning now to the compact operator H_ω, note first that the imaginary part of any of its eigenvalues is negative. For if $H_\omega(g) = \theta g$, where $\theta = a - ib$ and $g \neq 0, F = (g, \omega^{-1}\theta g)'$ is in D_+ and $\langle\langle F \rangle\rangle^2 = \langle F, F \rangle = 2b \int_\Sigma |g(x)|^2 \, dx = 2b\|g\|^2 > 0$. Note that since $\langle F_1, F_2 \rangle$ is positive definite on D_+, Minkowski's inequality is valid, that is for F_1 and F_2 in D_+,

(22) $\qquad \langle\langle F_1 \pm F_2 \rangle\rangle \leqq \langle\langle F_1 \rangle\rangle + \langle\langle F_2 \rangle\rangle.$

This inequality will be used to estimate F when F_1 and F_2 are taken as a boundary data for the single and double layer parts of the solution u having boundary data F. That is, in the notation of Section 2, $F_1 = (v_{N_+}, v)'$ and $F_2 = (w_{N_+}, w)'$, where $h = \omega^{-1}\theta g$ so that $F = F_1 - F_2$. The terms on the right in (22) will be estimated by replacing Σ in (4') by a large sphere S_R of radius R centered at the origin. Note that for y on Σ and x on S_R, $|Q(x, y)| = (4\pi R)^{-1} + 0(R^{-2})$. Then if A is the area of Σ the Schwarz inequality gives: for x on S_R:

$$|v(x)|^2 = \left| \int_\Sigma Q(x, y) g(y) \, dy \right|^2 \leqq A(4\pi R)^{-2}\|g\|^2 + 0(R^{-3}).$$

Therefore $\langle\langle F_1 \rangle\rangle^2 = i\omega \int_{S_R} (v\bar{v}_N - \bar{v}v_N) \, dx \leqq 2\omega^2 \int_{S_R} |v|^2 \, dx + 0(R^{-1}) \leqq 2\omega^2 A(4\pi)^{-1}\|g\|^2 + 0(R^{-1})$. Similarly if $\alpha(x, y)$ denotes the angle between $x - y$ and the normal to Σ at y, $|w(x)|^2 = |\int_\Sigma \omega^{-1}\theta g(y) K(x, y) \, dy|^2 \leqq (4\pi R)^{-2}|\theta|^2|\int_\Sigma g(y)\cos\alpha \, dy|^2 + 0(R^{-3}) \leqq (4\pi R)^{-2}|\theta|^2\|g\|^2 \times \int_\Sigma |\cos\alpha(x, y)|^2 \, dy + 0(R^{-3})$. But for any y on Σ, $\int_{S_R} \cos^2\alpha(x, y) \, dx = 4\pi R^2/3 + 0(R)$, so $\langle\langle F_2 \rangle\rangle^2 = \omega^2|\theta|^2 A\|g\|^2/6\pi$. Combining the above estimates in (22) gives

(23) $\qquad 0 < b \leqq A\omega^2(4\pi)^{-1}(1 + 3^{-1/2}|\theta|)^2.$

Therefore we have proved:

THEOREM 5.1. *If $\theta = a - ib$ is an eigenvalue of H_ω, (23) must be satisfied.*

6. Remarks on applications. The known methods of treating boundary value problems for (1) seek solutions u that are differences of single layer solutions v and double layer solutions w. Such a decomposition is not unique unless one requires, as we have here, that the single layer density g be identified with the boundary values of u_N and the double layer density h with those of u. There are advantages to insisting on such an identification in applications where the main interest is in the correspondence between u_N and u on Σ, that is, in the map H_ω. Helmholtz's own analysis of the operation of wind instruments is a good example of such an application.

A wind instrument works by replacing the energy that is lost through the radiation of sound with energy introduced by puffs of air admitted through the mouthpiece. Helmholtz sought to understand this process by isolating sound radiation from mouthpiece action by taking a cross-section S of the instrument bore near the mouthpiece and carrying out separate analyses on the parts of the divided instrument. Then the sounds that the instrument produces can be explained in terms of how the pressure and velocity can be made to match along S. (cf. [1], [2], [9], [10], [19]).

Acousticians still use methods stemming from Helmholtz's ideas. There are, to be sure, inessential differences such as the modern tendency to describe acoustical phenomena in the terminology of electrical engineering (cf. eg., [19]). For instance, it is usually assumed that at any given time pressure and velocity must be constant along the section S, and then the necessary conditions for sound production can be described in terms of "impedance match". More significantly, it is usual practice to avoid solving the exterior Neumann problem by replacing it with simpler, but less realistic, boundary value problems for the "horn" equation.

This paper has been motivated by the belief that now with the increased availability of computing facilities it should no longer be necessary to rely on simplifications such as those described in order to find practical solutions to acoustical problems. In particular, improvements in models of wind instrument operation could be expected from a better understanding of the map H_ω. Other work with similar purposes is underway in [4] and [6], where some of the smoothness requirements on Σ that have been employed here are eliminated (cf. [16], [17]).

CITY UNIVERSITY OF NEW YORK

REFERENCES

[1] Benade, A. H., *Fundamentals of Musical Acoustics*, Oxford University Press, (1976).

[2] Bixler, M. and Macksteder, R., "On the application and misapplication of acoustical theory to wind instruments", *The American Recorder*, **17** (1977), 136–142.

[3] Courant, R. and Hilbert, D., *Methods of Mathematical Physics*, vol. 2, Interscience Publishers (1962).

[4] Goldstein, J., Thesis, The City University of New York.

[5] Günter, N.M., *Potential Theory*, The Frederick Ungar Publishing Co., (1967).

[6] Harris, W.S., Jr., Thesis, The City University of New York.

[7] Hartman, P., "On the solutions of $\Delta V + V = 0$ in an exterior region," *Mathematische Zeitschrift*, **71** (1959), 251–257.

[8] Hartman, P., and Wilcox, C., "On the solutions of the Helmholtz equation in an exterior domain," *Mathematische Zeitschrift*, **75** (1961), 228–255.

[9] Helmholtz, H. von, *On the Sensations of Tone*, Appendix VII, Dover Publications Inc. (1954).

[10] ———, "Theorie der Luftschwingungen in Röhren mit offenen Enden", *Journal für Mathematik* **57** (1859), 1–72.

[11] Kupradse, W.D., *Randwertaufgaben der Schwingungstheorie und Integralgleichungen*, VEB Deutscher Verlag der Wissenschaften (1956).

[12] ———, "Über das Ausstrahlungsprinzip von A. Sommerfeld", *Doklady Akademiya Nauk*, U.S.S.R., **2** (1933), 55–58.

[13] Lax, P., and Phillips, R., *Scattering Theory*, Academic Press (1967).

[14] Leis, R., "Über das Neumannsche Randwertproblem für die Helmholtzsche Schwingungsgleichung", *Archive for Rational Mechanics and Analysis*, **2** (1959), 205–211.

[15] ———, "Zur Dirichletschen Randwertaufgabe des Aussenraumes der Schwingungsgleichung", *Mathematische Zeitschrift* **90** (1965), 205–211.

[16] ———, "Über die Randwertaufgabe des Aussenraumes zur Helmholtzschen Schwingungsgleichung", *Archive for Rational Mechanics and Analysis* **9** (1962), 21–44.

[17] ———, "Über die Dirichletsche Randwertaufgabe zur Schwingungsgleichung," *Zeitschrift für Angewandte Mathematik und Mechanik* **41** (1964), T45–T47.

[18] Levine, L., "A uniqueness theorem for the reduced wave equation", *Communications in Pure and Applied Mathematics* **17** (1964), 147–176.

[19] Morse, P. M., *Vibration and Sound*, McGraw-Hill Book Company, Inc., (1948).

[20] Müller, C., "Zur Methode der Strahlungskapazität von H. Weyl", *Mathematische Zeitschrift*, **56** (1952), 80–83.

[21] Rellich, F., "Über das asymptotische Verhalten der Lösungen von $\Delta u + \lambda u = 0$ in unendliche Gebieten", *Jahresbericht der Deutschen Mathematiker-Vereinigung* **53** (1943), 57–65.

[22] Schauder, J., "Potentialtheoretische Untersuchungen", *Mathematische Zeitschrift* **33** (1931), 602–640.

[23] Sommerfeld, A., "Die Greensche Funktion der Schwingungsgleichung", *Jahresbericht der Deutschen Mathematiker-Vereinigung*" **21** (1912), 309–353.

[24] ————, "Electromagnetiche Schwingungen", *Die Differential- und Integralgleichungen der Mechanik und Physik*, (ed. P. Frank and R. von Mises.) Dover Publications, Inc. (1961).

[25] Weyl, H., "Kapazität von Strahlungsfeldern", *Mathematische Zeitschrift* **55** (1952), 187–198.

[26] Wilcox, C.H., "Spherical means and radiation conditions", *Archive for Rational Mechanics and Analysis* **3** (1959), 133–148.

ON THE STRUCTURE OF TRANSLATION-INVARIANT CORE-MEMORIES*

By Juan Jorge Schäffer

In [1], Coffman and Schäffer examined the structure of the "memory" M that occurs in autonomous linear differential equations with delays of the form $u^{\cdot} + Mu = r$ under "natural Carathéodory conditions" in a Banach space E. In particular, M was a linear mapping from the space of all continuous functions on R with values in E to the space of all locally integrable functions on R with values in E. The autonomous nature of the equation was reflected in the translation-invariance of M. The memories considered in [1] were *short* memories; i.e., the values of Mu on an interval $[a, b]$ depended on u only through its values on $[a - l, b]$, where l is a fixed number; such memories are said to be *of length l* (in [1] the length l was normalized to be 1). It was shown in [2] that such a short memory is necessarily a continuous linear mapping. Among the results of [1] was a representation theorem for M when E is finite-dimensional.

The purpose of the work reported on here is to take a first step in a parallel investigation into the structure of *long* translation-invariant memories M, for which Mu on $]-\infty, b]$ may depend on u through all its values on $]-\infty, b]$. As usual when dealing with memories involving unbounded delays, an immediate problem is the choice of a suitable domain for the memory (for discussions of this matter from a different point of view, see [3], [4]). In this work we adopt the following attitude: no matter what the proper choice of domain for the memory M may be in any particular case, M will act at least on every continuous function that agrees with 0 on some left half-line. A first step in analysing M will therefore be to determine its action on such functions. We shall assume that this action is continuous in a mild sense, and this is sufficient to guarantee the validity of existence and uniqueness results for the equation. When M is restricted in this manner, we call it a *core-memory*.

---

*This research was supported in part by NSF Grants MCS 75-06667 and MCS 79-02546.

More specifically, let $\mathbf{K}(E)$ denote the space of all continuous functions from R to E, with the topology of uniform convergence on each compact interval. For each $t \in R$, $\mathbf{K}_{1t}(E)$ is the subspace of $\mathbf{K}(E)$ consisting of those functions that agree with 0 on $] - \infty, t]$. $\mathbf{K}_0(E)$ is the union of all the $\mathbf{K}_{1t}(E)$. The linear mapping M from $\mathbf{K}_0(E)$ to the space of locally integrable functions from R to E (with the local $\mathbf{L}^1$-topology) is a *core-memory* if $u \in \mathbf{K}_{1t}(E)$ implies that Mu agrees with 0 on $] - \infty, t]$, for all $t \in R$; it is *continuous* if the restriction of M to each $\mathbf{K}_{1t}(E)$ is continuous.

Since we have a fair understanding of the structure of short memories, it is natural to attempt to approximate the action of a core-memory by the action of suitable short memories. Because we are in "Carathéodory conditions", there are measurability problems that prevent us from carrying out this construction in a completely straight-forward manner. What we can achieve is recorded in the following theorem.

THEOREM A. *Let the translation-invariant continuous core-memory M and the numbers l, l' with $l' > l > 0$ be given. Then there exists a translation-invariant, constant-annihilating memory M' of length l' such that $M'f$ agrees with Mf on $] - \infty, t + l]$ for every $t \in R$ and all $f \in \mathbf{K}_{1t}(E)$.*

Using our knowledge of translation-invariant short memories, taken from [1], we are then able to prove the following representation theorem. (In this theorem, $[E \to E]$ denotes the space of bounded linear mappings from E to E.)

THEOREM B. *Assume that E is finite-dimensional.*

(a): *If M is a translation-invariant continuous core-memory, then there is exactly one locally square-integrable function G from $] - \infty, 0]$ to $[E \to E]$ such that*

(1) $$\int_{-\infty}^{t} Mf = \int_{-\infty}^{0} G(s)f(s + t)\,ds \quad \text{for all } f \in \mathbf{K}_0(E) \text{ and all } t \in R.$$

Moreover, for all numbers l, l' with $l' > l > 0$ there exists $G' \in \mathbf{L}_{[-l',0]}^2([E \to E])$ such that G agrees with G' on $[-l, 0]$ and

(2) *the function* $$t \mapsto \left\| t \int_{-l'}^{0} e^{its} G'(s)\,ds \right\| : R \to R \quad \text{is bounded.}$$

(b): *Conversely, if a locally square-integrable function G from* $]-\infty, 0]$ *to* $[E \to E]$ *is given, and if for every number* $l > 0$ *there exist a number* $l' > l$ *and* $G' \in \mathbf{L}^2_{[-l', 0]}([E \to E])$ *such that G agrees with G' on* $[-l, 0]$ *and G' satisfies* (2), *then there is exactly one translation-invariant continuous core-memory M satisfying* (1).

Proofs and a more detailed discussion will appear elsewhere [5].

CARNEGIE-MELLON UNIVERSITY

REFERENCES

[1] C. V. Coffman and J. J. Schäffer, The structure of translation-invariant memories. *J. Differential Equations* **16** (1974), 428–459.
[2] C. V. Coffman, B. E. Johnson, and J. J. Schäffer, The structure of translation-invariant memories: two addenda. *J. Differential Equations* **20** (1976), 270–281.
[3] B. D. Coleman and V. J. Mizel, On the general theory of fading memory. *Arch. Rational Mech. Anal.* **29** (1968), 18–31.
[4] J. Hale and J. Kato, Phase spaces for retarded equations with infinite delay. *Funkcial. Ekvac.* **21** (1978), 11–41.
[5] J. J. Schäffer, On the structure of translation-invariant core-memories. *J. Differential Equations* **37** (1980), 225–237.

DYONS EXIST

By MARTIN SCHECHTER and R. A. WEDER

---

1. Introduction. In 1968, J. Schwinger [1] formulated a dynamical interpretation of the subnuclear world on the basis of the hypothesis that electric and magnetic charge can reside on a single particle. He was led to a picture in which hadronic matter is viewed as a magnetically neutral composite of dual-charged particles. He called these particles dyons. Hitherto it has not been shown that such particles can exist according to the Yang-Mills-Higgs theory. Recently [2], the authors have given a rigorous proof of the existence of dyons according to the theory. The proof makes use of new, highly sophisticated technique generalizing work of Berger-Schechter [3]. In this paper we develop some of the mathematical methods that were used.

It was shown by Julia-Zee [4] that dyons will exist according to the Yang-Mills-Higgs theory if one can find functions $K(r)$, $J(r)$, $H(r)$ on $(0, \infty)$ such that

$$(1.1) \quad K'' = r^{-2}K(K^2 - 1) + (J^2 - H^2)K$$

$$(1.2) \quad J'' + 2r^{-1}J' = 2r^{-2}JK^2 + V'(J)$$

$$(1.3) \quad H'' + 2r^{-1}H' = 2r^{-2}HK^2$$

and

$$(1.4) \quad E = \int_0^\infty \left[K'^2 + \tfrac{1}{2}r^{-2}(K^2 - 1)^2 + \tfrac{1}{2}r^2H'^2 \right.$$
$$\left. + \tfrac{1}{2}r^2J'^2 + K^2(H^2 + J^2) + r^2V(J) \right] dr < \infty$$

Here $V(J)$ is a given non-negative function. (The expression in (1.4) represents the energy of the particle.) The requirement that rH' and rJ' are in $L^2 = L^2(0, \infty)$ implies that the limits

$$(1.5) \quad b = \lim_{r \to \infty} J(r), \quad c = \lim_{r \to \infty} H(r)$$

exist. We shall prove

THEOREM 1.1.   *If $V(J) \in C^1(\mathbb{R})$ and $V(b) = 0$ for some $b \neq 0$, then for every $c \in \mathbb{R}$ such that $c^2 < b^2$ there are functions $K \in C^4(0, \infty)$, $J \in C^2(0, \infty)$, $H \in C^6(0, \infty)$ satisfying (1.1)–(1.5). Moreover, $(1 + r)^{-\epsilon}(|J - b| + |H - c|) \in L^2$ for every $\epsilon > 0$. If $V \in C^m(\mathbb{R}), m \geqslant 1$, then $K \in C^{m+3}(0, \infty)$, $J \in C^{m+1}(0, \infty), H \in C^{m+5}(0, \infty)$. If V is real analytic, then K, J, H are real analytic on $(0, \infty)$. If $V \in C^\infty(\mathbb{R})$ and $V'(J) = JF(J)$ where $|F(J)| \leqslant C(1 + |J|^\alpha)$ for some constants C, α, $0 \leqslant \alpha < 4$, then K, J, H are infinitely differentiable at $r = 0$.*
(Note: Our notation differs from that of [4]. Our H is their J/r and our J is their H/r.)

We base the proof of this theorem on an abstract result in Banach space derived in Section 2. We cannot use the usual Frechet derivative because of the nature of our problem. Instead, we introduce a new definition of derivative in a Banach space which is appropriate for the application at hand. We anticipate that it will be useful in other applications as well. In Section 5 we describe how the abstract theory is used to prove Theorem 1.1.

2. The Generalized Derivative.   Let $F(x)$ be a mapping from a Banach space X to a Banach space Y. If $F(x)$ is defined everywhere in a neighborhood U of $x_0 \in X$ and there is a bounded linear operator A from X to Y such that

$$(2.1)\qquad \|F(x) - F(x_0) - A(x - x_0)\| = o(\|x - x_0\|), \qquad x \in U$$

then A is called the Frechet derivative of F at x_0 and designated by $F'(x_0)$. If the dependence of $F'(x_0)$ on x_0 is continuous, we say that $F \in C^1$.

In the application of the present paper we encounter a situation in which a mapping F is not defined everywhere and (2.1) does not hold. Yet the concept of derivative is appropriate nevertheless. One can easily find an alternate definition which is not so demanding and still preserves the concept of derivative. For example, suppose $G(u)$ is a map from D to Y where D is some subset of X. Let u_0 be an element of D and suppose there is a linear manifold $Q \subset X$ such that

(a) for each $q \in Q$ there is a $\delta > 0$ such that $u_0 + tq \in D$ for $|t| < \delta$.

(b) There is a linear operator A from X to Y with $D(A) = Q$ such that

(2.2) $\|G(u_0 + tq) - G(u_0) - tAq\| = o(t)$, $q \in Q$

In this case we can say that $G(u)$ has a derivative relative to Q at u_0.

This is the definition we wanted to use in the present application. However, it did not allow us to reach the conclusions that we desired. It was discovered that we needed a topology on Q as well. Our final definition is somewhere in between the two definitions given above.

Definition. Let $G(u)$ be a mapping from $D \subset X$ to Y, $u_0 \in D$ and Q a topological vector space $\subset X$. We shall say that a linear operator A from X to Y with $D(A) = Q$ is the derivative of $G(u)$ at u_0 with respect to Q if for every function $q(t) \in C^1 [(-1,1), Q]$ with $q(0) = 0$, $q'(0) = q$ there is a $\delta > 0$ such that $u_0 + q(t) \in D$ for $|t| < \delta$ and

(2.3) $\|G(u_0 + q(t)) - G(u_0) - tAq\| = o(t)$ as $t \to 0$

3. The Abstract Result. We were able to reduce the proof of Theorem 1.1 to that of the following abstract theorem in Banach space.

THEOREM 3.1. *Let N be a closed subspace of a Banach space X, and let $F(u)$ be a C^1 map from X to N. Let $G(u)$ be a real functional defined on $D \subset X$. Put $M = \{u \in X \mid F(u) = 0\}$, $S = D \cap M$. Suppose there exists a $u_0 \in S$ such that $G(u_0) = \min_S G(u)$ and $F'(u_0)$ is bijective on N. Assume further that there is a topological vector space $Q \subset X$ such that $N \subset Q$, $G'(u_0)$ exists with respect to Q and that topology of X on N is finer than that of Q on N. Assume finally that $x \in Q$, $u_0 + x \in M$ implies $u_0 + x \in D$. Then for each $x \in Q$ there is a $w \in N$ such that $G'(u_0)(x + w) = 0$.*

COROLLARY 3.2. *If, in addition, $G'(u_0)w = 0$ for all $w \in N$, then $G'(u_0)x = 0$ for all $x \in Q$.*

Proof. Let x be any element of Q. For $g \in N$ put $F(t, g) = F(u_0 + tx + g)$. Then $F(0,0) = 0$ since $u_0 \in M$, and $F_g(0,0) = F'(u_0)$, which is bijective on N. Thus by the implicit function theorem there is a C^1 function $g(t)$ in N for $|t| < \delta$, $\delta > 0$, such that $g(0) = 0$ and $F(t, g(t)) \equiv 0$. This means that $h(t) = u_0 + tx + g(t)$ is in M for

$|t| < \delta$, and consequently we see that $h(t) \in S$. Now $\| G(h(t)) - G(u_0) - t G'(u_0)(x + g'(0)) \| = o(t)$ since $g(t) \in C^1[(-\delta, \delta), Q]$. Since $G(u_0)$ is a minimum on S, we must have $G'(u_0)(x + g'(0)) = 0$. Since $g'(0) \in N$, the result follows.　　　　□

4. The Basic Technique. It is quite easy to show that equations (1.1)–(1.3) are the Euler-Lagrange equations associated with the functional

$$(4.1) \quad G = \int_0^\infty \left[K'^2 + \tfrac{1}{2} r^{-2}(K^2 - 1)^2 - \tfrac{1}{2} r^2 H'^2 + \tfrac{1}{2} r^2 J'^2 \right.$$

$$\left. + K^2(J^2 - H^2) + r^2 V(J) \right] dr$$

This functional is not bounded from below. Consequently, we cannot obtain a solution of (1.1)–(1.3) by minimizing (4.1). Our method of attack is to minimize it under certain constraints. The problem is that a minimum under constraints is almost never a stationary point without constraints. The method of the present paper is to choose the constraints in such a way that the minimum of (4.1) under the constraints is also an extremal point without constraints. This is a difficult problem; our solution requires very delicate abstract arguments as well as technical calculations. We describe the arguments here and sketch the proofs in the next section. Put $w = \{ K, J, H \}$

$$(4.2) \quad I(w) = \| K' \|^2 + \tfrac{1}{2} \| r^{-1}(K^2 - 1) \|^2 + (r, rV)$$

$$(4.3) \quad L(w) = \tfrac{1}{2} \| rJ' \|^2 + \| JK \|^2$$

$$(4.4) \quad M(w) = \tfrac{1}{2} \| rH' \|^2 + \| HK \|^2$$

Then

$$(4.5) \quad G = G(w) = I(w) + L(w) - M(w)$$

Let D be the set of those w such that (4.2)–(4.4) (and hence (4.1)) are finite. First we note

LEMMA 4.1. *For each K there is a unique H which satisfies (1.5) and minimizes $M(w)$. This function is a weak solution of (1.3).*

Let $H(K)$ be the function given by Lemma 4.1, and let T be the set of those w in D such that (1.5) holds and H is $H(K)$. Thus for w in T, equation (1.3) is satisfied in the weak sense. Moreover, G is bounded from below for such w. This follows from

LEMMA 4.2. *If $c^2 \leqslant b^2$ and H is $H(K)$, then $M(w) \leqslant L(w)$.*

Now we are ready to minimize G on T. Put $\alpha = \inf_T G$. If $\{w_n\}$ is a minimizing sequence, we have

LEMMA 4.3. *There is a constant C such that $I(w_n) \leqslant C$, $L(w_n) \leqslant C$, $M(w_n) \leqslant C$. There exists a subsequence (also denoted by $\{w_n\}$) which converges weakly to a limit w_0 in T such that*

$$M(w_n) - c^2\|K_n\|^2 \to M(w_0) - c^2\|K_0\|^2$$

and

$$b\|K_n\|^2 - (J_n K_n, K_n) \to b\|K_0\|^2 - (J_0 K_0, K_0)$$

Lemma 4.3 implies

$$I(w_n) + (b^2 - c^2)\|K_n\|^2 + \tfrac{1}{2}\|rJ_n'\|^2 + \|K_n(J_n - b)\|^2$$

$$= G(w_n) + M(w_n) - c^2\|K_n\|^2 + 2b^2\|K_n\|^2 - 2b(J_n K_n, K_n)$$

$$\to \alpha + M(w_0) - c^2\|K_0\|^2 + 2b^2\|K_0\|^2 - 2b(J_0 K_0, K_0)$$

The lower semicontinuity of the norms with respect to weak convergence now implies $G(w_0) \leqslant \alpha$. This shows that G assumes a minimum on T when $w = w_0$. The crucial step is to show that the function w_0 is not only a minimum point for G on T, but is also an extremal point of G without restrictions. This is where Theorem 3.1 comes into the picture. We describe the argument in the next section. Once we know this, we can conclude that w_0 is a weak solution of (1.1)–(1.3). It is then not difficult to show that w_0 has the stated degree of regularity. In particular, it is a regular solution.

5. The Reduction. We show that the function w_0 obtained in the preceding section is indeed a solution of (1.1)–(1.3). We do this by showing that it is an extremal of G on D as well as T. For convenience we introduce the following notation: $k = K$, $\beta = J - b$, $\gamma = H - c$,

$u = \{k, \beta, \gamma\},$

(5.1) $J_b(\beta) = \frac{1}{2}\|r\beta'\|^2 + \|k\beta\|^2 + 2b(k\beta, k)$

Then

(5.2) $G(u) = I(u) + J_b(\beta) - J_c(\gamma) + (b^2 - c^2)\|k\|^2$

Let $\delta > 0$ be fixed and let $\mathcal{H}$ be the Hilbert space of functions u such that

(5.3) $\beta(r) \to 0, \qquad \gamma(r) \to 0 \qquad$ as $r \to \infty$

with norm squared

(5.4) $\|u\|_{\mathcal{H}}^2 = \|k\|^2 + \|k'\|^2 + \|r\beta'\|^2 + \|r\gamma'\|^2 + \|(1 + r)^{-\delta}\gamma\|^2$

Let N be the closed subspace of $\mathcal{H}$ consisting of those u such that $k = \beta = 0$. We let B be the set of those u in $\mathcal{H}$ such that $r^{-1}(k^2 - 1)$, $k\beta, rV(\beta + b)^{1/2}$ are in L^2. Also let P be the set of those functions $y(r)$ such that $\{0, 0, y\}$ is in N. Note that P is a Hilbert space equivalent to N. Finally let Q be the set of those u in $\mathcal{H}$ for which k, β are in $\mathcal{D} = C_0^\infty(0, \infty)$. We give Q the topology of $\mathcal{D} \times \mathcal{D} \times P$ where $\mathcal{D}$ is given the usual topology. Note that Q contains N and its topology coincides with that of N on N. It is easily checked that $G(u)$ has a derivative $G'(u)$ on B with respect to Q and that

(5.5) $G'(u)\tilde{u} = 2(k', \tilde{k}') + 2(r^{-2}(k^2 - 1), \tilde{k})$

$\qquad\qquad + 2(b^2 - c^2)(k, \tilde{k}) + 2(k(\beta + 2b), \tilde{k}\beta) + (r\beta', r\tilde{\beta}')$

$\qquad\qquad + 2(k(\beta + b), k\tilde{\beta}) + (rV'(\beta + b), r\tilde{\beta}) - (r\gamma', r\tilde{\gamma}')$

$\qquad\qquad - 2(k(\gamma + c), k\tilde{\gamma}) - 2(k(\gamma + 2c), \tilde{k}\gamma)$

where $\tilde{u} = \{\tilde{k}, \tilde{\beta}, \tilde{\gamma}\}$. Next define the map $F(u)$ from $\mathcal{H}$ to N as follows. If $v = \{0, 0, y\}$ is in N, put

(5.6) $\Phi(u, v) = (r\gamma', ry') + 2(k(\gamma + c), ky)$

We have

(5.7) $|\Phi(u,v)| \leqslant c(\|u\|_{\mathcal{K}} + \|u\|_{\mathcal{K}}^3)\|v\|_N$

This follows from the first part of

LEMMA 5.1. *We have*

(5.8) $\|ky\|^2 \leqslant 2^{1+2\delta}\|k\|_{\infty}^2\|(1+r)^{-\delta}y\|^2 + \|k\|^2\|ry'\|^2$

Moreover, if $k(0)^2 = 1$, *there is a constant* C *depending only on* $\|k'\|$ *and* δ *such that*

(5.9) $\|(1+r)^{-\delta}y\|^2 \leqslant c(\|ry'\|^2 + \|ky\|^2)$

From (5.7) we see that there is an element $F(u)$ in N such that

(5.10) $(F(u),v)_N = \Phi(u,v), \qquad u \in \mathcal{K}, \quad v \in N$

It is easily checked that $F(u) \in C^1$ with

(5.11) $(F'(u)\tilde{u},v)_N = (r\tilde{\gamma}',ry') + 2(k\tilde{\gamma},ky) + 4(\tilde{k}(\gamma + c),ky)$

In particular, we have

(5.12) $(F'(u)v,v) = \|ry'\|^2 + 2\|ky\|^2, \qquad v \in N$

Moreover, if u is in B, then $k(0)^2 = 1$. Thus (5.9) implies

$$\|v\|_N^2 \leqslant C(F'(u)v,v), \qquad v \in N$$

where the constant C depends on u. This implies that $F'(u)$ is bijective on N for each u in B. Let M be the set of those u in $\mathcal{K}$ such that $F(u) = 0$. We note

LEMMA 5.2. $H = H(K)$ *iff* u *is in* M.

Let S be the intersection of M and B. It is easily shown that S is not empty. Moreover, the argument in Section 4 shows us that there is an element u_0 in S such that $G(u_0) = \min_S G(u)$. It is easily checked that $u \in S$, $\tilde{u} \in Q$, $u + \tilde{u} \in M$ imply $u + \tilde{u} \in S$. Finally we note that

$G'(u_0)\tilde{u} = 0$ for $\tilde{u} \in N$. Thus $G'(u_0) = 0$ by Corollary 3.2. This shows that u_0 is a solution of (1.1)–(1.3). The regularity of u_0 follows from standard arguments.

We now describe some of the proofs of the lemmas.

Proof of Lemma 4.1. Put $\alpha = \inf_P M_c(\gamma)$, where $M_c(\gamma) = \frac{1}{2}\|r\gamma'\|^2 + \|k(\gamma + c)\|^2$, and let $\{\gamma_n\}$ be a minimizing sequence. By (5.9) this sequence is bounded in norm. Thus there is a subsequence (also denoted by $\{\gamma_n\}$) which converges weakly in P to some element γ_c. The usual argument shows that $M_c(\gamma_c) = \alpha$. This fact gives

$$(5.13) \quad (r\gamma_c', ry') + 2(k(\gamma_c + c), ky) = 0, \quad y \in P$$

There can be only one function $\gamma_c \in P$ satisfying (5.13). For the difference γ between any two satisfies

$$(r\gamma', ry') + 2(k\gamma, ky) = 0, \quad y \in P$$

which implies $\gamma = 0$. We take $H(K) = \gamma_c + c$.

Proof of Lemma 4.2. Note that $c\gamma_1$ is a solution of (5.13). Thus $\gamma_c = c\gamma_1$. Hence we have

$$M_c(\gamma_c) = M_c(c\gamma_1) = c^2 M_1(\gamma_1) \leqslant b^2 M_1(\gamma_1)$$
$$= M_b(b\gamma_1) = M_b(\gamma_b) \leqslant M_b(y), \quad y \in P$$

The proofs of Lemmas 4.3 and 5.1 are based on the following inequalities.

$$(5.14) \quad r|\gamma(r)|^2 \leqslant \|r\gamma'\|^2$$

$$(5.15) \quad |k(r)| \geqslant 1 - r^{1/2}\|k'\|, \quad k(0)^2 = 1$$

which are easily proved. The arguments are a bit tedious and can be found in [2]. To prove Lemma 5.2 we note that $F(u) = 0$ iff $\Phi(u, v) = 0$ for all v in N. Thus γ is a solution of (5.13) and consequently $\gamma = \gamma_c$.

YESHIVA UNIVERSITY

UNIVERSIDAD NACIONAL AUTONOMA DE MEXICO

REFERENCES

[1] Julian Schwinger, "A magnetic model of matter" *Science*, **165** (1969), 757–761.
[2] Martin Schechter and R. A. Weder, "A theorem on the existence of dyon solutions," *Comunicaciones Tecnicas, serie Naranja: Investigaciones* No. 227, Instituto de Investigaciones en Matematicas Aplicadas y en Sistemas, Universidad Nacional Autonoma de Mexico, 1980.
[3] M. S. Berger and Martin Schechter, "On the solvability of semilinear gradient operator equations," *Advances in Mathematics* **25** (1977), 97–132.
[4] B. Julia and A. Zee, "Poles with both magnetic and electric charges in non-Abelian gauge theory," *Physical Review D*, **11** (1975), 2227–2232.

THE KINETIC THEORY OF GASES, A CHALLENGE TO ANALYSTS*

By C. TRUESDELL

---

1. Basic concepts and variables of the kinetic theory. I speak of Maxwell's second kinetic theory, which is based on a quadratic integro-differential equation Boltzmann promulged later as a quintessential expression of Maxwell's treatment of what can be recognized as the infinite differential system satisfied by the moments of a solution of the integro-differential equation. Though mathematically expressed, the kinetic theory has so far resisted all attempts even to formulate mathematically its fundamental problems. Some subsidiary problems have been set, and some of these have been solved recently by analysts, mostly by appeal to powerful modern methods and even so only at the cost of great labor.

The kinetic theory is a field theory like hydrodynamics. It differs from hydrodynamics in resting its field equations upon appeal to properties of a numerous assembly of point molecules. The basic variables that we use in discussing molecular motions are a *time t*, a *place* **x**, and a *velocity* **v**:

$t \in \mathfrak{T}$, a 1-dimensional Euclidean space,

$\mathbf{x} \in \mathcal{E}$, a 3-dimensional Euclidean space,

$\mathbf{v} \in \mathcal{V}$, a 3-dimensional inner-product space.

We do not follow the motions of individual molecules; rather, we suppose that their positions and velocities are distributed randomly, according to a specific rule. This rule is described by means of a *molecular density F*, a real-valued function defined and measurable over $\mathfrak{T} \times \mathcal{E} \times \mathcal{V}$, integrable over $\mathcal{V}$ and non-negative.

The *number density* $n(t, \mathbf{x})$ is the value of the scalar field defined as

---

*Lecture for the Philip Hartman Symposium, Johns Hopkins University. April 25, 1980. The research reported here was supported in part by grants of the U.S. National Science Foundation's Programs in Applied Mathematics and Solid Mechanics.

321

follows:

$$n \equiv \int F.$$

Here and henceforth $\int$ means $\int_{\mathcal{V}} \cdots d\mathbf{v}$.

If g is a real-valued function defined on $\mathcal{E} \times \mathcal{V}$, and if $\mathbf{v} \mapsto F(t, \mathbf{x}, \mathbf{v})$ $g(t, \mathbf{x}, \mathbf{v})$ is integrable over $\mathcal{V}$, then the *average* or *expectation* $\bar{g}$ of the function g at the place $\mathbf{x}$ and at the time t is defined as follows:

$$\bar{g} \equiv \frac{1}{n} \int Fg.$$

The expectation $\bar{g}$ is field over $\mathcal{E}$, and the field $\bar{g}$ is to be interpreted in the same way as are the fields introduced directly when the gas is regarded as a continuum. Our concern is with just a few expectations, certain fields that have immediate and useful counterparts in continuum thermomechanics. Generally we assume that these particular fields are smooth, though such an assumption cannot justly be imposed until more specific knowledge about the kinetic theory than is now available shall have been established.

The first expectation of direct interest is the *mass density* ρ, namely, the expected total mass of the molecules per unit volume at $\mathbf{x}$ and t, whatever be their velocities:

$$p \equiv mn.$$

Here m is an assigned constant which represents the mass of one molecule. Other important expectations are defined as follows:

The *velocity field* $\mathbf{u}$ of the gas: $\mathbf{u} \equiv \bar{\mathbf{v}}$
the *random velocity* $\mathbf{c}$ of a molecule: $\mathbf{c} \equiv \mathbf{v} - \mathbf{u}$
the *pressure tensor field* $\mathbf{M}$: $\mathbf{M} \equiv \rho \overline{\mathbf{c} \otimes \mathbf{c}}$ $(\mathbf{M}^{\mathrm{T}} = \mathbf{M})$
the *pressure deviator field* $\mathbf{P}$: $\mathbf{P} \equiv \mathbf{M} - \frac{1}{3}(\mathrm{tr}\,\mathbf{M})\mathbf{1}$.
the *energy flux field* $\mathbf{q}$: $\mathbf{q} \equiv \frac{1}{2}\rho \overline{c^2 \mathbf{c}}$

The entire purpose of the kinetic theory is to relate the 13 scalar fields ρ, u_k, M_{pm}, *and* q_r *to various circumstances of the kinetic gas.* All these fields are determined by the molecular density F, so if we know F, we know in principle all that we wish to know. Conversely, it is plain from the definitions that these few fields cannot determine F. To any one particular set ρ, $\mathbf{u}$, $\mathbf{M}$, and $\mathbf{q}$ correspond infinitely many F. Thus F contains and is capable of expressing vastly more information than we shall ever wish to have about the motion of the kinetic gas. In order to answer a particular question, we may use any one of the infinitely many appropriate F; we need not find any of the others. The particular beauty of the kinetic theory lies in this superabundance of solutions and the consequent problem of selecting some one F that is good enough for a given task. In many cases F itself is never found, but only some useful relations among some of the quantities it determines. Viewed in the light of its desired applications, the kinetic theory can never be reduced to problems of boundary values or initial values of any classical type, though the study of such problems may provide handy tools for probing its deeper reaches.

Some scalars derived from the 13 basic fields are important for their interpretation. One of these is the *mean normal pressure p*:

$$p \equiv \tfrac{1}{3} \operatorname{tr} \mathbf{M}.$$

Another is the specific internal energy or *energetic* ε, the expected random kinetic energy per unit mass:

$$\varepsilon \equiv \tfrac{1}{2} \overline{c^2} .$$

Thus

$$p = \tfrac{2}{3} \rho \varepsilon.$$

We may interpret ε as being proportional to the temperature field.

While the fields I have just introduced come near to exhausting the interest the molecular density F offers us, that density embodies a great deal more information about the molecular motions. Some of this information is expressed by

the *relative moment* $^n\mathbf{M}$ *of order n*: $\quad ^n\mathbf{M} \equiv \rho \overline{\mathbf{c} \otimes \cdots \otimes \mathbf{c}} .$

Here we encounter the first central difficulty of the kinetic theory: Its protagonist, the molecular density F, would tell us too much, if ever we could get hold of it. Nobody desires F or its general moment. Only the *principal moment*

$$\rho \equiv (\rho, \mathbf{u}, \varepsilon)$$

and the further moments

$$\mathbf{P}, \ \mathbf{q}$$

and two other expectations have any physical interpretation. We desire to determine at most 15 scalar fields on $\mathfrak{T} \times \mathfrak{E}$, but we must do so in terms of a single scalar F on $\mathfrak{T} \times \mathfrak{E} \times \mathfrak{V}$. It is not just a matter of the difficulty in determining F. Infinitely many different F can correspond to the 15 fields we wish. The differences among these F are useless. One F will do. But if we have one F, it contains a lot of superfluous detail, which we must somehow discard, the detail which manifests itself only in the way the various F compatible with the 15 desired fields differ from each other. The matter is not so simple as that, anyway. At present we don't have any F to work with. The first non-trivial F were delivered by an existence theorem (very restricted at that) proved by Glikson in 1972. The first non-trivial explicit F was found by Muncaster in 1977. Glikson's existence theorem shares the defects of all existence theorems so far proved in the kinetic theory, defects I will come back to. Muncaster's F is a jolly fellow, but it happens to suffer from the very same defects.

2. The Maxwell-Boltzmann equation. I am getting ahead of what I have presented. Of course, F must satisfy the integro-differential equation I mentioned. That integro-differential equation, which Hilbert called *the Maxwell-Boltzmann equation*, may be written as

$$\partial_t F + \mathbf{v} \cdot \partial_x F + \mathbf{b} \cdot \partial_v F = \int \int_{\mathfrak{S}} w \left(F' F'_* - F F_* \right).$$

To explain this equation it is easiest to revert to an equation of evolution, of which the Maxwell-Boltzmann equation is the local and differential expression. We suppose given the set of curves that the molecules would follow, did they never collide with each other or with walls. The only force acting would be none at all or gravity or something of the kind, say some time-independent field $\mathbf{b}$. If all of $\mathfrak{E}$ were empty except for a single molecule subject to the extrinsic force $\mathbf{b}$, then should the molecule at time t occupy the place $\mathbf{x}$ and have the velocity $\mathbf{v}$, its position at another time s would be given by a function $s \mapsto \chi[t, \mathbf{x}, \mathbf{v}](s)$, and χ would be determined as the solution of the single Newtonian equation

$$\chi'' = \mathbf{b}(\chi)$$

such as to satisfy the initial conditions

$$\chi[t, \mathbf{x}, \mathbf{v}](t) = \mathbf{x}, \qquad \chi'[t, \mathbf{x}, \mathbf{v}](t) = \mathbf{v}.$$

The resulting *extrinsic trajectory* in $\mathfrak{I} \times \mathcal{E} \times \mathcal{V}$ would be the curve defined parametrically as follows: $s \mapsto (s, \chi[t, \mathbf{x}, \mathbf{v}](s), \chi'[t, \mathbf{x}, \mathbf{v}](s))$. We assume **b** smooth enough to let us call upon the theorems on differential equations that deliver the mapping χ and make it unique in such intervals of time as we shall consider. The *retrogressor* is an operator $\mathfrak{r}_s$ on $\mathfrak{I} \times \mathcal{E} \times \mathcal{V}$ defined as follows for each real number s:

$$\mathfrak{r}_s(t, \mathbf{x}, \mathbf{v}) \equiv \left(t - s, \chi[t, \mathbf{x}, \mathbf{v}](t - s), \chi'[t, \mathbf{x}, \mathbf{v}](t - s)\right).$$

The *retrogression* $\mathfrak{R}_s G$ of a function G on some subset of $\mathfrak{I} \times \mathcal{E} \times \mathcal{V}_i$ is defined by

$$\mathfrak{R}_s G \equiv G \circ \mathfrak{r}_s.$$

If the operator $\mathbb{C}$ is such that $\mathbb{C}F$ is the rate of increase of $\mathbb{C}$ through collision with other molecules, then the basic *Equation of Evolution* is

$$F(t, \mathbf{x}, \mathbf{v}) = \mathfrak{R}_{t - t_0} F(t, \mathbf{x}, \mathbf{v}) + \int_{t_0}^{t} \mathfrak{R}_{t - s}(\mathbb{C}F)(t, \mathbf{x}, \mathbf{v}) \, ds.$$

For the *collisions operator* $\mathbb{C}$ Maxwell motivated and proposed the form

$$\mathbb{C}f \equiv \int_{\mathcal{V}} \int_{\mathcal{S}} w\left(F'F'_* - FF_*\right) dS \, d\mathbf{v}_*.$$

The notations are defined as follows:

$$\mathbf{w} \equiv \mathbf{v}_* - \mathbf{v}, \qquad w \equiv |\mathbf{w}|,$$

g stands for $g(t, \mathbf{x}, \mathbf{v})$,

g_* stands for $g(t, \mathbf{x}, \mathbf{v}_*)$,

g' stands for $g(t, \mathbf{x}, \mathbf{v}')$,

g'_* stands for $g(t, \mathbf{x}, \mathbf{v}'_*)$.

The *cross-section* $\mathcal{S}$ is the portion of the plane $\mathcal{P}$ through $\mathbf{x}$ normal to $\mathbf{w}$ that corresponds to possible collisions of the molecule at $\mathbf{x}$ with a molecule having asymptotic velocity $\mathbf{v}_*$. In the physical literature the asymptotic velocities $\mathbf{v}'$ and $\mathbf{v}'_*$ after encounter of two molecules whose asymptotic velocities before encounter were $\mathbf{v}$ and $\mathbf{v}_*$ is calculated by

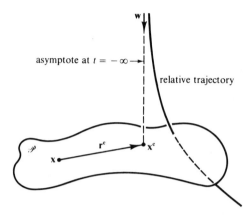

solving explicitly the problem of analytical dynamics that arises when a particular intermolecular force is laid down. The simplest such laws are those for smooth spheres and for molecules whose intermolecular force is a repulsion varying as the inverse kth power of the intervening distance. If $k = 5$, the molecules are called *Maxwellian*; as Maxwell himself discovered, many otherwise complicated formulae become relatively simple for these very special molecules; accordingly, many explorations begin with Maxwellian molecules, and such is the difficulty of the kinetic theory that most of them end there. Mathematicians may prefer to use a collective formulation in terms of an abstract *encounter operator* :

An *Encounter Operator* $\mathbb{E}$ is any piecewise continuously differentiable function

$$\mathbb{E} : \mathcal{V} \times \mathcal{V} \times \mathcal{S} \to \mathcal{V} \times \mathcal{V} \times \mathcal{S}$$

which satisfies

$$\mathbb{E} \circ \mathbb{E} = \text{identity} ;$$

$$\mathbb{M} \circ \mathbb{E} = \mathbb{M},$$

$$\mathbb{K} \circ \mathbb{E} = \mathbb{K};$$

$$\mathbb{X} \circ \mathbb{E} = \mathbb{E} \circ \mathbb{X};$$

$$\mathbf{Q} \circ \mathbb{E} = \mathbb{E} \circ \mathbf{Q} \qquad \text{for all orthogonal } \mathbf{Q};$$

$$\frac{r'}{r} \left| \frac{\partial(\mathbf{v}', \mathbf{v}'_*, r', \zeta')}{\partial(\mathbf{v}, \mathbf{v}_*, r, \zeta)} \right| = 1;$$

the operators $\mathbb{M}, \mathbb{K}$, and $\mathbb{X}$ are defined as follows:

$$\mathbb{M}(\mathbf{v}, \mathbf{v}_*, r, \zeta) \equiv \mathbf{v} + \mathbf{v}_*,$$

$$\mathbb{K}(\mathbf{v}, \mathbf{v}_*, r, \zeta) \equiv v^2 + v^2_*;$$

$$\mathbb{X}(\mathbf{v}, \mathbf{v}_*, r, \zeta) \equiv (\mathbf{v}_*, \mathbf{v}, r_*, \zeta_*);$$

r, ζ are polar co-ordinates of $\mathbf{x}^e$ in the cross-section $\mathcal{S}$; and the action of an orthogonal tensor $\mathbf{Q}$ on points of $\mathcal{V} \times \mathcal{V} \times \mathcal{S}$ is denoted also by $\mathbf{Q}$:

$$\mathbf{Q}(\mathbf{v}, \mathbf{v}_*, \mathbf{r}^e) \equiv (\mathbf{Q}\mathbf{v}, \mathbf{Q}\mathbf{v}_*, \mathbf{Q}\mathbf{r}^e).$$

Then

$$(\mathbf{v}', \mathbf{v}'_*, r', \zeta') \equiv \mathbb{E}(\mathbf{v}, \mathbf{v}_*, r, \zeta).$$

The constitutive variety of gases is modelled in the kinetic theory by three quantities:

> molecular mass m
> cross-section $\mathcal{S}$
> encounter operator $\mathbb{E}$

Only these may be chosen. Once a choice has been made, every further step must be mathematical. But that fact does not tell us what problems to pose.

Whatever will be our choice of problem, the choice of the cross-section $\mathcal{S}$ will make a great difference in any attempt to solve that problem. If the intermolecular force drops to zero at a finite distance, we say that the model refers to *molecules with a cut-off*. For molecules with a cut-off $\mathcal{S}$ is a bounded part of the plane $\mathcal{P}$, and the integral defining the collisions operator $\mathbb{C}$ may be absolutely convergent. In that case $\mathbb{C}F$ is the difference of two convergent integrals. Various techniques of iteration seem feasible for molecular models of this kind, and for that reason the few analysts who have studied the kinetic theory have mainly treated only molecules with a cut-off. Physicists and chemists, on the contrary, regard intermolecular forces of infinite range as better models of nature. Except for occasional treatments of ideal spheres, most work by physicists has considered the far more difficult problems that arise when the integral used to define the collisions operator is only conditionally convergent. This is one of several points at which analysts and physicists treating the kinetic theory have turned into different roads.

3. Monotonicity of the collisions operator.

A word regarding the collisions operator itself is necessary here. As Boltzmann showed, that operator is monotone in the following sense:

$$\int \log F \mathbb{C} F \leq 0$$

$$\int \log F \mathbb{C} F = 0 \quad \Leftrightarrow \quad \mathbb{C} F = 0 \quad \Leftrightarrow \quad F = ae^{-bc^2}$$

Thus a certain trend in time is built into the kinetic theory through Maxwell's proposal for the operator $\mathbb{C}$. The only molecular density exempt from change due to collisions is ae^{-bc^2}, the *Maxwellian density*; if a, b, and $\mathbf{u}$ are suitably chosen, it is appropriate to equilibrium. Since a Maxwellian density is determined uniquely by its own principal moment fields, which may be selected arbitrarily, a Maxwellian F is by no means limited to equilibrium. It is possible to find all those particular Maxwellian F that are solutions of the Maxwell-Boltzmann equation; while these are its trivial solutions, even they are not limited to gross conditions associated with equilibrium. Finally, the fact that $\mathbb{C} F = 0$ for a Maxwellian F enables us to show that Maxwell's kinetic theory is strictly compatible with classical continuum mechanics in the sense that *for every solution of the Maxwell-Boltzmann equation the corresponding moment fields of the kinetic gas satisfy the standard, familiar field equations expressing the balance of mass, momentum, and energy.* For all molecular densities other than ae^{-bc^2}, the effect of collisions is to make a certain time rate negative. This fact indicates some irreversibility; it suggests that an arbitrary molecular density will exhibit some trend in time. While the physicists usually speak of "a trend to equilibrium", we know from counter-examples that states other than equilibrium may be approached and that there need be no discernible trend of any kind. We now have a good deal of information, in part from analytical theorems and in part from special solutions, but it remains a major open problem to discover *the asymptotic behavior of solutions of the Maxwell-Boltzmann equation as time increases.* Note that I do not say as $t \to \infty$, for we have examples to show that just as in ordinary gas dynamics, classical solutions may cease to exist after a finite time.

4. Viscosity and Maxwell numbers. I have stated that only part of the information embodied in F is of any interest. An example is furnished by the asymptotic trend just mentioned. Another, that for which Maxwell proposed his kinetic theory in the first place, is to determine "transport coefficients" like viscosity and thermal conductivity in terms of a particular molecular structure giving rise to one or another particular choice of the kinetic constitutive quantities $m, \mathbb{S}$, and $\mathbb{E}$. The *viscosity* and the *Maxwell number* are the coefficients μ and M that occur in the simplest continuum theory of dissipative gas dynamics:

$$\mathbf{P} = -2\mu\mathbf{E}; \qquad \mathbf{E} \equiv \tfrac{1}{2}\big(\operatorname{grad}\mathbf{u} + (\operatorname{grad}\mathbf{u})^{T}\big) - \tfrac{1}{3}E\mathbf{1}, \qquad E \equiv \operatorname{div}\mathbf{u}.$$

$$\mathbf{q} = -M\mu\operatorname{grad}\varepsilon.$$

The product $M\mu$ is proportional to the *heat conductivity*. In continuum gas dynamics these "transport coefficients" are adjustable; they are determined by experiment on actual gases or set out by hypotheses. A principal aim of the kinetic theory is to calculate counterparts of these coefficients mathematically from the molecular structure as reflected by the choice of $m, \mathbb{S}$, and $\mathbb{E}$. To do any such thing, we must first derive from the kinetic theory relations among the moments of F and their gradients having just the foregoing forms. The coefficients can then be evaluated by trivial comparison of the two statements: The *assumption* in gas dynamics and the *theorem* in the kinetic theory.

5. Role of special solutions. But no such theorem can hold for general solutions in the kinetic theory. To see that, we have only to look back at the basic equation of evolution:

$$F(t, \mathbf{x}, \mathbf{v}) = \Re_{t-t_0}F(t, \mathbf{x}, \mathbf{v}) + \int_{t_0}^{t}\Re_{t-s}(\mathbb{C}F)(t, \mathbf{x}, \mathbf{v})\,ds.$$

If in this equation we put $t = t_0$, the result is $F(t_0, \mathbf{x}, \mathbf{v}) = F(t_0, \mathbf{x}, \mathbf{v})$. The equation of evolution imposes *no condition at all* on F at the arbitrarily selected initial time. The initial value of F is arbitrary. Thus at an arbitrarily selected time there can be no restriction on the moments of F. Therefore a general solution cannot have moments and gradients of moments which satisfy any time-independent relation like a constitutive relation of continuum mechanics. *A fortiori*, the constitutive relations of classical gas dynamics cannot be satisfied in the kinetic theory. In other

words, a general solution of the Maxwell-Boltzmann equation has no
viscosity, no heat conductivity, no Maxwell number. Yet everyone
knows that kinetic calculations of these quantities abound. This fact is
easily justified if the following two statements are true:

1. Solutions in a special class have a viscosity and a Maxwell
   number.
2. The solutions in this special class correspond to the facts of
   nature, while the others ones do not generally do so.

Experience with special cases has taught us to narrow Statement 2:

2 *bis*. Any solution is asymptotic in time to some solution in this
   special class.

Here we see another difference between researches by analysts and
researches by physicists or chemists. An analyst looks at the equation of
evolution and says Aha! An initial-value problem. I will find a function
space in which the problem has a unique solution. Most analytical
researches so far have been of just this kind and have restricted attention
to place-independent solutions for molecules with a cut-off. These circumstances are trivial for application: The gas is grossly at rest, nothing
at all is happening, any viscosity and heat conductivity the gas may have
cannot manifest themselves. For these degenerate circumstances it is
natural to formulate and prove a trend to kinetic equilibrium. A measure
of the difficulty of the kinetic theory may be inferred from the heavy
artillery, intricate constructions of sequences, and involved estimates
that have been brought to bear in these analytical researches on the
aspect of the kinetic theory that would seem to be the simplest of all.
Corresponding results for intermolecular forces of infinite range are
scant. Many years ago I constructed the general solution for the system
of moment equations for a gas of Maxwellian molecules in a grossly
homogeneous state. In a paper now in press Arkeryd proves that for a
broad class of molecules with mutual forces of infinite range, Maxwellian molecules being included, spatially homogenous solutions exist.
    Even though an existence theorem for the initial-value problem does
not offer much interest in itself, it is of course a necessary first step in
approach to the fundamental questions, but still this step is altogether
lacking for a gas with intermolecular forces of infinite range subject to
initial conditions that are not close to equilibrium. There are half a

dozen different senses of "weak solution"—for example, a solution of the equation of evolution which need not be spatially differentiable—yet all applications by physicists and engineers so far have employed classical solutions and at present do not seem to follow for more general kinds of solutions.

The table below summarizes the various differences in approach I have mentioned.

Table

Analysts (HILBERT, CARLEMAN, MORGENSTERN, POVZNER, KANIEL)	Physicists, Chemists, Engineers (CHAPMAN, ENSKOG, and numerous later persons)
Molecules with a cut-off ($\mathcal{S}$ bounded)	Intermolecular forces of infinite range ($\mathcal{S}$ unbounded)
Existence of general weak solutions for the above	Properties of special classical solutions giving rise to "transport coefficients" for the above
Properties independent of the choice of encounter opeator $\mathbb{E}$	Differences arising from different choices of $\mathbb{E}$
Spatially homogeneous solutions	Non-trivial gas flows
Proof of a trend to equilibrium for the above	MISSING but desired opposite: characterization of the class of asymptotic solutions for the above

EXCEPTIONS (Students who have crossed the double line at least once): MAXWELL, BOLTZMANN, GRAD, TRUESDELL, IKENBERRY, CARLESON, GILKSON, GUIRAUD, Y.-P. PAO, MUNCASTER, UKAI, ARKERYD.

If the exceptions seem to outnumber the conforming examples, please note the word "numerous" in the right-hand column. Sometimes a paper by an entry in one column could scarcely be perceived to concern the same theory as one by an entry in the other column, were it not for the respective titles.

6. Maxwell's relaxation theorem. Hint of principal solutions. A hint toward the position of special solutions is provided by Maxwell's own work. Thus far I have shielded you from the ugly glare of the enormous and unsymmetrical equations that the kinetic theory provides, but they have been and still are ineluctable. Even Hilbert's paper on the kinetic theory is full of them. Here are three systems of partial differential equations derived in principle by Maxwell:

$$\dot{P}_{km} + P_{km}E - \tfrac{2}{3}P_{ab}E_{ab}\delta_{km} + P_{ak}u_{m,a}$$
$$+ P_{am}u_{k,a} + (M_{kma} - \tfrac{2}{3}q_a\delta_{km})_{,a} + 2pE_{km} = -\tau^{-1}P_{km},$$

$$\dot{q}_k + \tfrac{5}{3}q_k E + P_{0\,|\,kab}E_{ab} + q_a u_{k,a} + \tfrac{4}{5}q_a E_{ka} - \frac{5p}{2\rho}P_{,k} - \frac{5p}{2\rho}P_{ka,a}$$
$$- \frac{1}{\rho}P_{ka}P_{,a} - \frac{1}{\rho}P_{ka}P_{ab,b} + \tfrac{1}{2}M_{kaab,b} = -\tfrac{2}{3}\tau^{-1}q_k,$$

$$\dot{P}_{0\,|\,kmr} + P_{0\,|\,kmr}E + 3P_{0\,|\,a(km}u_{r),a} - \tfrac{6}{5}P_{0\,|\,ab(k}\delta_{mr)}E_{ab} + \tfrac{12}{5}q_{(k}E_{mr)}$$
$$- \tfrac{24}{25}q_a E_{a(k}\delta_{mr)} - \frac{3}{\rho}P_{(km}P_{,r)} - \frac{3}{\rho}P_{(km}P_{r)a,a} + \frac{6}{5\rho}P_{a(k}\delta_{mr)}(p_{,a}$$
$$+ P_{ab,b}) + M_{kmra,a} - \tfrac{3}{5}\delta_{(km}M_{r)aab,b} = -\tfrac{3}{2}\tau^{-1}P_{0\,|\,kmr}.$$

These are the first thirteen members of the infinite system of *equations satisfied by the moments* of a smooth solution F for a gas of Maxwellian molecules. The superimposed dot denotes a substantial time derivative in the sense of hydrodynamics; parentheses around a set of subscripts denote the symmetric part; τ is defined as follows:

$$\tau \equiv \frac{1}{3a}\sqrt{\frac{2m^3}{g}}\,\frac{1}{\rho},$$

in which

$$a = 1.3703 \ldots;$$

g is the magnitude of the force of intermolecular repulsion at unit distance exerted by a pair of molecules of unit mass; and $P_{0\,|\,kmr}$ is defined as follows from the third moments M_{kmr}:

$$P_{0\,|\,kmr} \equiv M_{kmr} - \tfrac{1}{5}(M_{aak}\delta_{mr} + M_{aar}\delta_{km} + M_{aam}\delta_{kr}).$$

These are the simplest examples of the dreaded *equations for moments*. They are typical in that the divergence of moments of order $n + 1$ appears in the equations for the moments of order n; that is, the infinite system is linked forward. Stated here for a gas of Maxwellian molecules, they are atypical in that on the right-hand sides stand linear functions of the moments of order n rather than the infinite series in bilinear combinations of the moments of lower and higher orders which appear there when moments of order greater than 3 and general molecules are considered. Most studies of the kinetic theory by physicists and chemists have rested upon the system of equations for moments rather than on the Maxwell-Boltzmann equation. The infinite series I have just mentioned—those that arise when molecules other than Maxwellian are considered—have never been shown to converge in any sense; the appliers truncate these series in one way or another and hope for the best.

If the prinicipal moment is constant in space and time, as is the case for a spatially homogeneous molecular density, the gross condition is said to be *gross rest*. We expect infinitely many different molecular densities to correspond to gross rest, but only one of these, the uniform Maxwellian density, corresponds to kinetic equilibrium. For gross rest all terms but the first on the left-hand sides of the equations of second and third moments for Maxwellian molecules vanish. Each member of the system of moments becomes an ordinary differential equation for a single quantity. The general solutions of the three systems are

$$P_{km} = P_{km}(0)e^{-t/\tau},$$

$$q_k = q_k(0)e^{-t/\frac{3}{2}\tau}.$$

$$P_{0|kmr} = P_{0|kmr}(0)e^{-t/\frac{2}{3}\tau}.$$

This is Maxwell's *relaxation theorem*. It shows that the pressure deviator and all third moments may assume arbitrary initial values but tend to 0 as $t \to \infty$. Note that only one solution, that which vanishes identically, can have a viscosity, but that viscosity may be arbitrary as far as this solution is concerned. For the logarithmic decrement of the general solution the pressure deviator is the time τ, called the *time of relaxation*; it is determined by the density and by molecular constants:

$$\tau \equiv \frac{1}{3a} \sqrt{\frac{2m^3}{g}} \frac{1}{\rho}$$

The logarithmic decrements of the third moments are $\frac{3}{2}/\tau$ and $\frac{2}{3}/\tau$. In 1956, using results obtained in collaboration with Ikenberry, I showed that for a gas of Maxwellian molecules the moments of all orders if spatially homogeneous would approach exponentially as $t \to \infty$ values appropriate to a Maxwellian equilibrium density, and that none of them approached that limit any more slowly than the heat flux. It is natural to expect that the molecular density itself approaches a uniform Maxwellian density, but that conjecture remains unproved today for all intermolecular forces of infinite range, though a single, explicit, and very simple example discovered by Krook & Wu in 1976 supports it. For molecules with a cut-off Carleman in a celebrated paper of 1933 demonstrated for an extremely special instance a strict trend to equilibrium in the sense of uniform convergence. Arkeryd in 1972 proved such a trend for general molecules with a cut-off, but only in the sense of weak convergence. It is natural to expect that similar results will be found for all molecular models. We can summarize these results and conjectures by stating that for a condition of gross rest there is a special solution, namely the unique uniform Maxwellian density, which all the infinitely many other solutions approach with time. We may call this special solution the *principal solution* corresponding to an assigned constant principal moment.

7. Maxwell's evaluation of viscosity and Maxwell number. This is still a long way from evaluating the viscosity μ and the Maxwell number M. Maxwell himself obtained those quantities for a gas of Maxwellian molecules by considering the same equations:

$$\dot{P}_{km} + P_{km}E - \tfrac{2}{3}P_{ab}E_{ab}\delta_{km} + P_{ak}u_{m,a}$$
$$+ P_{am}u_{k,a} + (M_{kma} - \tfrac{2}{3}q_a\delta_{km})_{,a} + 2pE_{km} = -\tau^{-1}P_{km},$$

$$\dot{q}_k + \tfrac{5}{3}q_kE + P_{0\,|\,kab}E_{ab} + q_au_{k,a} + \tfrac{4}{5}q_aE_{ka} - \frac{5p}{2\rho}P_{,k} - \frac{5p}{2\rho}P_{ka,a}$$
$$- \frac{1}{\rho}P_{ka}P_{,a} - \frac{1}{\rho}P_{ka}P_{ab,b} + \tfrac{1}{2}M_{kaab,b} = -\tfrac{2}{3}\tau^{-1}q_k.$$

Inspecting them, he gave reasons for discarding most of the terms. In effect he supposed that on the left-hand sides all moments of orders greater than 1 could be replaced by the functions of ρ and ε to which they would reduce, were the molecular density Maxwellian. The first equation then becomes

$$2pE_{km} = -\tau^{-1}P_{km}.$$

This is of·the same form as the phenomenological equation

$$\mathbf{P} = -2\mu\mathbf{E},$$

and the viscosity μ is evaluated:

$$\mu = p\tau = \frac{1}{3a} \sqrt{\frac{2m^3}{g}} \frac{2}{3} \varepsilon.$$

Similar manhandling of the second equation yields

$$\mathbf{M} = \tfrac{5}{2}.$$

The treatment is crude, but nobody has ever doubted the results. The questions have been, what do they mean about the kinetic theory, and how we can derive them rigorously?

Surveying Maxwell's work nearly twenty years after his death, Boltzmann wrote:

> Many terms of the equations must be neglected in order to obtain the hydrodynamical equations in their usual form. Even if this course in most cases is justifiable, it cannot be rigorously proved that such is the case, and the mathematician is not satisfied. The following question arises: is this a defect of the theory of gases, or is it rather a defect of hydrodynamics? Are these terms required by the theory of gases not an essential correction of the equations of hydrodynamics?

8. Hilbert's work. The problem was taken up by Hilbert in 1912. In a celebrated memoir, in its importance and influence upon later research into the kinetic theory second only to the works of Maxwell and Boltzmann themselves, Hilbert was the first to recognize and assert that "the most general task" of the kinetic theory was to determine solutions of a *special kind*. He chose to define those special solutions by a power-series expansion in terms of a mysterious parameter. He showed that the successive terms in that expansion, treated purely formally, had to satisfy linear integral equations. He claimed that by evaluating the terms in the expansion it would be possible to determine the viscosity and heat conductivity. While he did not substantiate that assertion, he clearly saw that *only special solutions* would have transport coefficients. He asserted that his special class of molecular densities would correspond to "a gas in a stable state of motion", but his definition of "stable state" is untenable. His special molecular densities have the altogether remarkable property that they satisfy formally an initial-value problem restricted to their own principal moments. If these solutions exist, they are uniquely determined by the initial values of their own fields ρ, $\mathbf{u}$, and ε. These are the very same initial values as are prescribed in ordinary gas dynamics, so it is no wonder Hilbert was pleased with what he found: solutions determined by gas-dynamic data! Among the strange properties of his alleged solutions discovered later is that if the defining power series is convergent for all values of his parameter, that parameter is a dummy which may be given any non-zero value without affecting the results. Hilbert's paper has been widely misunderstood. Physicists, seeing a great mathematician's name on it, have presumed that it dealt with existence theorems and have dismissed it accordingly, unread, as being devoid of physical interest. In fact the paper is systematic but almost purely formal, and it is aimed straight at the major physical problem. Mathematicians, especially German ones and again because of Hilbert's name, until recently sometimes claimed that Hilbert had solved all the fundamental problems of the kinetic theory, so only "applications" remained. Hilbert himself made various sweeping assertions, none of which he proved true, and some of which we now know are not true, but certainly he made no attempt to prove an existence theorem. Equally certainly, he settled no fundamental question whatever; rather, he opened a new conceptual domain, one that has not yet been fully explored. I cannot here enter into the details of this complicated, fascinating, and still incompletely developed study.

Although, as far as proof is concerned, only one of the several major gaps in Hilbert's reasoning has been closed by later researches, Hilbert's paper remains a monument of the subject. It was Hilbert who first stated outright that *only a special class of solutions* can describe gas flows in the terms appropriate to gas dynamics; who first specified it as his objective to calculate directly *just those solutions*; who first conceived a *systematic procedure*, if only a formal one, for approximating such solutions; and who first showed, though again only formally, that a suitable limiting process led to a *solution determined uniquely by the initial value, which may be assigned at will, of its own principal moment*.

9. Simple shearing and its principal solution.   In 1956 I sought to extend Maxwell's relaxation theorem to some simple gas flows, like Maxwell himself considering only a gas of Maxwellian molecules. The example I choose to work out was simple shearing, a gas flow with a linear velocity profile:

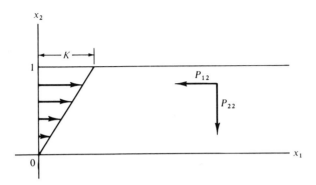

$$u_1 = Kx_2, \quad u_2 = 0, \quad u_3 = 0, \quad K = \text{const}.$$

It happened to be easy to do so. Maxwell's system of equations for the second moments, namely

$$\dot{P}_{km} + P_{km}E - \tfrac{2}{3}P_{ab}E_{ab}\delta_{km} + P_{ak}u_{m,a}$$
$$+ P_{am}u_{k,a} + (M_{kma} - \tfrac{2}{3}q_a\delta_{km})_{,a} + 2pE_{km} = -\tau^{-1}P_{km},$$

can be solved in general for this velocity field. The non-trivial part of the solution is

$$\frac{p}{p(0)} = Ae^{Rt/\tau} + e^{-(1+\frac{1}{2}R)t/\tau}\left[B\cos JKt + C\sin JKt\right],$$

$$\frac{P_{12}}{p(0)} = \frac{1}{\frac{2}{3}T}\left\{-ARe^{Rt/\tau} + e^{-(1+\frac{1}{2}R)t/\tau}\left[\left(\{1 + \tfrac{1}{2}R\}B - JTC\right)\cos JKt\right.\right.$$

$$\left.\left. + \left(\{1 + \tfrac{1}{2}R\}C + JTB\right)\sin JKt\right]\right\},$$

$$\frac{P_{22}}{p(0)} = -\frac{1}{R+1}\left\{ARe^{Rt/\tau} + e^{-(1+\frac{1}{2}R)t/\tau}\right.$$

$$\times\left[\left(\{\tfrac{3}{2} + R\}B + \frac{JT}{R}C\right)\cos JKt\right.$$

$$\left.\left. + \left(\{\tfrac{3}{2} + R\}C - \frac{JT}{R}B\right)\sin JKt\right]\right\},$$

in which the constants A, B and C satisfy the following system of linear equations:

$$1 = A + B,$$

$$\frac{\frac{2}{3}\mathsf{T}P_{12}(0)}{p(0)} = -AR + (1 + \tfrac{1}{2}R)B - J\mathsf{T}C,$$

$$-\frac{(1 + R)P_{22}(0)}{p(0)} = AR + (\tfrac{3}{2} + R)B + \frac{J\mathsf{T}}{R}C.$$

The parameter $\mathsf{T} \equiv \tau K$; the factors R and J are easily calculated as functions of it:

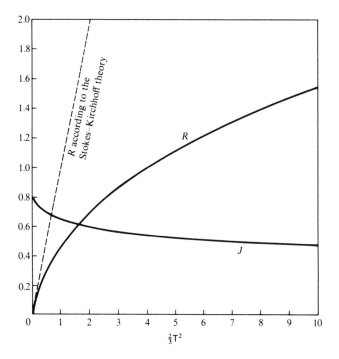

Again there is a *principal solution*, namely, the unique solution corresponding to initial values such that $A = 1$, $B = C = 0$:

$$\varepsilon^P = \varepsilon^P(0)e^{Rt/\tau},$$

$$P_{11}^P = \frac{4}{3}\frac{R}{1 + R}\rho\varepsilon^P(0)e^{Rt/\tau}.$$

$$P_{22}^{P} = -\frac{2}{3}\frac{R}{1+R}\rho\varepsilon^{P}(0)e^{Rt/\tau},$$

$$P_{33}^{P} = -\frac{2}{3}\frac{R}{1+R}\rho\varepsilon^{P}(0)e^{Rt/\tau},$$

$$P_{12}^{P} = -\frac{R}{T}\rho\varepsilon^{P}(0)e^{Rt/\tau},$$

$$P_{13}^{P} = 0, \qquad P_{23}^{P} = 0.$$

This solution is uniquely determined by its own principal moment; such a solution we may call *grossly determined*. For a fixed value of the parameter T, each solution is exponentially asymptotic as $t \to \infty$ to a multiple of the principal solution. The solution is not the same as that given by classical gas dynamics for this flow. There is what Boltzmann called "an essential correction of the equations of hydrodynamics". Nevertheless, when t is fixed, in the limit as $T \to 0$ a viscosity emerges asymptotically, and it has the value Maxwell discovered. The term "solution" in these phrases refers only to the pressure deviator, not to the molecular density, which has not been found.

10. Unidirectional expansion and its principal solution. The solution for shearing was discovered independently by Galkin in the same year, 1956. In 1964 he worked out a similar solution for unidirectional expansion:

$$u_1 = \frac{1}{t+T}(x_1 - x_0), \quad u_2 = 0, \quad u_3 = 0,$$

namely

$$\frac{p}{p(0)} = A\left(1 + \frac{t}{T}\right)^{R} + (1 - A)\left(1 + \frac{t}{T}\right)^{S},$$

$$\frac{P_{11}}{p(0)} = -\frac{3R+5}{2}A\left(1 + \frac{t}{T}\right)^{R} - \frac{3S+5}{2}(1 - A)\left(1 + \frac{t}{T}\right)^{S},$$

in which T is an arbitary constant time. The quantity τ is defined as before, but because ρ now varies with the time, τ is no longer constant. We define a dimensionless parameter T as follows: $T = \frac{4}{3}\tau(0)/T$. This T, which determines R and S, we may regard as the parameter that governs the problem. If $T > 0$, the solution exists for all time, but if $T < 0$, the

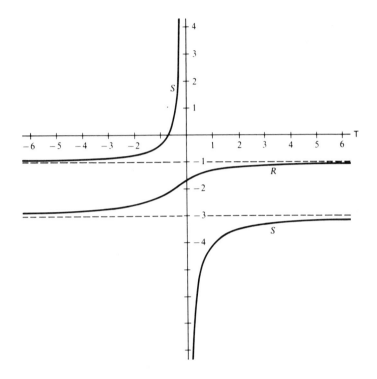

solution ceases to exist when $t = -T$. Muncaster & I completed Galkin's work by showing that his class of solutions includes a *principal solution* which is grossly determined and has a viscosity, which has the value Maxwell found. Our solutions for extension tend asymptotically to a multiple of the principal solution if $T > 0$ or $0 > T > -\frac{2}{3}$, but the rate of approach follows a power law instead of exponential decrease. If $T \leqq -\frac{2}{3}$, there is no approach to a principal solution. Again we recover the classical solution and Maxwell's formula for the viscosity in an asymptotic form as $T \to 0$ when t is fixed.

In all these special cases the problem treated reduces to the solution of ordinary differential equations. I expect that all of our results and Galkin's are subsumed under an appropriate center-manifold theorem. No doubt more results of the same kind can be gotten by further applications of that theorem, but still only a few special examples would come out. What is needed is a center-manifold theorem for the Maxwell-Boltzmann equation—a quadratic integro-differential equation—or, if you prefer generalized solutions, a center-manifold theorem for the underlying equation of evolution. If such a theorem can be found, it may well deliver only weak solutions. The analyst then will face a second problem: either prove that the solutions are indeed classical and smooth, or go over all the non-trivial applications and show that they can be obtained or possibly modified by using weak solutions only.

10. Enskog's procedure. Let me step backward to the evaluation of transport coefficients and to Boltzmann's suggestion, which reflects some partial results obtained by Maxwell in 1879, that the hydrodynamical equations themselves should be corrected. So far I have spoken in this regard only of a gas of Maxwellian molecules and for it only in very special examples, and I have mentioned that Hilbert's procedure does not deliver transport coefficients effectively. A formal method applicable to all molecular models was invented by Enskog in 1917. He attempted to modify Hilbert's iterative procedure. There is no doubt that he obtained final and correct forms for the viscosity and the Maxwell number, and that subsequent authors have successfully applied his method to obtaining approximate corrections to the hydrodynamical equations appropriate to moderately rarefied flows. The details are not so happy. By mysterious juggling, Enskog tailored his procedure to get results of the type he sought. The results themselves suggest that he considered only solutions in a special class, but, unlike Hilbert, he neither specified that class nor formulated his iterative method unequivocally. Thanks to C.-Y. Pao, Hilbert's method has at last been proved effective in the sense that the steps he set up formally can in fact be carried out, to within a still lacking existence theorem for his successive systems of partial differential equations. For Enskog's procedure we cannot approach a proof of even that much, for beyond one or two steps, we do not know what his procedure really is. It rests upon a strange secondary iteration in which time derivatives are approximated by space derivatives.

11. Muncaster's theory of gross determiners. In a thesis accepted at Johns Hopkins in 1975 and published in improved form in 1980, R. G. Muncaster, who studied analysis in courses given by Hartman and Menikoff, obviated the formal obscurities in Enskog's process by introducing a new approach to the problem of calculating special solutions.

In contrast with Hilbert's and Enskog's immediate resort to infinite series for defining and approximating their special solutions, Muncaster introduced the class of *grossly determined solutions*. These are the molecular densities that are determined uniquely at every time by their own fields of principal moments at that time. The time is eliminated from the problem once and for all, at the start, and exactly. Obviously solutions in this class, if they exist, are exempt from relaxation and from the objections raised against solutions of the general initial-value problem. The mappings which determine these molecular densities are called *gross determiners*. Muncaster characterized them as solutions of a time-independent functional equation. A mapping satisfying that equation delivers a solution of the Maxwell-Boltzmann equation if and only if its moments satisfy the field equations of continuum mechanics. Thus gross determiners serve a use similar to that of constitutive relations in continuum theories. This part of Muncaster's work, like Hilbert's, is purely formal. An existence theorem for the gross determiners is lacking. It would provide exactly the apparatus that is needed to study rigorously the type of solution that will correct hydrodynamics.

12. Muncaster's method of stretched fields. Next Muncaster introduced a scheme of functional approximation, which he called the *method of stretched fields*. The idea is similar to Coleman & Noll's method of retardation, but it is applied to space rather than to time. Space in the neighborhood of a point is stretched uniformly by a parameter, and a mapping whose domain is a set of spatial fields is then expanded in powers of that parameter. Muncaster established the validity of this method in certain function spaces, provided the stretching parameter be small enough. This is a matter of pure analysis, not kinetic theory.

Applying his method of stretched fields to his functional equation for gross determiners, Muncaster obtained and completed results previously obtained by Enskog's method. In contrast, Muncaster's method is explicit. He also proved it effective in the sense that the nth step can be carried out. If the gross determiner can be proved to exist and to have the required smoothness, the expansion will be justified. Thus the aspect of approximate calculation retreats, as it should, to a subsidiary role. The concepts are now clear, and the mathematical problem still to be solved is definite. For the first time, the main problem of the kinetic theory has been reduced to a problem of analysis.

If grossly determined solutions exist, what is their status with respect to the much larger class of solutions of the general initial-value problem? Except for special cases such as those I mentioned, no answer to this question is known. For the same special cases a different process of approximation, one introduced by Ikenberry & me in 1956 and called *atemporal Maxwellian iteration*, has been proved to converge to a principal solution if the defining parameter is sufficiently small. It is an open problem to relate atemporal Maxwellian iteration to the method of stretched fields. In fact we do not presently know even how to formulate atemporal Maxwellian iteration as a general procedure of analysis, independent of the kinetic theory. It is a very strange method in that it rests upon repeated differentiation and thus is appropriate only to very smooth functions.

13. Main open problems. Principal conjectures. I close by listing what Muncaster & I consider to be the main open problems of the kinetic theory:

1. Completion of existence theory on the general initial-value problem.
2. Proof of an asymptotic trend to a grossly determined state.
3. Interpretation of Boltzmann's *H*-theorem (which I have not mentioned before now and will not go into here) and its bearing on the trend to equilibrium.
4. Asymptotic status of solutions according to classical gas dynamics and replacement for that theory when it is not valid.

I add also our

Principal Conjectures. If the fields of density, gross velocity, and energetic corresponding to a solution F of the Maxwell-Boltzmann equation lie in sufficiently small neighborhoods of constant fields, and if the body force is sufficiently close to **0**, then

1. For large times the fields **P**, **q**, h, and **s** corresponding to F have asymptotic forms which are grossly determined.
2. To at least one of the sets of four asymptotic forms which Conjecture 1 asserts to exist correspond infinitely many solutions F which have asymptotically a common gross condition ρ, **u**, ε, and **b**.
3. Among the infinitely many solutions to which Conjecture 2 refers there is exactly one, a grossly determined one, for which the asymptotic approximations that Conjecture 1 asserts to exist are also the exact values of **P**, **q**, h, and s, respectively, for every t in the interval on which the particular solution exists.

Of course the analyst who works with the kinetic theory will be led by the analytic structure of the theory itself as he unfolds it. As in any other kind of analysis, he will frame appropriate definitions and in terms of them prove what he can, step by step. But he must not think the task of analysis done until he can discover and establish the status of viscosity and Maxwell number with respect to solutions of the general initial-value problem.

THE JOHNS HOPKINS UNIVERSITY

REFERENCE

C. Truesdell & R. G. Muncaster, *Fundamentals of Maxwell's Kinetic Theory of a Simple Monatomic Gas treated as a Branch of Rational Mechanics*, New York etc., Academic Press, 1980. This book contains full attributions and references to the sources, except for one paper now in press: L Arkeryd, "Intermolecular forces of infinite range and the Boltzmann equation," *Archive for Rational Mechanics and Analysis* **77**, No. 1.

GEOMETRY INDUCED BY MATERIAL STRUCTURE*

By C.-C. Wang

1. Introduction. This is an expository article on the mathematical structure of a topic in continuum mechanics, known as theory of continuous dislocations. We review its basic mathematical formulation and present its central problems and results. Its background in continuum mechanics may be found in reference [1], and that in differential geometry, in reference [2]. Contributions towards its mathematical formulation are given principally by Noll in reference [3] and by Wang in reference [4]. A survey of its recent development may be found in reference [5].

2. Constitutive Equation. In continuum mechanics the physical nature of a material point is characterized by a constitutive equation. Let $\mathscr{B}$ denote a body manifold, and let p be a typical point in $\mathscr{B}$. We define a *configuration* of $\mathscr{B}$ as an (orientation preserving) diffeomorphism $\varphi : \mathscr{B} \to \mathscr{R}^3$, where $\mathscr{R}^3$ represents the physical Euclidean space, and a *local configuration* of p as an (orientation preserving) linear isomorphism $\kappa : \mathscr{B}_p \to \mathscr{R}^3$, where $\mathscr{B}_p$ denotes the tangent space of $\mathscr{B}$ at p. Then a configuration φ gives rise to an induced local configuration φ_{*p} at p. We call p a *simple material point* if its local configuration determines—in a sense that we shall make precise—its constitutive equation. A body manifold $\mathscr{B}$ made up of simple material points is then called a *simple material body*.

A possible constitutive equation of p, determined by the local configuration κ, is of the form

$$(1) \qquad \mathbf{T} = \mathbf{G}_\kappa(\mathbf{F}),$$

where $\mathbf{F}$ denotes the deformation gradient at p relative to κ, and where $\mathbf{T}$ denotes the stress tensor at p in the deformed local configuration $\mathbf{F}\kappa$.

*This paper was supported by a grant from the U.S. National Science Foundation to the author through Rice University.

The constitutive equation (1) gives $\mathbf{T}$ as a function of $\mathbf{F}$; that function, $\mathbf{G}_\kappa$, is called the *response function* of p relative to κ.

In the physical interpretation κ is just a representation of $\mathscr{B}_p$ by $\mathscr{R}^3$. Once such a representation is taken, the stress tensor $\mathbf{T}$ may be expressed as a function of the deformation gradient $\mathbf{F}$. Each deformation gradient $\mathbf{F}$ is an (orientation preserving) linear isomorphism $\mathbf{F} : \mathscr{R}^3 \to \mathscr{R}^3$; it transforms the local configuration κ into the (deformed) local configuration $\mathbf{F}\kappa$. We can regard the stress tensor $\mathbf{T}$ as a response of p towards the deformation $\mathbf{F}$, so that the function $\mathbf{G}_\kappa$ that gives $\mathbf{T}$ in terms of $\mathbf{F}$ is called the response function.

A material point p whose physical nature is characterized by (1) is called an *elastic material point*, since (1) asserts that the response of p is instantaneous, *i.e.*, the stress tensor $\mathbf{T}(t)$ of p at any instant t depends only on the deformation gradient $\mathbf{F}(t)$ relative to κ at the same instant t. An elastic material point is a simple material point, since (1) is determined by the local configuration κ in the sense that $\mathbf{G}_\kappa$ is a function of κ; the value of $\mathbf{G}_\kappa$ is itself also a function, which transforms $\mathcal{L}(\mathscr{R}^3)$—the proper linear group of $\mathscr{R}^3$ where $\mathbf{F}$ belongs—to $\mathcal{S}(\mathscr{R}^3)$—the space of symmetric 3×3 matrices where $\mathbf{T}$ belongs.

The general form of constitutive equation for a simple material point is

$$(2) \qquad \mathbf{T}(t) = \mathbf{G}_\kappa(\mathbf{F}(t - s), s \in [0, \infty)),$$

where $\mathbf{F}(t - s)$ denotes the deformation gradient of p relative to κ at time $t - s$. Since we require $s \in [0, \infty)$, the function $\mathbf{F} : [0, \infty) \to \mathcal{L}(\mathscr{R}^3)$ represents a deformation history of p relative to κ up to the (present) time t. As we have pointed out, (1) asserts that $\mathbf{T}(t) = \mathbf{G}_\kappa(\mathbf{F}(t))$, which is clearly a special case of (2). Unlike an elastic material point, a simple material point generally may have memory effects, since (2) asserts that $\mathbf{T}(t)$ may depend on the deformation gradients $\mathbf{F}(\tau)$ at past time $\tau < t$. We say that the local configuration κ determines the constitutive equation (2) in the same sense as before: $\mathbf{G}_\kappa$ is a function of κ. However, in (2) the value of $\mathbf{G}_\kappa$ is a function which transforms a deformation history —a curve $\{\mathbf{F}(t - s), s \in [0, \infty)\}$ in $\mathcal{L}(\mathscr{R}^3)$—into a tensor $\mathbf{T}(t)$ in $\mathcal{S}(\mathscr{R}^3)$. Since a deformation history is itself a function, we call $\mathbf{G}_\kappa$ the *response functional* of p relative to κ.

In this article we are not interested in the explicit form of the response function in (1) or the response functional in (2). The fact that

(1) or (2) are determined by κ suffices. For simplicity of writing we assume that p is elastic and may be characterized by a constitutive equation of the form (1). Since we are primarily concerned with the dependence of $\mathbf{G}_\kappa$ on κ, we now write $\mathbf{G}(\kappa)$ for $\mathbf{G}_\kappa$; it is important to remember that $\mathbf{G}(\kappa)$ is itself a function, $\mathbf{G}(\kappa) : \mathcal{L}(\mathcal{R}^3) \to \mathcal{S}(\mathcal{R}^3)$.

We denote the set of all local configurations of p by $\mathcal{L}(\mathcal{B}_p, \mathcal{R}^3)$ and the set of all functions from $\mathcal{L}(\mathcal{R}^3)$ to $\mathcal{S}(\mathcal{R}^3)$ by Φ. Then there is a mapping $\mathbf{G} : \mathcal{L}(\mathcal{B}_p, \mathcal{R}^3) \to \Phi$ whose value $\mathbf{G}(\kappa) \in \Phi$ is the response function of p relative to κ for each $\kappa \in \mathcal{L}(\mathcal{B}_p, \mathcal{R}^3)$. We denote the proper linear group of $\mathcal{B}_p$ by $\mathcal{L}(\mathcal{B}_p)$. Then the set $\mathcal{L}(\mathcal{B}_p, \mathcal{R}^3)$ may be acted on the right by $\mathcal{L}(\mathcal{B}_p)$ and on the left by $\mathcal{L}(\mathcal{R}^3)$ as transformation groups. Of course, the group $\mathcal{L}(\mathcal{R}^3)$ may be acted both on the right and on the left by itself. Furthermore, the actions may be lifted to the set Φ, *viz*, for any $\mathbf{H} \in \mathcal{L}(\mathcal{R}^3)$ we define $L_\mathbf{H} : \Phi \to \Phi$ and $R_\mathbf{H} : \Phi \to \Phi$ by

$$(3) \qquad L_\mathbf{H} \mathbf{A}(\mathbf{F}) = \mathbf{A}(\mathbf{HF}), \qquad R_\mathbf{H} \mathbf{A}(\mathbf{F}) = \mathbf{A}(\mathbf{FH}) \qquad \forall \mathbf{F} \in \mathcal{L}(\mathcal{R}^3)$$

for any $\mathbf{A} \in \Phi$.

Now from the very definition of the constitutive equation (1) the mapping $\mathbf{G}$ obeys the following transformation rule:

$$(4) \qquad \mathbf{G}(\mathbf{H}\kappa) = R_\mathbf{H} \mathbf{G}(\kappa) \qquad \forall \kappa \in \mathcal{L}(\mathcal{B}_p, \mathcal{R}^3)$$

for any $\mathbf{H} \in \mathcal{L}(\mathcal{R}^3)$. Thus the response function $\mathbf{G}(\eta)$ relative to any local configuration η may be obtained from the response function $\mathbf{G}(\kappa)$ relative to a particular local configuration κ through the transformation (4) with $\mathbf{H} = \eta \kappa^{-1}$.

3. Material Isomorphism. In the preceding section we defined the constitutive equation of a typical point p in a body manifold $\mathcal{B}$. We now apply our formulation to all points in $\mathcal{B}$. In order to make our notations precise, we denote a local configuration of p by κ_p and the response function of p relative to κ_p by $\mathbf{G}(\kappa_p, p)$. Notice, however, that the group of deformation gradients, $\mathcal{L}(\mathcal{R}^3)$, and the space of stress tensors, $\mathcal{S}(\mathcal{R}^3)$, are based on the physical space $\mathcal{R}^3$ and, therefore, are independent of the points in $\mathcal{B}$.

Let p and q be any two material points in $\mathcal{B}$ characterized by response functions $\mathbf{G}(\kappa_p, p)$ and $\mathbf{G}(\kappa_q, q)$ relative to κ_p and κ_q, respectively. Then we say that κ_p and κ_q are (*materially*) *isomorphic* if their corresponding response functions $\mathbf{G}(\kappa_p, p)$ and $\mathbf{G}(\kappa_q, q)$ are the same function in Φ. If p and q possess isomorphic local configurations, then they are called (*materially*) *isomorphic points*.

In view of the transformation rule (4), if κ_p is isomorphic to κ_q, then $\mathbf{H}\kappa_p$ is isomorphic to $\mathbf{H}\kappa_q$ for all $\mathbf{H} \in \mathcal{L}(\mathcal{R}^3)$. Hence there is an (orientation preserving) linear isomorphism $\boldsymbol{\alpha}(p,q): \mathcal{B}_p \to \mathcal{B}_q$ such that $\boldsymbol{\eta}_q$ is isomorphic to $\boldsymbol{\eta}_q \boldsymbol{\alpha}(p,q)$ for all $\boldsymbol{\eta}_q \in \mathcal{L}(\mathcal{B}_q, \mathcal{R}^3)$, viz,

$$(5) \qquad \mathbf{G}(\boldsymbol{\eta}_q \boldsymbol{\alpha}(p,q), p) = \mathbf{G}(\boldsymbol{\eta}_q, q).$$

Indeed, we may take

$$(6) \qquad \boldsymbol{\alpha}(p,q) = \kappa_q^{-1} \kappa_p$$

for any isomorphic κ_p and κ_q. Then $\boldsymbol{\eta}_q \boldsymbol{\alpha}(p,q) = (\boldsymbol{\eta}_q \kappa_q^{-1}) \kappa_p$ while $\boldsymbol{\eta}_q = (\boldsymbol{\eta}_q \kappa_q^{-1}) \kappa_q$. In other words, if (5) is satisfied by any one local configuration $\boldsymbol{\eta}_q$, say by $\boldsymbol{\eta}_q = \kappa_q$, then it is satisfied by all local configurations $\boldsymbol{\eta}_q \in \mathcal{L}(\mathcal{B}_q, \mathcal{R}^3)$ by virtue of the transformation rule (4).

We denote the set of all (orientation preserving) linear isomorphisms from $\mathcal{B}_p$ to $\mathcal{B}_q$ by $\mathcal{L}(\mathcal{B}_p, \mathcal{B}_q)$. Then we call $\boldsymbol{\alpha}(p,q) \in \mathcal{L}(\mathcal{B}_p, \mathcal{B}_q)$ a *material isomorphism* if it satisfies (5), which means that $\boldsymbol{\eta}_q \boldsymbol{\alpha}(p,q)$ and $\boldsymbol{\eta}_q$ are isomorphic. From (6) we see that p and q are isomorphic if and only if there is a material isomorphism $\boldsymbol{\alpha}(p,q)$ from p to q.

We assume now that all material points in $\mathcal{B}$ are pairwise isomorphic, so that $\mathcal{B}$ is (materially) uniform. This assumption characterizes the physical condition that $\mathcal{B}$ is made up of the same material. Under this assumption there is a material isomorphism between each pair of points in $\mathcal{B}$. If those material isomorphisms are unique, then they give rise to a distant parallelism on $\mathcal{B}$. Such a simple structure, however, does not correspond to the material structure of most material bodies of interest in continuum mechanics. Indeed, uniqueness of material isomorphisms corresponds to total lack of material symmetry, which is appropriate only for a triclinic crystal body.

From (5) we see that the set of all material isomorphisms on $\mathcal{B}$ obeys the following conditions:

i) The identity map ι_p of $\mathcal{B}_p$ is a material automorphism of p.
ii) If $\boldsymbol{\alpha}(p,q)$ is a material isomorphism from p to q, then $\boldsymbol{\alpha}(p,q)^{-1}$ is a material isomorphism from q to p.

iii) If $\alpha(p,q)$ and $\alpha(q,r)$ are material ismorphisms from p to q and from q to r, respectively, then their composition $\alpha(q,r)\alpha(p,q)$ is a material isomorphism from p to r.

These conditions imply that the set of all material automorphisms of any point p form a group $g(p) \in \mathcal{L}(\mathcal{B}_p)$, called the (material) symmetry group of p. The group $g(p)$ characterizes the set of all material isomorphisms from p in the following way: Let $\alpha(p,q)$ be a particular material isomorphism from p to q. Then $\beta(p,q) \in \mathcal{L}(\mathcal{B}_p, \mathcal{B}_q)$ is a material isomorphism if and only if $\alpha(p,q)^{-1}\beta(p,q) \in g(p)$. In other words, given any one particular material isomorphism $\alpha(p,q)$ we can express the family of all material isomorphisms from p to q as $\alpha(p,q)g(p)$. Thus material isomorphisms from p to q are unique if and only if $g(p)$ is the trivial group consisting of the identity map ι_p of $\mathcal{B}_p$ only.

As remarked before, the special case with $g(p) = \{\iota_p\}$ is appropriate for a triclinic crystal. While our theory does not exclude such a trivial symmetry group, we are more interested in the general case with a non trivial symmetry group. For instance $g(p)$ may be a discrete finite group (corresponding to other types of crystals), a special orthogonal group (corresponding to an isotrophic solid), a special linear group (corresponding to a fluid), or others (corresponding to various types of fluid crystals). In all these cases material isomorphisms from p to q are not unique.

Physical experience does restrict $g(p)$ to be a subgroup of the special linear group $\mathcal{SL}(\mathcal{B}_p)$ of $\mathcal{B}_p$. Setting $p = q$ in (5), we see that $\alpha_p \in g(p)$ if and only if

(7) $\mathbf{G}(\eta_p\alpha_p, p) = \mathbf{G}(\eta_p, p)$.

There are no known materials for which (7) is satisfied by some α_p with $\det \alpha_p \neq 1$.

It should be noted that, if the symmetry group $g(p)$ of p and a material isomorphism $\alpha(p,q)$ from p to q are given, then the symmetry group $g(q)$ of q is determined by

(8) $g(q) = \alpha(p,q)g(p)\alpha(p,q)^{-1}$.

Indeed, Conditions ii) and iii) imply that $g(q) \supset \alpha(p,q)g(p)\alpha(p,q)^{-1}$ for arbitrary p and q. Using this result and Conditions ii) and iii) again, we then have $\mathfrak{g}(q) \subset \boldsymbol{\alpha}(p,q)\mathfrak{g}(p)\boldsymbol{\alpha}(p,q)^{-1}$. Hence (8) holds. It follows from (8) that the set of all material isomorphisms form p to q may be expressed as $\mathfrak{g}(q)\boldsymbol{\alpha}(p,q)$ also.

4. Geometry Induced by Material Isomorphisms. In differential geometry an (orientable) manifold $\mathfrak{M}$ has an (oriented) tangent space $\mathfrak{M}_x$ at each $x \in \mathfrak{M}$. Since the dimension of $\mathfrak{M}_x$ is independent of $x \in \mathfrak{M}$, there are (orientation preserving) linear isomorphisms $\gamma(x, y):$ $\mathfrak{M}_x \to \mathfrak{M}_y$ for any pair of points x and y in $\mathfrak{M}$. The family of all such linear isomorphisms obviously satisfies Conditions i), ii), and iii) similar to those satisfied by the family of all material isomorphisms of a materially uniform body. Within the family of all linear isomorphisms the proper linear group $\mathfrak{L}(\mathfrak{M}_x)$ plays the role similar to that of the symmetry group $\mathfrak{g}(p)$ within the family of all material isomorphisms.

If $\mathfrak{M}$ is a Riemannian manifold with metric m, then there are (orientation preserving) linear isomorphisms $\gamma = \gamma(x, y): \mathfrak{M}_x \to \mathfrak{M}_y$ which transform m_x into m_y in the usual sense: $m_x(\mathbf{u}, \mathbf{v}) = m_y(\gamma\mathbf{u}, \gamma\mathbf{v})$ $\forall \mathbf{u}, \mathbf{v} \in \mathfrak{M}_x$. The family of all such linear isomorphisms also satisfies Conditions i), ii), and iii). For this family the special orthogonal group $\mathfrak{SO}(\mathfrak{M}_x)$ plays the role similar to that of the symmetry group (p).

The families of linear isomorphisms just mentioned for a manifold or for a Riemannian manifold satisfy also the following condition of smoothness: iv) Each point $x \in \mathfrak{M}$ has a neighborhood $\mathfrak{M}_x$ in which there is a smooth field γ whose values $\gamma(x, y)$, $y \in \mathfrak{M}_x$, are linear isomorphisms belonging to the specific family.

By using Conditions i)–iv), we can characterize each aforesaid family of linear isomorphisms by a principal bundle over the manifold $\mathfrak{M}$ (which is the base space of the bundle). For the family of all linear isomorphisms among tangent spaces of $\mathfrak{M}$ the corresponding principal bundle is known as the *bundle of (linear) frames*, while for the family of all linear isomorphisms which preserve the Riemannian metric the corresponding principal bundle is known as the *bundle of orthonormal frames*. For the family of material isomorphisms on a uniform body a similar principal bundle may be defined provided that Condition iv) is satisfied. Specifically, we require that each point $p \in \mathfrak{B}$ has a neighborhood (a subbody containing p) in which there is a smooth field of material isomorphisms from the point p.

A principal bundle corresponding to the family of material isomorphisms is called a *bundle of (materially) isomorphic frames*, defined as follows: Conditions i)–iii) imply that material isomorphism is an equivalence relation on the set of all local configurations of points in $\mathscr{B}$. We claim that any one equivalence class of isomorphic local configurations has the structure of a principal bundle over the body manifold $\mathscr{B}$.

First, we remark that a local configuration κ_p may be visualized as a frame $\{\mathbf{e}_1, \mathbf{e}_2, \mathbf{e}_3\}$ at p (a frame at p is just a basis of $\mathscr{B}_p$) by setting

$$(9) \qquad \kappa_p(\mathbf{e}_a) = \mathbf{i}_a, \qquad a = 1, 2, 3,$$

where $\{\mathbf{i}_1, \mathbf{i}_2, \mathbf{i}_3\}$ denotes the standard basis of $\mathscr{R}^3$. We call two frames (at the same point or at different points in $\mathscr{B}$) isomorphic if their corresponding local configurations are. Then an equivalence class of isomorphic local configurations corresponds to a *bundle of isomorphic frames*. The structure group of this bundle is a subgroup $\mathfrak{g} \subset \mathcal{SL}(\mathscr{R}^3)$, defined as follows:

Let $\{\mathbf{e}_a\}$ be a frame at p belonging to the aforesaid bundle. Then any other frame $\{\mathbf{f}_a\}$ at p is isomorphic to $\{\mathbf{e}_a\}$ if and only if there is $\alpha_p \in \mathfrak{g}(p)$ such that $\mathbf{f}_a = \alpha(\mathbf{e}_a)$, $a = 1, 2, 3$. As usual, we can express the change of basis from $\{\mathbf{e}_a\}$ to $\{\mathbf{f}_a\}$ by a right-multiplication, *viz*,

$$(10) \qquad \mathbf{f}_b = \mathbf{e}_a A_b^a, \qquad b = 1, 2, 3,$$

where the repeated index a is summed from 1 to 3. The set of components matrices $\mathbf{A} = [A_b^a]$ for all $\{\mathbf{f}_a\}$ isomorphic to $\{\mathbf{e}_a\}$ form a group $\mathfrak{g} \subset \mathcal{SL}(\mathscr{R}^3)$. We claim that $\mathfrak{g}$ is independent of $\{\mathbf{e}_a\}$ as well as the point $p \in \mathscr{B}$.

This result may be proved from the following condition for $\mathbf{A} \in \mathfrak{g}$:

$$(11) \qquad \mathbf{G}_{\kappa_p}(\mathbf{F}\mathbf{A}^{-1}, p) = \mathbf{G}_{\kappa_p}(\mathbf{F}, p) \qquad \forall \mathbf{F} \in \mathcal{L}(\mathscr{R}^3).$$

Indeed, from (10) the local configuration $\boldsymbol{\eta}_p$ corresponding to $\{\mathbf{f}_a\}$ is related to κ_p by $\boldsymbol{\eta}_p = \mathbf{A}^{-1}\kappa_p$. Hence (11) follows from the transformation rule (4). Now since the response function is invariant within an equivalence class of local configurations, (11) is independent of κ_p and p and, therefore, $\mathfrak{g}$ is also.

From (9) Condition iv) is equivalent to the requirement that each point $p \in \mathscr{B}$ has a neighborhood $\mathfrak{N}_p$ in which there is a smooth field $\{\mathbf{e}_a\}$ of isomorphic frames. Then from (10) all other smooth fields $\{\mathbf{f}_a\}$ of isomorphic frames over $\mathfrak{N}_p$ may be obtained from $\{\mathbf{e}_a\}$ by smooth fields $\mathbf{A}$ of right multiplications in the group $\mathfrak{g}$. In other words, we can regard $\{\mathbf{e}_a\}$ as the section of identity over $\mathfrak{N}_p$ in a bundle chart and define other sections over $\mathfrak{N}_p$ by right multiplications of $\mathfrak{g}$. Such bundle charts cover the body manifold $\mathfrak{B}$ and define a bundle atlas for the bundle of isomorphic frames.

The bundle of isomorphic frames just defined clearly characterizes the family of material isomorphisms. In fact, the linear extension of any mapping from a frame to an isomorphic frame is a material isomorphism and, conversely, a material isomorphism maps any frame into an isomorphic frame.

As remarked before, $\mathfrak{B}$ is called an isotropic solid if there is a Riemannian metric m on $\mathfrak{B}$ such that $\mathfrak{g}(p) = \mathfrak{SO}(\mathfrak{B}_p)$ with respect to m_p. For such a body the bundle of orthonormal frames considered in differential geometry is a bundle of isomorphic frames. (Other bundles of isomorphic frames may be obtained from the bundle of orthonormal frames by right multiplications of non rotation matrices.) The Riemannian geometry is the intrinsic geometry on the body manifold $\mathfrak{B}$ of an isotropic solid.

We call $\mathfrak{B}$ a fluid if $\mathfrak{g}(p) = \mathfrak{SL}(\mathfrak{B}_p)$. For such a body the bundle of isochoric frames with respect to a volume tensor field is a bundle of isomorphic frames. The geometry of an oriented manifold is the intrinsic geometry on the body manifold $\mathfrak{B}$ of a fluid.

In general, the intrinsic geometry on the body manifold $\mathfrak{B}$ of a smooth uniform body is the geometry associated with a bundle of isomorphic frames, which may be viewed as a field of geometric objects invariant under the family of material isomorphisms. The bundle of linear frames is the underlying space of all bundles of isomorphic frames which are equivalence classes with respect to material isomorphism.

5. Material Connection. A useful tool for studying the intrinsic geometry on the body manifold of a smooth uniform body is a *material connection*, defined as follows: It is an affine connection whose parallel transports along all curves are material isomorphisms. For example, if $\mathfrak{B}$ is an isotropic solid, then the Riemannian connection associated with the intrinsic Riemannian metric m is a material connection.

We can visualize the parallel transports induced by an affine connection as mappings among frames along curves. In order that mappings among frames may be extended linearly into the same linear isomorphisms among tangent spaces, they must commute with right multiplications by $\mathfrak{L}(\mathfrak{R}^3)$. For a material connection frames mapped by the parallel transports must be isomorphic to one another. Then the mappings among isomorphic frames commute with right multiplications by elements of the structure group $\mathfrak{g}$. In differential geometry such a connection is called a $\mathfrak{g}$-connection. Hence a material connection is just a $\mathfrak{g}$-connection on a bundle of isomorphic frames.

Existence of $\mathfrak{g}$-connections on principal bundle in general may be proved by using a partition of unity subordinate to a covering of the base manifold by bundle charts. Thus material connections exist for any smooth uniform body. A criterion for the connection symbols of a material connection may be derived as follows:

Let $\boldsymbol{\varphi} : \mathfrak{B} \to \mathfrak{R}^3$ be a configuration of $\mathfrak{B}$. We denote the coordinates of a typical point $p \in \mathfrak{B}$ by $\boldsymbol{\varphi}(p) = (x^a, a = 1, 2, 3)$. The natrual frame $\{\partial_p x_a\}$ of (x^a) at p corresponds to the induced local configuration $\boldsymbol{\varphi}_{*p}$, viz, $\boldsymbol{\varphi}_{*p}(\partial_p x_a) = \boldsymbol{i}_a$ as defined by (9). Then a field of isomorphic frames $\{\mathbf{e}_a\}$ may be expressed in component form relative to $\{\partial x_a\}$,

(12) $\mathbf{e}_b = e_b^a \partial x_a,$ $b = 1, 2, 3,$

which means that $\{\mathbf{e}_a\}$ differs from $\{\partial x_a\}$ by the field of right multiplications of component matrices $\mathbf{e} = [e_b^a]$. Let Γ_{bc}^a be the connection symbols of an affine connection relative to the coordinate system (x^a), and suppose that $\lambda(t)$ is a curve passing through the point p at $t = 0$. Then a field of frames $\{\mathbf{f}_a\}$ on λ, parallel with respect to the affine connection, satisfies the equations of parallel transport:

(13) $\dfrac{df_a^b}{dt} + \Gamma_{ch}^b f_a^c \dfrac{d\lambda^h}{dt} = 0,$ $a, b = 1, 2, 3.$

The affine connection is a material connection if and only if the field $\{\mathbf{f}_a\}$ is related to the field $\{\mathbf{e}_a\}$ by (10) with $\mathbf{A} = \mathbf{A}(t) \in \mathfrak{g}$.

Since the parallel transport commutes with right multiplications by $\mathcal{L}(\mathfrak{R}^3)$, we can choose the value of $\{\mathbf{e}_a\}$ at p as the initial value $\{\mathbf{f}_a(0)\}$. Then $\mathbf{A}(0) = [\delta_b^a]$. Substituting (10) into (13) and evaluating the result at $t = 0$, where $p = \lambda(0)$, we get

(14) $\dfrac{dA_a^b(0)}{dt} = -\left[(e^{-1})_c^b \left(\dfrac{\partial e_a^c}{\partial x^h} + \Gamma_{ih}^c e_a^i \right) \right]_p \dfrac{d\lambda^h(0)}{dt} .$

Now since $\mathbf{A}(t)$ is a curve in $\mathfrak{g}$ passing through the identity matrix at $t = 0$, and since $\lambda(t)$ is an arbitrary curve passing through the point p at

$t = 0$, (14) implies that the matrices

(15) $\qquad \left[(e^{-1})_c^b \left(\dfrac{\partial e_a^c}{\partial x^h} + \Gamma_{ih}^c e_a^i \right) \right]_p, h = 1, 2, 3,$

are contained in the Lie algebra of $\mathfrak{g}$.

Notice that the matrices in (15) are just $[(e^{-1})_c^b e_{a,h}^c]$, $h = 1, 2, 3$, where $e_{a,h}^c$ denotes the covariant derivative of the vectors $\mathbf{e}_a$ with respect to the connection. In particular, if the coordinate system (x^a) is chosen in such a way that $\{\partial_p x_a\} = \{\mathbf{e}_a(p)\}$, i.e., the induced local configuration φ_{*p} coincides with the local configuration κ_p in the equivalence class under consideration, then (15) are just the matrices $[e_{a,h}^c]$, $h = 1, 2$, 3. Hence the condition that the matrices (15) are contained in the Lie algebra of $\mathfrak{g}$ simply means that the connection form takes its values in the fundamental fields of the bundle of isomorphic frames.

The preceding condition is both necessary and sufficient for the connection symbols of a material connection. We have proved necessity of the condition in the derivation of (15). Conversely, if the condition is satisfied at all points $p \in \mathfrak{B}$ (we may have to use different fields of isomorphic frames to cover different points in $\mathfrak{B}$, of course), then the tangent vector field of the curve $\mathbf{A}(t)$ is contained in the Lie algebra of $\mathfrak{g}$ for all t. Consequently, the curve $\mathbf{A}(t)$ is contained in $\mathfrak{g}$, since the Lie algebra is integrable. ($\mathfrak{g}$ is an integral manifold of its Lie algebra at the identity matrix $[\delta_b^a]$.)

If $\mathfrak{g}$ is discrete, then the matrices in (15) must vanish. Hence

(16) $\qquad \Gamma_{ih}^c = -(e^{-1})_i^a \dfrac{\partial e_a^c}{\partial x^h}.$

In this case material connection is unique; it corresponds to a distant parallelism if the body manifold is simply connected. Material isomorphisms between any pair of points in $\mathfrak{B}$ need not be unique, however, since $\mathfrak{g}$ need not be the trivial group. This remark shows clearly that parallel transports induced by material connections do not necessarily give all material isomorphisms. In other words, the holonomy group associated with a material connection may be only a proper subgroup of $\mathfrak{g}$.

As usual integrability conditions (vanishing of torsion tensor and curvature tensor) for a material connection tell us whether or not the intrinsic geometry on the body manifold $\mathfrak{B}$ is an Euclidean geometry. Such conditions are very important in mechanics, since an Euclidean intrinsic geometry corresponds precisely to the physical notion that $\mathfrak{B}$ is a homogeneous body.

Finally, we remark that a material connection plays an important role in the formulation of the equations of motion for a smooth uniform elastic body.

6. Symmetry of Material Structure. A diffeomorphism $\psi : \mathfrak{B} \to \mathfrak{B}$ such that its induced linear isomorphisms $\psi_{*p} : \mathfrak{B}_p \to \mathfrak{B}_q$, where $q = \psi(p)$, are material isomorphisms for all $p \in \mathfrak{B}$ is called a (*global*) *symmetry transformation* of $\mathfrak{B}$. Such a transformation preserves the material structure on $\mathfrak{B}$. The set of all symmetry transformations of $\mathfrak{B}$ form a group $\mathcal{G}$, called the (*global*) *symmetry group* of $\mathfrak{B}$.

If $\mathfrak{B}$ is an isotropic solid, its material structure is characterized by an intrinsic Riemannian metric m, then a symmetry transformation is just an isometry of $\mathfrak{B}$ with itself. Even in this special case, the problem of characterizing the group $\mathcal{G}$ completely is quite difficult in general. However, one-parameter groups in $\mathcal{G}$ correspond to flows generated by Killing vector fields whose integrability conditions are known. For a material structure in general we may introduce similar vector fields on $\mathfrak{B}$ by requiring that the flows preserve a bundle of isomorphic frames.

Like the parallel transports induced by an affine connection, the flow generated by a vector field commutes with right multiplications by $\mathcal{L}(\mathfrak{R}^3)$. Consequently, we may use any convenient bundle of isomorphic frames to test the flow of a vector field.

Let $\mathbf{v}$ be a vector field on $\mathfrak{B}$. Then a field of frames $\{\mathbf{f}_a\}$ is parallel with respect to the flow generated by $\mathbf{v}$ if

$$(17) \qquad \frac{\partial f_a^b}{\partial x^c} v^c - \frac{\partial v^b}{\partial x^c} f_a^c = 0, \qquad a, b = 1, 2, 3,$$

where we have used a coordinate system $\varphi(p) = (x^a)$ as before. The condition that the flow preserves the material structure simply means that a solution $\{\mathbf{f}_a\}$ of (17) is a field of isomorphic frames. We can test this condition by (10) with respect to any convenient field of isomorphic frames $\{\mathbf{e}_a\}$.

Without loss of generality, we choose $\{\mathbf{f}_a\}$ to coincide with $\{\mathbf{e}_a\}$ at a point $p \in \mathfrak{B}$, so that $\mathbf{A}(p) = [\delta_a^b]$. Then substituting (10) into (17) and evaluating the result at p, we get

$$(18) \qquad \left[\frac{\partial A_a^b}{\partial x^c} v^c \right]_p = - \left[(e^{-1})_h^b \left(\frac{\partial e_a^h}{\partial x^c} v^c - \frac{\partial v^h}{\partial x^c} e_a^c \right) \right]_p.$$

Notice that (18) has the same form as (14) except that the covariant derivative in (14) is replaced by the Lie derivative with respect to the vector v in (18). Since the left hand side of (18) is just the derivative of the field of matrices $A = [A_a^b]$ along the orbit of v passing through p, we see that the matrix

$$(19) \qquad \left[(e^{-1})_h^b \left(\frac{\partial e_a^h}{\partial x^c} v^c - \frac{\partial v^h}{\partial x^c} e_a^c \right) \right]_p$$

must be contained in the Lie algebra of $\mathcal{g}$.

By the same argument as that for a material connection the condition that the matrix (19) is contained in the Lie algebra of $\mathcal{g}$ for all $p \in \mathcal{B}$ is both necessary and sufficient for v to generate a 1-parameter group in the symmetry group $\mathcal{G}$ of $\mathcal{B}$.

In the special case $\mathcal{B}$ is an isotropic solid, the preceding condition is equivalent to the classical equations of Killing, which require that the Lie derivative of the intrinsic Riemannian metric m vanishes, so that the flow of an orthonormal basis remains an orthonormal basis. If $\mathcal{g}$ is discrete, the preceding condition reduces to

$$(20) \qquad [e_a, v] = 0, \qquad a = 1, 2, 3,$$

which means that the flow of v commutes with the distant parallelism induced by the field $\{e_a\}$. Since $[v, v] = 0$, (20) implies that v must be a parallel field with respect to $\{e_a\}$.

Like the classical equations of Killing, the preceding condition on v need not have any non zero solutions. Certain integrability conditions must be satisfied for the existence of non zero solutions v.

RICE UNIVERSITY

REFERENCES

[1] Truesdell, C. & Noll, W., The non-linear field theories of mechanics, *Handbuch der Physik*, III/3, Berlin-Heidelberg-New York: Springer, 1965.

[2] Kobayashi, S. & Nomizu, K., *Foundations of Differential Geometry*, Vol. 1, Interscience Publ., New York-London, 1963.

[3] Noll, W., Materially uniform simple bodies with inhomogeneities, *Arch. Rational Mech. Anal.* **27** 1967, 1–32.

[4] Wang, C.-C., On the geometric structures of simple bodies, a mathematical foundation for the theory of continuous distributions of dislocations, *Arch. Rational Mech. Anal.* **27** 1967, 33–92.

[5] Bloom, F., Modern differential geometric techniques in the theory of continuous distributions of dislocations, *Lecture Notes in Math.* Vol. 733, Berlin-Heidelberg-New York: Springer, 1979.